CLASSIC MUSTANG

Restoration, Repair & Upgrades

From the Editors of

Mustang Monthly

HPBooks

HPBooks
Published by the Penguin Group
Penguin Group (USA) Inc.
375 Hudson Street, New York, New York 10014, USA
Penguin Group (Canada), 90 Eglinton Avenue East, Suite 700, Toronto, Ontario M4P 2Y3, Canada
(a division of Pearson Penguin Canada Inc.)
Penguin Books Ltd., 80 Strand, London WC2R 0RL, England
Penguin Group Ireland, 25 St. Stephen's Green, Dublin 2, Ireland (a division of Penguin Books Ltd.)
Penguin Group (Australia), 250 Camberwell Road, Camberwell, Victoria 3124, Australia
(a division of Pearson Australia Group Pty. Ltd.)
Penguin Books India Pvt. Ltd., 11 Community Centre, Panchsheel Park, New Delhi—110 017, India
Penguin Group (NZ), 67 Apollo Drive, Rosedale, North Shore 0632, New Zealand
(a division of Pearson New Zealand Ltd.)
Penguin Books (South Africa) (Pty.) Ltd., 24 Sturdee Avenue, Rosebank, Johannesburg 2196, South Africa

Penguin Books Ltd., Registered Offices: 80 Strand, London WC2R 0RL, England

CLASSIC MUSTANG

Cover design by Bird Studios
Cover photos courtesy *Mustang Monthly* magazine
Contributors: Miles Cook, Jim Smart, Donald Farr, Mark Houlahan

First edition: March 2011

ISBN: 978-1-55788-556-2

PRINTED IN THE UNITED STATES OF AMERICA

5th Printing

CONTENTS

CONTENTS

INTRODUCTION

It started as little more than a restyled Falcon. With its long-nose, short-tail styling and standard equipment like bucket seats and floor shifter, the new 1965 Mustang appealed to the emerging Baby Boom generation that was just entering the work force. Ford hoped to sell 200,000; they sold over one million in the first year, launching a phenomenon that continues to this day. From 1965 to 1973, the "first generation" Mustang was a sports car for everyone. Today, untold thousands continue to be coveted by their owners.

Since 1978, *Mustang Monthly* magazine has been guiding '65–'73 Mustang owners in their ongoing quests to improve their cars. Through how-to and technical articles, the editors have tackled everything from basic maintenance chores to the most elaborate restoration procedures. With the recent trend toward modern upgrades, we've covered many of the ways to add performance, comfort, and convenience with late-model electronic fuel injection, power windows, and rack-and-pinion steering. Our goal is simple—help readers improve their vintage Mustangs for however they plan to use them, whether it's a weekend driver, show car, or even everyday transportation.

Classic Mustang: Restoration, Repair & Upgrades is a compilation of popular and helpful tech stories from the editors of *Mustang Monthly* magazine. Inside, you'll find informative articles covering engine rebuilds, suspension tips, basic maintenance repairs, sheetmetal replacement, and performance upgrades. Three decades ago, *Mustang Monthly* pioneered the easy-to-follow how-to format, and that's what you'll find here. Each technical article will guide you along step-by-step as you repair or upgrade your Mustang, whether it's a basic six-cylinder hardtop or a fire-breathing Shelby.

Inside you'll find a guide for aligning your Mustang's front end, replacing an old exhaust system with a brand-new concours version, installing rack-and-pinion steering, optimizing your cooling system, restoring exhaust manifolds, and more.

Improving your vintage Mustang is the goal, and that's what this book is all about. —Donald Farr

Editor, *Mustang Monthly*

ENGINE
&
DRIVETRAIN

Chapter 1

How to Build a Budget 289

Text and Photos by Jim Smart

Mark Jeffrey of TransAm Racing took on the challenge of freshening up our Mexican block 289 within a strict budget.

This Mexican block 289 has been hanging around our garage for at least six years, collecting dust and rust. When a Southern California machine shop built this engine six years ago, they started with a 4.040-inch-bore block we thought we could save. The machine shop, no longer in business, was able to hone the bores at 4.040-inches and fit them with forged pistons. The crank was turned 0.010/0.010-inch undersize. The C3OE 289 connecting rods were reconditioned and fitted with new ARP bolts.

When we began planning a project to show how to convert a six-cylinder Mustang to a V-8, we weren't sure where we would source our small-block engine until we remembered this old dust bunny. Perhaps a gut feeling was the reason we bagged this engine and never touched it again. The irregular appearance of the cylinder walls bothered us, so there it sat inside a plastic bag for six years.

When we blew the dust off and took the engine to Mark Jeffrey at TransAm Racing, he looked it over and knocked it down for a freshening up. Mark didn't like the way the bores looked either. He hauled the block, crank, rods, and pistons out to L&R Automotive Supply Company in Santa Fe Springs, California, for inspection, machine work, and a recheck of the dynamic balancing. Then, he brought it all back to his shop for assembly.

Now we're going to confess one of our more memorable screw-ups with the hope you don't make the same mistake. When we found this Mexican block 289 eight years ago, it was in a '66 Mustang that had been sitting for years. We tore the engine down to see what was inside and learned that machine shops and engine builders can make mistakes by assuring you the castings and forgings you've brought them will "probably" work fine. We accept their word because we don't want to search for another engine block or spend the money to get it.

In the world of engine building and car restoring, there are no free lunches. If something is on the ragged edge of "acceptable," it's a throwaway unless you want trouble later on. This logic applies to driveline components, body panels, seat upholstery, electrical system, suspension and braking systems, engines, and personal computers. If something is marginal, it's best to cut your losses and start over with better stuff. This is the main lesson of this budget Mexican block 289 build.

When a Simi Valley machine shop accepted this engine for a freshening up years ago, it didn't exhibit proper judgment when deciding to clean up and hone a 0.040-inch-overbore block. As engines power our Mustangs, the cylinder walls experience the extreme cycling of heat, cold, and friction. Pistons travel millions of times back and forth in the bores for thousands of miles at extremely high temperatures. This cycling process alters the molecular structure of the iron, which considerably changes a perfectly dimensioned cylinder bore. When we refer to a seasoned block, this is exactly what we mean. A seasoned block or cylinder head has been cycled thousands of times through heat and cold, helping the iron castings settle into relative normalcy. On the down side, cylinder bores tend to taper and warp with operation. When cylinder-bore taper exceeds 0.011-inch from top to bottom,

This is our just-machined '68 Mexican 302 block, easily identified by the casting number (C8AM) and the hump in each side of the front of the block. L&R Automotive Supply has re-honed the bores to 4.040-inches with room for piston growth. The line bore has been checked. Decks didn't need to be cut. It's ready for assembly.

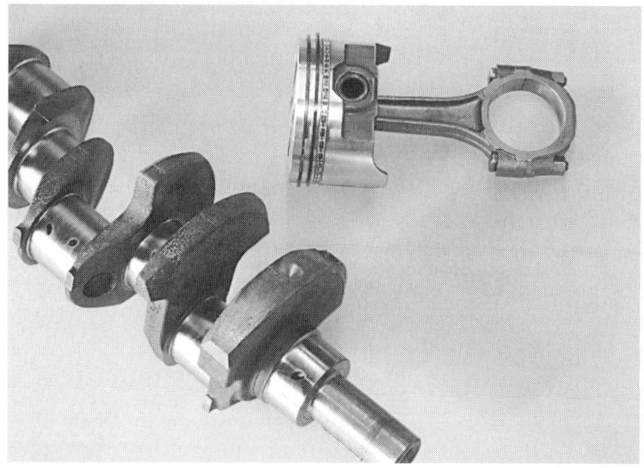

Our bottom-end rotating assembly consists of a 1m crank with a 2.87-inch stroke, reconditioned C30E rods, and Speed Pro-forged flat-top pistons from Summit Racing Equipment. Cast or hypereutectic pistons would have been enough for cruising. Forged pistons are overkill for a weekend driver. Ductile iron rings are ideal for this application.

it's time to cut a fresh bore to the next oversize, typically 0.020, 0.030, 0.040, and 0.060 inch. This creates a perfect cylinder devoid of irregularities.

From the factory, 289/302/351ci Fords had a 4.000-inch bore—exactly four inches. The 221ci and 260ci Fords ('62-'64) had smaller bores. To make the bores perfect again after thousands of miles of use, they can be bored to 4.020, 4.030, 4.040, or 4.060 inches in diameter. A good rule of thumb with small-block Fords is to bore 0.020- or 0.030-inch oversize on a standard 4.000-inch bore, then 4.040-inches on a 4.030-inch bore, but no more than 0.040-inch oversize. If racing, it's best to limit yourself to a 0.030-inch oversize in the interest of safety and security. Some like to punch the bores out to 4.060-inches, 0.060-inch oversize, but this is strongly discouraged. If you're determined to bore your block to 4.060-inches, have it sonic-checked first to ascertain cylinder-wall thickness.

Our Mexican block 289 had already been bored to 4.040-inches, but bore taper wasn't too bad, well under the mandated 0.011-inch taper. Our original machine shop concluded the cylinders could be honed and fitted with new 4.040-inch forged pistons, but the visible irregularities in the cylinder walls when the engine was assembled disturbed us. We could see low and high spots in the walls amid the fresh crosshatch pattern. The machine shop assured us the engine would perform well.

When Mark Jeffrey saw our cylinder bores, he had the same concerns about the irregularity's impact on reliability. Because we were building a budget 289 for the Six to Eight project, we both

Mark sets the 0.010-inch oversize Speed Pro bearings from Summit Racing Equipment. He sets them dry, with no lubricant between the bearing and block. He uses assembly lube on the bearing surfaces before setting the crank. Here's one example of where overkill can cost you. These are race bearings, also known as H-bearings, installed by a Simi Valley machine shop years ago. Hard-surfaced race bearings are designed for steel cranks and hard use, not cast cranks in street use. Whenever you set the bearings, check oil-galley passages for proper alignment.

Two-piece rear main seals do not have to leak. Mark uses good old-fashioned Permatex Form-A-Gasket between the seals and rear main grooves. He staggers the seal end gaps away from the main bearing-cap parting lines to minimize the risk of leakage. He also lays down Form-A-Gasket between the main cap and the block.

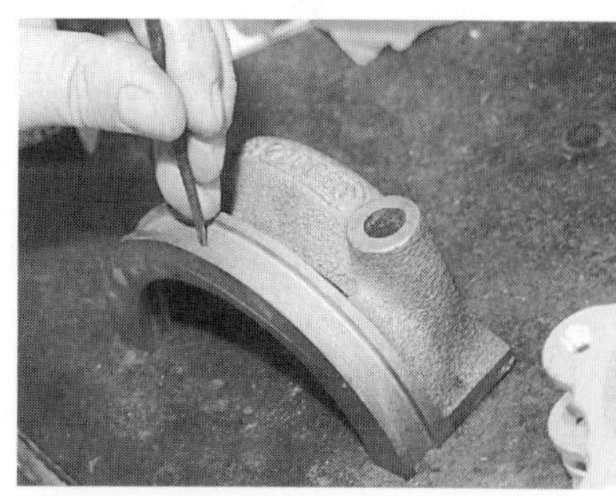

Older generation small-block Fords were equipped with rope rear main seals from the factory that used this pin to keep the rope in place. Rebuilders tend to forget the pin during the rebuild process, leaving it in place and damaging the new rubber seal. Our Simi Valley machine shop made this mistake and left the rope-retaining pin installed, which damaged the rear main seal. Mark caught this error and removed the pin.

Mark fills the pinhole with high-temp RTV silicone sealer. You can also use JB Weld. Make sure the surface is smooth before installing the seal.

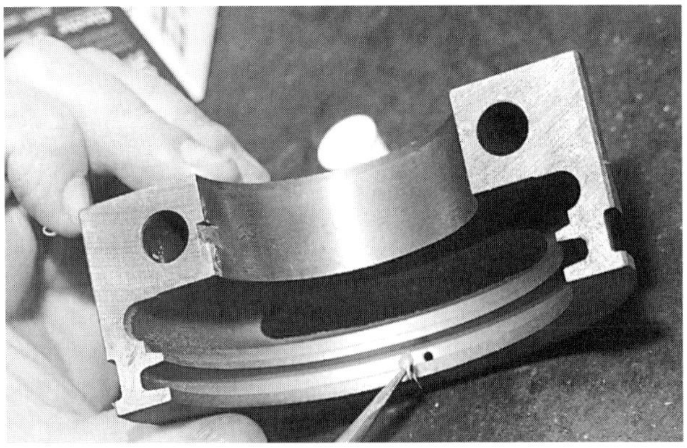

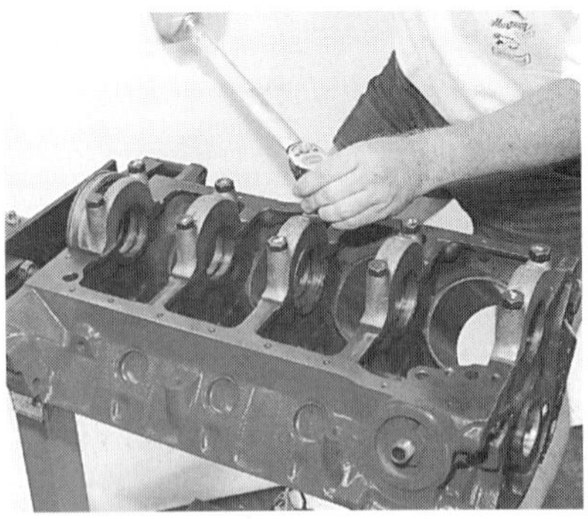

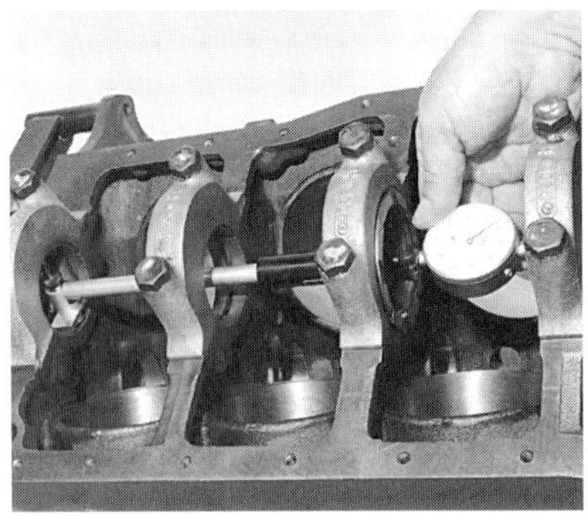

Before permanently installing the crankshaft, measure the main-bearing clearances. Mark installs the main-bearing caps and torques each of them to 60–70 lb-ft. This sets the main cap up with the bearing at installed sizing. He checks the main-bearing diameters with a dial-bore gauge. Then he checks the crank journals with a micrometer. Clearances should be 0.0008 to 0.0015 inch. Our 1m crank fits these main bearings perfectly.

concluded it was probably a passable block, and we were just worrying too much. When the block came back from L&R Automotive Supply, it had a fresh 4.040-inch hone and appeared ready for assembly. As Mark prepared the block, gut feelings continued to hound him, as well as us. He was disturbed by the cylinder wall irregularities he could see with the naked eye. We decided to go for broke and assemble the short-block anyway. When Mark assembled our short-block, he learned a couple of disappointing facts. One connecting rod was hung on the piston backwards, a formula for disaster. Connecting rod chamfers always go toward the crank journal radius. The machine shop made the error, and mistakes like this happen to even the most talented professionals. This is why you must always double and triple check everything before closing up.

REAR MAIN SEAL TRICK

It's so easy to make a mistake at the rear main seal and wind up with an oil leak. Again, you want the two-piece seal end gaps positioned away from the main-cap parting lines. Apply Permatex Form-A-Gasket between the main cap and block. Coat the seal tips with Form-A-Gasket. Carefully fit the #5 main cap in place and observe the seal installation. Make sure the seal tips groove into the cap smoothly.

Notice liberal doses of engine assembly lube on the crank-journal seal-mating surface. Also note the seal lip is correctly positioned toward the inside. Apply Form-A-Gasket to the main cap mating surfaces and seal the ends for added protection against leakage.

As Mark hand-cranked and checked the engine with each piston/rod installation, he couldn't help but notice resistance in one narrow segment of rotation. He removed the pistons one at a time and checked each bore with a dial bore gauge. Some of the cylinders were not true despite L&R's best efforts to get them straight, and regardless of the 0.011-inch taper rule, piston binding is piston binding. Our block was unacceptable for use.

We did have one option: to bore the block to

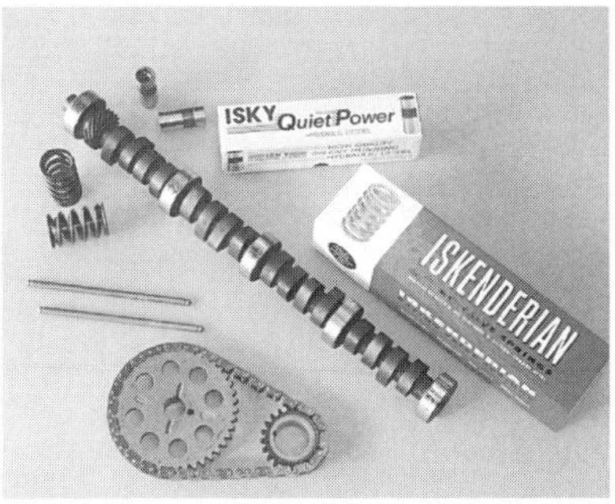

Originally, we considered a hydraulic roller camshaft for our Mexican block 289. Because a roller cam and all of the related components can run upwards of $500, we decided to build a conventional small-block with a flat-tappet camshaft. This is an Isky Super Cam flat-tappet hydraulic camshaft with 0.278/0.278-inch cam lift, 0.445/0.445 valve lift, and 208/208 duration on 108 lobe centers. This means a smooth idle with good off-idle throttle response.

Mark coats the cam journals with assembly lube. The lobes get molybdenum grease for proper break-in.

4.060-inches and order new 4.060-inch pistons. Because we wanted a fiercely reliable 289-inch small-block, we elected to pass on our Mexican block and order a replacement. Derek Ranney of L&R Automotive Supply set us up with a fresh vintage 302 block that was machined, bored, and honed to 4.040-inches.

We could have taken a chance with the C8AM 302 Mexican block with its distorted cylinder walls, and may have had a fine engine that could have gone 100,000 miles, or we could have wound up with an engine that had to come out of the car at 5,000 miles because the rings didn't seat properly. It's cheaper to replace the block before it goes in the car than doing it all again later. This is where a decision has to be made on the side of caution. Our greatest failing was not recognizing the distorted cylinder walls during the machining process before assembly began.

Our formula here is designed for a mild, reliable, street small-block for those who want to get into V-8 power without breaking the bank. You can save money by taking a common-sense approach. Rule

Mark gives the crankshaft ends a whack with a mallet to help center the crank on the bearings.

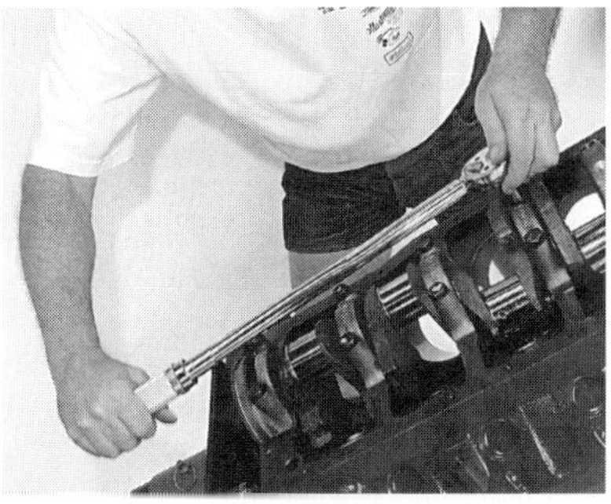

The main caps are torqued from center outward to 60–70 lb-ft of in three increments. First torque is 20 lb-ft, then check rotation. Next, 40 lb-ft for freedom of rotation. Finally, 60–70 lb-ft and check rotation. Our main saddles are happy and the crank turns freely. Double check all main cap bolts.

Next, Mark checks crankshaft endplay with a dial indicator. Endplay is how much the crank moves back and forth in the main saddles and should be 0.004–0.008 inch.

The cylinder walls are lubed with engine assembly lube, but you can also use SAE 30-weight engine oil.

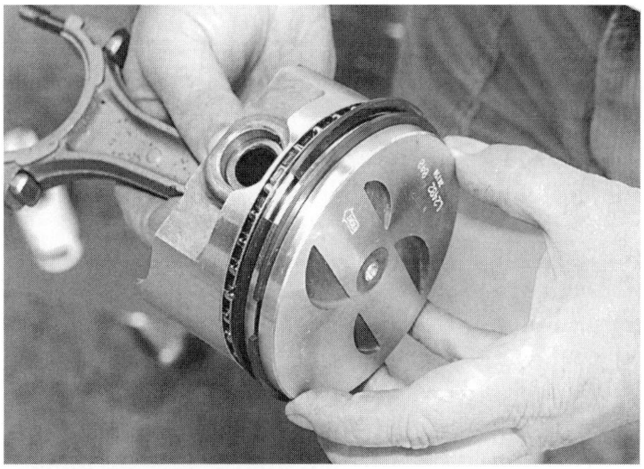

Check piston rings for proper gap locations. Ring end gaps should be positioned at 9, 12, 3, and 6 o'clock.

one is don't plan more for your engine than it needs. Weekend cruisers and commuter cars do not need forged pistons, big valves, roller cams, big carburetors, headers, or single-plane manifolds.

Mild street engines need what it takes to be reliable, which is mostly just good engine-building technique and extra pains to ensure the engine is built properly. Most of those pains won't cost you a dime. If you want to improve on your budget small-block and have extra cash, here's where you should spend your money:

• Roller hydraulic cam with under 0.500-inch lift for reduced friction and better performance
• Base-level roller rocker arms, like the Magnum from Comp Cams, to reduce friction and improve performance
• One-piece pushrods for reliability
• Autolite 4100 four-barrel carburetor from Pony Carburetors
• New generation dual-plane manifold, like the Edelbrock Performer RPM or Weiand Stealth
• Autolite or Motorcraft distributor with PerTronix Ignitor II electronic conversion
• Heavy-duty ignition wires
• Cast-aluminum valve covers with thick steel-reinforced gaskets
• Closed crankcase ventilation system for cleaner air
• High-flow water pump
• Ceramic-coated shorty or long-tube headers.

Hit the Swap Meets and eBay

We saved money on our budget rebuild by thinking "used" on the things that didn't affect engine integrity. Intake manifolds, carburetors, valve covers, distributors, and other items can be found for a song in the classifieds, at swap meets, and on eBay. For our Budget 289 effort, we unearthed an old Ford Motorsport dual-plane manifold and a Holley 600cfm carburetor for next to nothing from someone who decided to go with something else years ago and never installed the stuff. Ditto for the Scott Drake Cobra valve covers that you can't get new anymore. There's always something out there someone didn't want. Take advantage and save.

Another great way to save money is buying old speed equipment from the '60s. There's nothing like a set of old Cal Custom finned valve covers or an Edelbrock F4B high-rise manifold to get your motor running. We're beginning to see these pieces on period restomods and retro-rides in our travels. We thought most of this stuff was gone, but pieces surface periodically at the shows to help us remember a golden era.

Sources

Comp Cams
www.compcams.com

Ed Iskenderian Racing Cams
www.iskycams.com

Holley Performance Products
www.holley.com

L&R Automotive Supply Co.
www.lnrengine.com

Summit Racing Equipment
www.summitracing.com

TransAm Racing
www.transamracing.net

We found this old Ford Motorsport dual-plane manifold for a bargain on eBay. Items that you may think had long ago disappeared may very well be available online.

Budget Building Tips

Mark uses a custom billet piston-ring compressor to install each piston and rod assembly. He has already checked the ring end gaps.

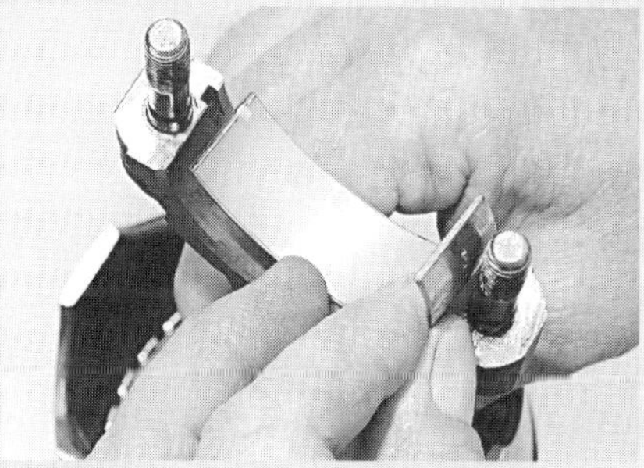

There are a lot of misconceptions when it comes to bearing installation. One basic rule is to install main and rod bearings dry—no lubrication between the bearing and block. A dry surface enables the bearing to remain anchored to the block. Keep in mind there's a tiny amount of crush going on between the bearing and block which keeps the bearing secure. Always use engine assembly lube or engine oil between the bearings and crank.

When building a small-block Ford or any other Ford with press-in oil-galley plugs, replace the press-ins with screw-in oil-galley plugs to keep the oiling system safe. There are three passages in front.

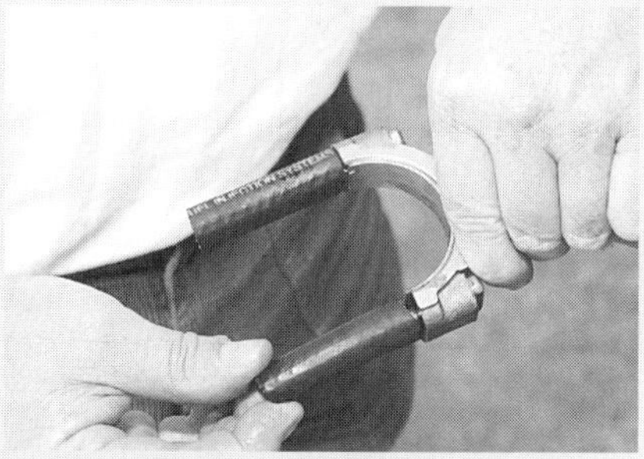

Because rod bolts are harder than cylinder walls and crankshaft journals, it's easy to nick a cylinder or chip a crank with the rod bolts during installation. Rod bolt protectors are available at machine shops and performance parts stores. If on a tight budget, use fuel line or transmission cooler line hose—as we did—for protection.

Fel-Pro gaskets are the gasket of choice for any engines we build. This is the performance set with Print-O-Seal gaskets for added security from Summit Racing Equipment.

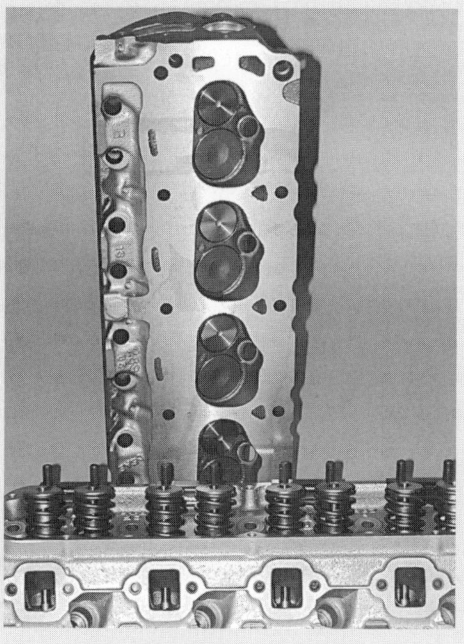

Clearance is important where the piston meets the cylinder head. Our 289 has approximately 0.020-inch of deck height, which is the distance from the top of the piston to the top of the block deck. We will check this with a dial indicator. When we install the Fel-Pro head gasket, this clearance will increase by at least 0.010-inch or more. This affects compression.

Ford buffs like Mexican blocks for the high nickel content. But the real gold in a Mexican 289/302 block isn't the block at all, but the wider main bearing caps like those used on '63–'67 289 high-performance blocks. A 289 Hi-Po block isn't all that unique. It's the same casting as a 289-2V/4V block. The wider main bearing caps make it special, and the same can be said for a Mexican block. If you have a bad Mexican block, don't throw it all away. Save the main bearing caps for another block.

L&R Automotive Supply provided us with these C5OE cylinder-head castings. These '65 289 heads have 53-57cc chambers and very little quench area, which minimizes detonation, perfect for the Speed Pro flat-top pistons we're using. We're going with stock valve sizing at 1.78/1.45-inches. For a bit more power, opt for larger 1.94/1.60-inch Chevrolet valves. L&R also machined these heads for screw-in rocker-arm studs and pushrod guide plates. Stock valve springs with dampers were installed to go with our mild Isky cam.

Chapter 2

A Basic Small-Block Rebuild

Text and Photos by Jim Smart

This 289 is proof that you don't have to decide between your kids' college education or a rebuild for your Mustang.

Over the years, we've shown our readers how to build lots of engines. In many cases, we've opted for forged pistons, roller cams, aftermarket induction systems, striking dress-up goodies, and more. But this stuff costs a lot of money. So how do you get into a fresh, reliable engine without having to mortgage the farm? We're going to show you with help from Jim Grubbs Motorsports and Federal-Mogul. We'll also show you options that will make your basic engine build even more productive, depending on your budget.

In addition, there's the option of purchasing a low-buck crate engine from a national discount auto-parts chain. But the problem is, you don't always know what you're getting. Mass-production rebuilders tend to have a "one size fits all" mentality. Your '65 Mustang may have come with a 289 V-8. However, the chances of getting a 289 from any of the national chains are slim. You're going to get a 302, with a smattering of components from four decades of small-block Ford production. The block might be an '81 302 casting, while the heads might be mismatched, one from 1978 and another from 1983. The problem is, combustion chamber sizes can be different, making compression ratios different on each cylinder bank. You probably won't notice it, but it does affect engine longevity and performance.

Tips for a Successful Engine Rebuild

When we rebuild an engine, we go inside and make it like new. But what does that mean? An engine rebuild consists of the teardown, inspection of the permanent parts we're going to machine and reuse, and assembly of the engine using new parts as necessary. Here's a good blueprint to follow for a successful engine build:

• Disassemble and inspect all components, and thoroughly clean the ones you're going to reuse. All block and head plugs should be removed and the oil passages should be wire-brushed and checked for blockages. Water jackets and passages should be inspected and cleaned. Chunks of rust should be removed.

• Check the block and heads for cracking and inspect the crankshaft for irregularities and cracks. Inspect the connecting rods for irregularities. Bore and hone the cylinders to the next oversize (but never over .040 inch).

• Check the line bore, and hone or bore as necessary, although boring is rarely necessary.

• Check the block deck surfaces and mill only as necessary. Measure crankshaft journals and machine to the next undersize only if necessary. Otherwise, polish the journals for good oil control and bearing mating.

• Recondition the connecting rods and install new bolts.

• Install new cam bearings.

• Install screw-in oil gallery plugs anytime there are press-in types.

• If equipped with a manual transmission, resurface the flywheel.

• Thoroughly wash the block with soap and water. Wire-brush the oil gallery passages to ensure removal of metal shavings, rust, and other debris. Do the same with the crankshaft oil passages.

• For the heads, go with a three-angle valve job with new valves and guides. If you're going to drive your Mustang daily, install hardened exhaust-valve seats. If you're building a show car or an occasional driver, you can pass on the hardened valve seats.

• Invest in a good, high-volume oil pump. Have your machine shop disassemble the pump and do an inspection, and make sure they check the pump rotor side clearances and machine as necessary. Have them check the pressure relief valve for proper operation because not all oil pumps leave the factory in correct operating condition. Invest in a new oil pickup as well.

• Be sure to dynamic-balance the rotating assembly for smooth operation and longevity.

• Check ring end gaps, even if you're using pre-gapped rings. Assume nothing.

• Opt for the best gaskets money can buy, such as Fel-Pro Print-O-Seal. And use the best silicone gasket sealer while you're at it.

• Go with a roller hydraulic camshaft if you can afford it. Install a dual-roller timing set for less internal friction.

• Opt for a high-flow water pump for best cooling efficiency.

• Paint oil-flow areas with a high-heat paint to improve oil-return flow.

• Use brass freeze plugs because steel plugs will ultimately rust out.

• Install a new timing cover from California Pony Cars. They are the best reproductions out there.

• Go with a new harmonic balancer from Mustangs Etc. They are identical to original equipment.

Getting Started

When disassembling an engine, be mindful of its problems and ready to move in the right direction when they're discovered. Our 289 was removed from a '65 Mustang convertible. The original owner passed on and left the car to his nephew, Jeff Fischbach. Jeff was just getting off the freeway when the engine developed a knock. He knew his uncle had the engine rebuilt a few years earlier, but he didn't know the specifics.

Disassembling an engine requires detective work. Our job was to determine what happened to make the engine inoperative. We knew by the sharp metallic knock that Jeff's 289 was ill. We began our investigation while the engine was still in the car, starting it to try to determine the source of the knock. It sounded like piston-skirt noise or a rod bearing. To determine which cylinder, we disconnected the spark-plug wires one at a time. At No. 4 on the right bank, the knock stopped, eliminating all doubt. At Jim Grubbs Motorsports, disassembly confirmed our fears: a broken piston in a 0.040-inch oversize block.

A Budget Building Plan—When it comes to doing a successful budget-engine rebuild, we look to the same folks we always have: Federal-Mogul, the name behind Speed-Pro, Fel-Pro gaskets, Sealed Power, and more. Barry Rabotnick at Federal-Mogul formulated a plan for reliable, affordable power. Here's what he came up with for our 289.

Because the 289 will be used mainly for weekend pleasure cruising, Barry believes we don't need much more than a stock engine, with nice refinements that will improve reliability. Stock applications do not need forged pistons and H-beam connecting rods. They don't need a steel crank. And they really don't need a roller camshaft. Roller cams are a nice upgrade if you can afford them because they reduce internal friction and improve performance at the same time. For Jeff's weekend driver, we're going with a flat-tappet hydraulic camshaft. We'll opt for a high-volume oil pump with a heavy-duty oil-pump shaft to keep moving parts flooded with oil for longer engine life.

Sources

California Pony Cars
www.calponycars.com

Federal-Mogul
www.federal-mogul.com

Jim Grubbs Motorsports
661/257-0101

Mustangs Etc.
www.mustangsetc.com

Virginia Classic Mustang
www.vamustang.com

1. Our subject engine is a '65 289 small-block with at least one rebuild in its history. Casting numbers and date codes confirms its status as the car's original engine. Ryan Peart of JGM removes the flywheel and mounts the engine on a stand for teardown.

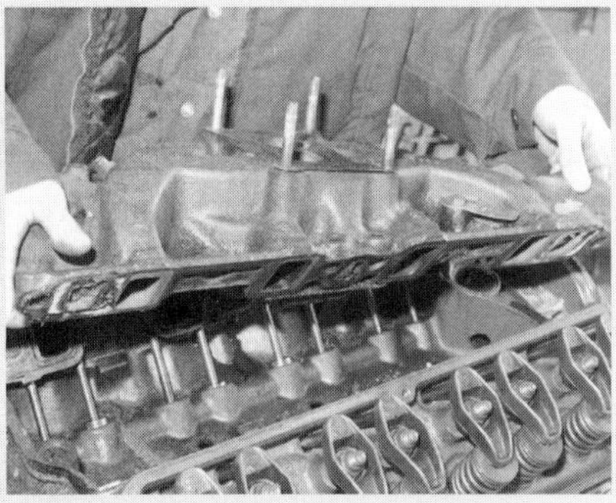

2. Engine disassembly begins topside with valve covers and intake manifold.

3. Valvetrain removal is next. Closely inspect rocker arms and pushrods for unusual wear patterns. Abnormal patterns can indicate casting and machining flaws.

4. Both cylinder heads are unbolted and removed.

5. At first glance, we have a blown head gasket at the No. 5 cylinder. Had we torn down the engine immediately upon hearing the knock, we might not have known. The car sat for nearly a year, while coolant leaked into the No. 5 bore. The aluminum piston and iron cylinder wall didn't have a happy marriage, with the coolant providing the electrolyte to generate corrosion.

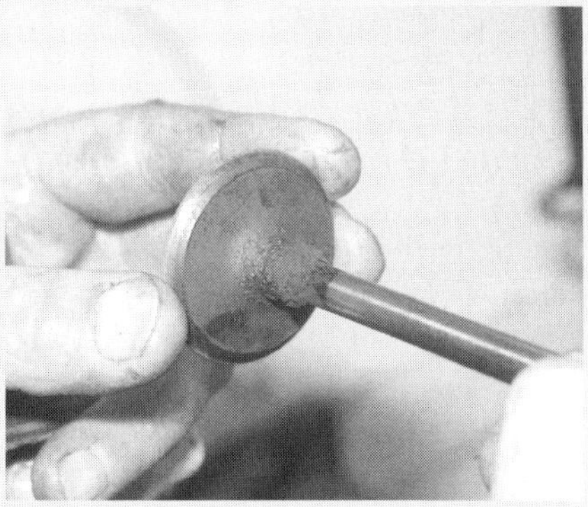

6. When the cylinder heads were torn down, we discovered a smattering of different valves. One head was fitted with valves designed for rail-style rocker arms (from mid-'66 through '77) with a longer stem above the keeper grooves for use with rail-style rocker arms. Prior to May 8, 1966, these engines used shorter valve stems, with pushrod guides cast into the cylinder head. Drew Whaley of JGM tells us these cylinder heads are victims of a mass-production rebuilder. They need a complete rework, installing all new valves, guides, and seats.

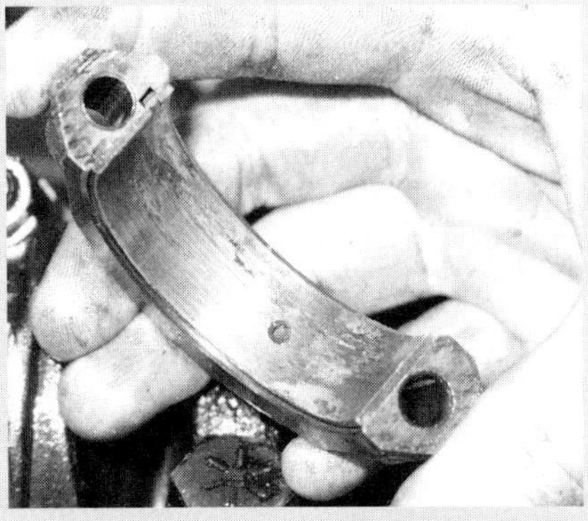

7. Piston and connecting-rod removal is next. Each rod cap is removed and its bearing inspected. Look at this rod bearing, which is worn down to the copper. Dirty oil, poor assembly procedures, and a dirty environment are reasons bearings fail like this. Overheating may have caused oil to break down where it was needed most.

8. This is our No. 4 cylinder knock—a cracked piston. Reason? A right-bank cylinder-head gasket was installed backwards, which caused the right bank to run hot. Piston destruction was the outcome on a California freeway.

9. Main bearing caps are removed next to free the crankshaft.

10. The same reasons rod bearings were scored led to the same kind of damage at the main bearings and journals. The crankshaft journals are miked and checked for wear. This crank will have to be machined 0.020-inch oversize.

11. Lifter and camshaft removal is next. It's a good idea to inspect camshaft and lifter wear patterns at this time. The same problems that damaged this camshaft could easily trouble the new parts.

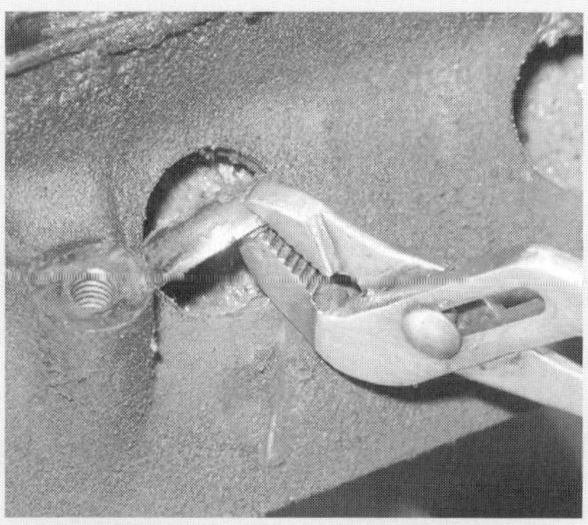

12. All block, head, and manifold plugs should be removed for cleaning and machining.

13. The block is bored 0.027-inch oversize. This means each cylinder bore is bored out to the next oversize, which gets the cylinder true and straight. Later, honing will cover the other 0.003 inch, arriving at 0.030-inch oversize.

14. Block deck surfaces are milled 0.010 inch in three passes to achieve perfect surfaces.

15. Jim Grubbs sets up the block for honing. Before honing the cylinder bores, he hones the lifter bores, which will improve oil control.

16. All bolt holes in the block get a thread chasing for smooth fastening and assembly.

17. With a torque plate installed to simulate cylinder-head installation, the cylinders are honed 0.003 inch for a total bore diameter of 4.030 inches. The torque plate alters the block casting to the point it will be when the cylinder heads are installed.

18. Finally, the main saddles are honed and trued. This is called line-honing. The main caps are removed and milled to square them up. Then they're installed, torqued to specifications, and honed to achieve a perfect line bore.

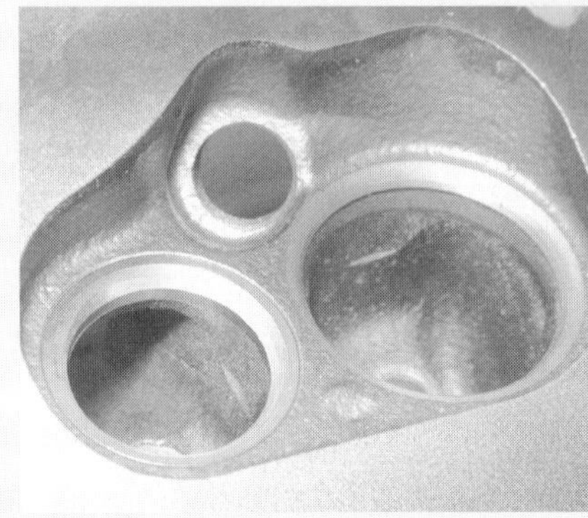

19. Because Jeff's cylinder heads had suffered from the unconventional treatment of a mass-production rebuild with a smattering of incorrect parts, they needed a lot of work. All valve guides received new bronze bushings for improved sealing and oil control. One guide had to be replaced. Hardened exhaust valve seats were installed for reliable operation.

20. Each cylinder head is milled the minimum amount to true up the mating surfaces. Because the right cylinder head suffered from extremely hot operation (head gasket installed backwards), it was warped significantly.

21. We asked JGM cylinder-head expert Drew Whaley about screw-in rocker-arm studs. He showed us something pretty clever: screw-in studs that don't mandate pedestal machining. He removed the factory press-in studs, tapped in the threads, and installed screw-in studs. Ask your engine builder about this and save money.

22. The new Sealed Power valves are ground to match the new seats, then installed employing proper spring pressures and installed height. JGM uses special valve seats that will virtually never wear out. Throw away the umbrella seals included in your gasket set because they do not last.

23. We had to throw our 0.040-inch-over 289 block away. Fortunately, we found a standard bore C5OE block at Mustangs Etc. Forged-aluminum TRW pistons from the '60s were in this short-block. They had to be pressed off the perfect C3OE connecting rods for this rebuild.

24. Because connecting rods stretch over time and use, any engine rebuild worth its salt includes reconditioning the connecting rods. The large ends and caps are machined, new ARP bolts installed, and the large ends honed for a perfect fit.

25. New Federal-Mogul cast-aluminum pistons are pressed onto the connecting rods after being dynamically balanced.

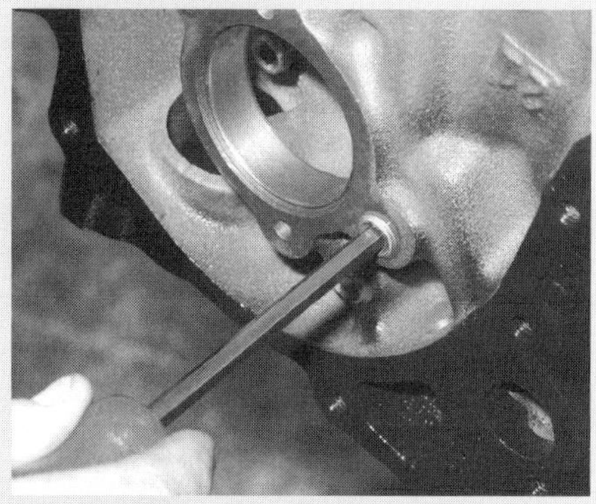

26. New plugs are installed in the block, including screw-in oil gallery plugs in all locations and brass freeze plugs that resist corrosion. Sealer must be used on all plugs.

27. Because we're running a Speed-Pro flat-tappet hydraulic camshaft, special care is used during installation. Cam lobes should be coated with moly coat, which will lubricate the lobes and help them wear in properly during break-in. Journals should be coated with assembly lube for protection during initial fire-up.

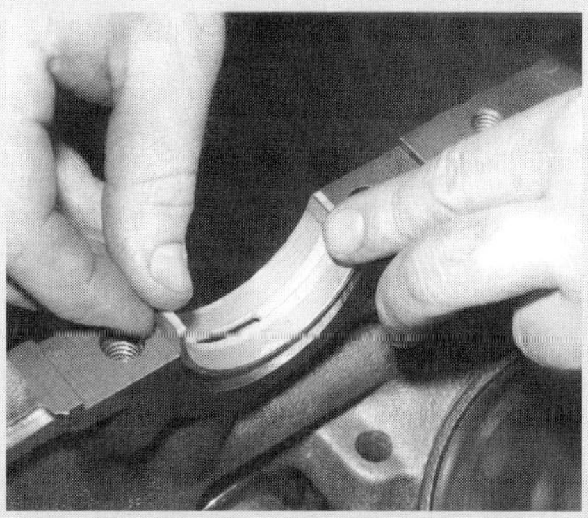

28. Sealed Power main bearings from Federal-Mogul are installed dry in the block and main caps. "Dry" means between the bearings and block only. Bearings should be lubricated with assembly lube before crankshaft installation.

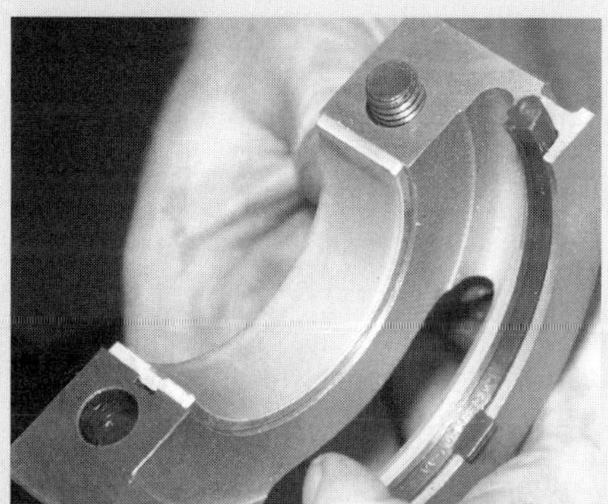

29. Because rear-main-seal leakage can be a chronic problem with older Ford V-8s, extra care must be taken installing new seals. Andy Hairfield of JGM has a method that minimizes the risk of leakage. He staggers the rear-main-seal end gaps as shown inside the No. 5 main cap. This keeps the seal gaps away from the main cap parting lines.

30. Always check to see if there's a rope-seal retainer pin in the No. 5 main bearing cap. If this pin remains, remove it before installing the new rubber seal.

31. Lubricate the main bearings with assembly lube and lay in the crankshaft. Ascertain absolute cleanliness of the bearings first. Install and torque the main caps.

32. Although we do have pregapped piston rings, it's always good to check the ring end gaps. The compression-ring end gaps should be 0.010 to 0.020 inch.

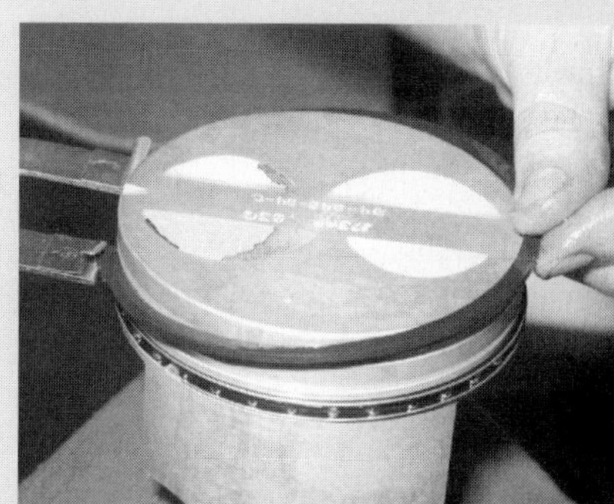

33. Piston rings are rolled onto the pistons as shown. The oil-ring expander is laid in first, followed by each of the two oil rings. Oil-ring gaps should be 180 degrees opposite. Compression rings roll in the same way, with ring end gaps 180 degrees opposite, 45 degrees from the oil-ring end gaps.

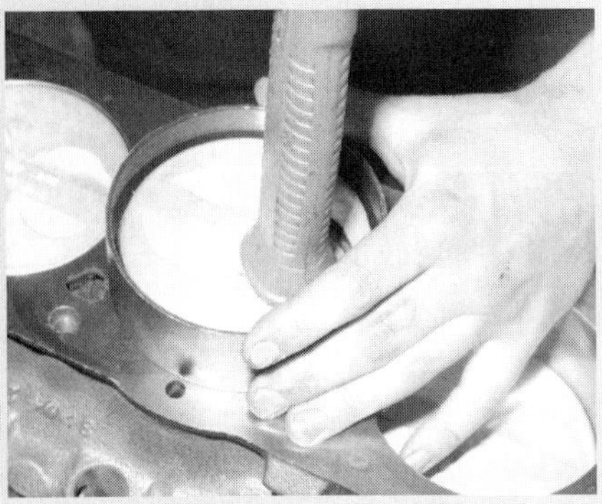

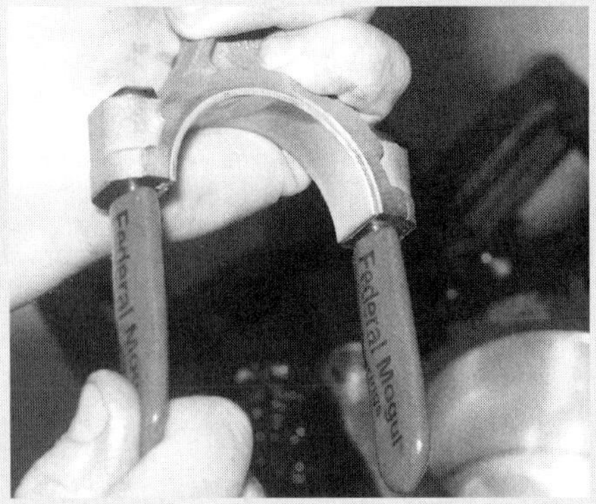

34. Cylinder walls get a wipe-down, then they are coated with SAE 30 weight engine oil. Rod bearings are installed dry between the bearing and rod. Then, use plenty of assembly lube for good protection during initial fire-up.

35. Connecting-rod bolts are torqued to 19 to 24 lb-ft. Check side clearances while you're at it, making sure they are 0.010 to 0.020 inch. Maximum allowable is 0.023-inch.

36. Andy performs a critical check mandatory with every engine build: degreeing the camshaft. This checks two important issues, crank and camshaft indexing, ensuring that valve and piston timing events are correct. Although rare, we learned the camshaft indexing wasn't per the camshaft specifications. Camshaft timing was a few degrees retarded, which would have adversely affected performance. Andy had to exchange the standard timing set for a dual-roller set where the timing could be advanced the appropriate number of degrees.

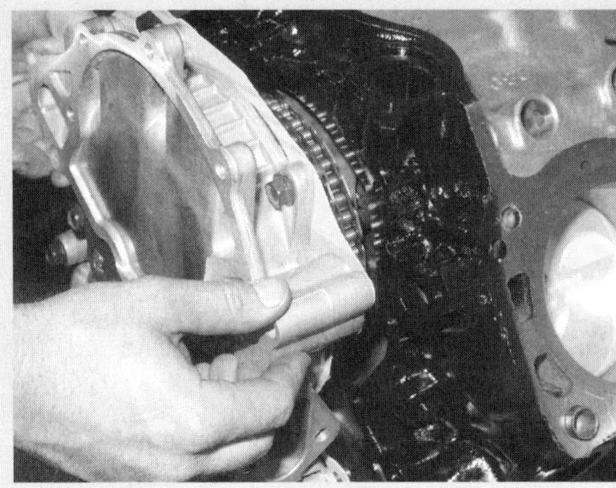

37. We're installing a new timing cover from California Pony Cars, available from your favorite Mustang parts vendor. Andy has applied silicone sealer between the gasket and timing cover, paying special attention to coolant passages.

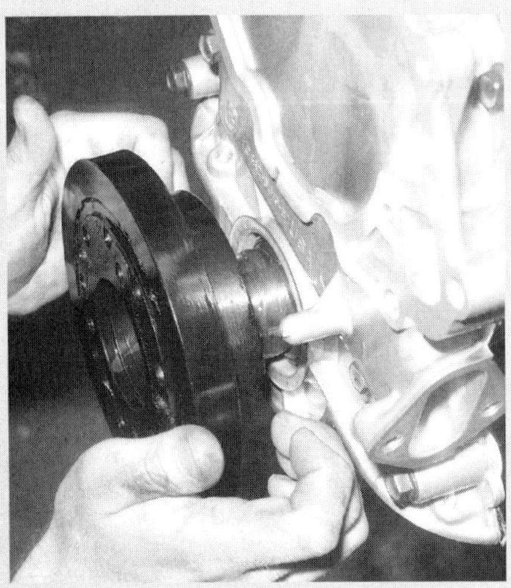

38. Mustangs Etc. provided us with a new period-correct harmonic balancer that's easy to read with a timing light. Andy has coated the seal-contact area with plenty of assembly lube.

39. Cylinder-head installation is next. Remember, cylinder-head gaskets are installed only one correct way: with the coolant passages at the rear of the block. "FRONT" means front on both sides. Lubricate the head bolts and torque them to 65–75 lb-ft from center outward.

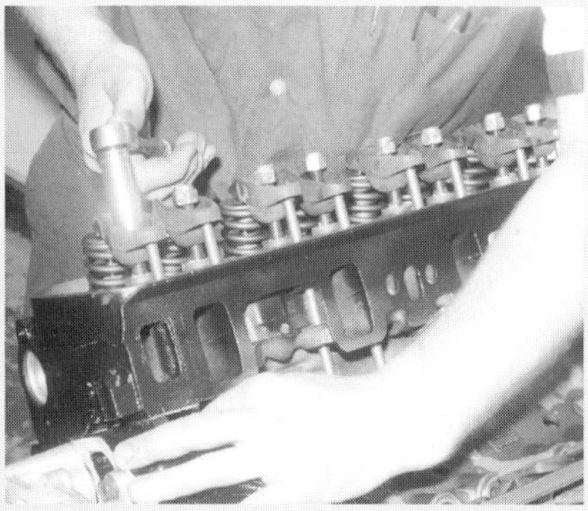

40. Lifters are lubricated and installed, followed by pushrods and rocker arms. Andy adjusts the rocker arms as follows: Crank the engine slowly following the 289's firing order of 1-5-4-2-6-3-7-8. At each compression/ignition stroke, adjust the rocker-arm lock nuts to the point the pushrod cannot be turned. Then tighten the locknut one-half turn.

41. Intake-manifold installation is next. Use a modest amount of sealer around each of the ports and passages as shown. Throw away the cork end gaskets supplied with the gasket set and run a bead of silicone gasket sealer along the end rails. Install the intake manifold and torque the bolts in a criss-cross pattern.

42. Roll over the engine and install the oil-pump shaft and pump. Install the shaft with the stamped-steel lock ring at the pump. Use the gasket between the pump and block.

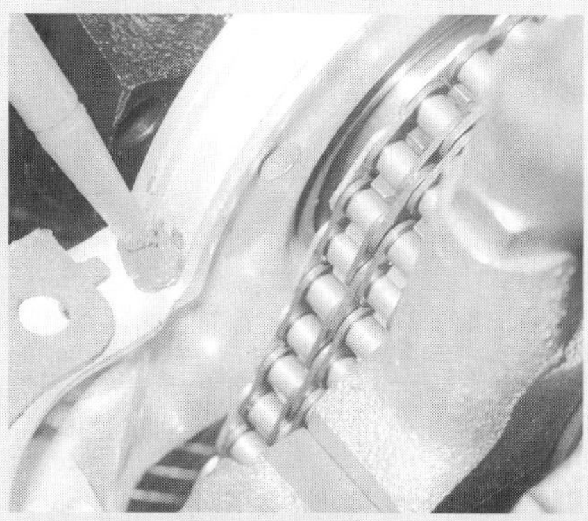

43. Proper oil-pan installation is crucial to leak-free performance. Run a light bead of silicone sealer between the block pan rails and gaskets. Lubricate and seat the end gaskets. They must be square with the timing cover and No. 5 main bearing cap. Apply sealer where the rail gaskets meet the end seals. Install the new oil pan from Virginia Classic Mustang.

44. We've decided to dress up this 289 with a period-look Cobra Dress-Up Kit from Virginia Classic Mustang. The hollow-letter, cast-aluminum Cobra valve covers look terrific.

Chapter 3

How to Install an Aluminum Radiator

Text by Jim Smart
Photos by Jim Smart and Laurie Slawson

Improve your cooling capacity with a Griffin aluminum radiator.

Classic Mustangs have always struggled with cooling issues, especially big-block cars. This is mainly due to small engine compartments and limited flow through the grille, coupled with under-capacity and vertical-flow radiators. It wasn't until the late '70s that Ford began to address Mustang cooling-system woes with good cross-flow radiators and greater capacity. Those with '79–'04 Mustangs have rarely been concerned with overheating because Ford did a pretty good job in those years.

The automotive aftermarket has taken care of classic Mustang enthusiasts with better radiators, high-flow water pumps, improved engine-driven and electric cooling fans, and water-wetter chemical treatments to virtually eliminate overheating problems. Griffin is undoubtedly front and center with high-quality aluminum radiators available from Mustangs Plus.

Choosing the right radiator depends on how you intend to drive the car and how much power you have under the hood. High-horsepower strokers, for example, need heavy-duty cooling capacity to carry away all that heat. Even stockers need four-row core desert-cooler radiators, which offer ample cooling capacity for just about any Mustang application.

We installed a Griffin aluminum radiator in a '69 Mustang GT convertible with a 390 High Performance V-8. The car was in Tucson, Arizona, a place legendary for its searing heat and summertime monsoons that bring unbearable humidity. It's not uncommon for the mercury to reach 120 degrees F with a blazing sun unmatched anywhere this side of the Middle East.

This is the 24-inch Griffin aluminum radiator for '67–'70 Mustangs with FE big-block power. The radiator provides unequalled capacity, pulling more heat out of your engine than a stock 2–3–row copper/brass radiator. You still need a fan shroud for proper cooling with the fan positioned so half of it shows.

Overheating Issues

A new radiator will only solve part of the problem if you have other issues in the cooling system. Belt slippage is a sneaky overheating gremlin because it isn't always apparent. Make sure your belts are solid and tight.

Another overheating issue belongs to a lower radiator hose that collapses on the open highway. As the engine demands more coolant flow at higher revs, the lower hose tends to collapse from the heavy rush of coolant flow into the block. Make sure you have an anti-collapse coil spring in the lower hose.

Removing the thermostat to solve an overheating problem is not only foolish, it's stupid. Do not remove the thermostat. Run a 180-degree thermostat and check it for proper operation. Inspect coolant flow through the radiator when the engine reaches operating temperature. Use a 7-lb cap on your classic Mustang, not a 12–14-lb cap.

Coolant recovery systems are a great idea because they allow room for coolant expansion when hot, then release coolant back to the engine when cool. If originality is important to you, hide the reservoir in a location you can reach.

Cooling Additives

There are plenty of additives advertised to eliminate cooling-system problems. The only additives you should use are water wetters, which improve the cooling capacity of your coolant mix. Never use stop-leak additives. They may stop leaks, but they can also clog your cooling system and create hot spots in the engine.

Thoroughly flush and service your cooling system every two years regardless of how often you drive the car. Antifreeze has a service life of about two years at the most. After that, the corrosion inhibitors start to degrade. What's more, heat and combustion by-products get into the coolant and create acids and other contaminants. Flush and fill every two years.

Evans non-aqueous coolant is expensive, but it never has to be replaced. Once installed, you never have to sweat cooling-system service again except to replace hoses every 4–5 years. The installation begins on the next page.

Source
Mustang Plus
www.mustangplus.com

Griffin delivers all the things you expect and need from a Mustang radiator, including an automatic-transmission cooler and drain petcock. All you have to do is transfer automatic-transmission–cooler line fittings.

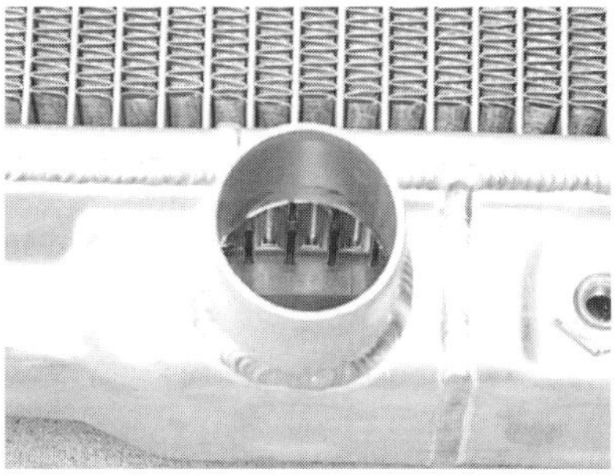

Aluminum radiators have a different tube configuration than copper/brass radiators, and fewer rows and wider tubes are the norm. The result is greater heat transfer. This is a two-row core, but it yields the cooling capacity of a standard four-row core.

Belt slippage is a common cause of overheating, so if you're still experiencing problems, check they are not slipping, in good condition, and adjusted properly.

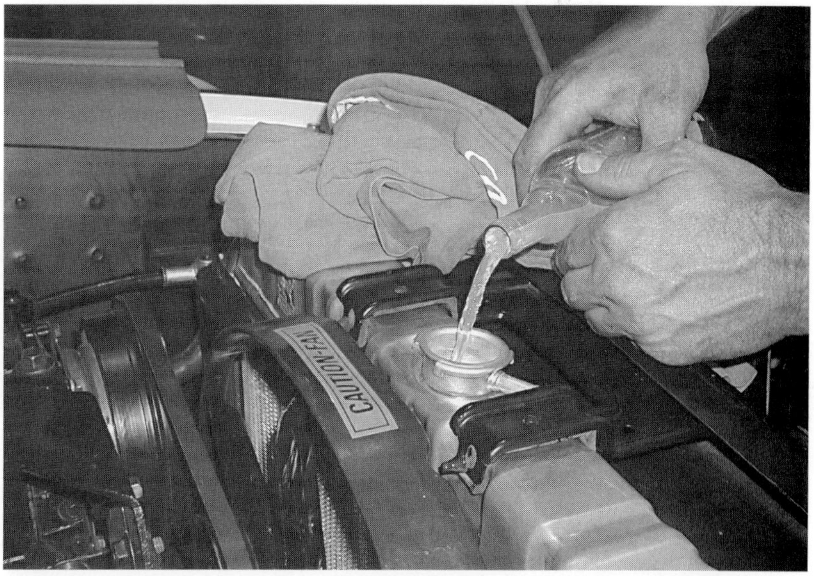

The cooling system should be flushed every two years, even if it is only occasionally driven. After about two years, the corrosion inhibitors in the antifreeze begin to break down. Evans makes a synthetic coolant that never has to be replaced, however, so you may feel it is worth the added expense.

1. Drain the radiator at its lowest point via the petcock or by removing the lower hose. Always recycle antifreeze, and never allow it to leak on the floor. Because animals love the sweet taste, they will drink it. However, antifreeze can kill your pet.

2. Automatic-transmission–cooler lines are next. Disconnect the lines as shown, and protect the fluid by capping the lines.

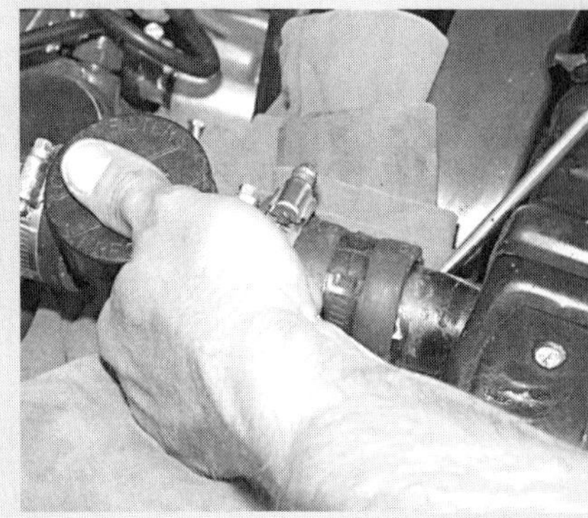

3. Disconnect the upper hose. This is a good time to consider new hoses for your cooling system. Remember, you need an anti-collapse coil spring in the lower radiator hose. Without it, the lower hose can collapse at higher revs, closing off coolant supply from the radiator, causing overheating.

4. Remove the fan shroud next, using a 3/8-inch socket in most applications. This one, however, used bolts with 7/16-inch heads.

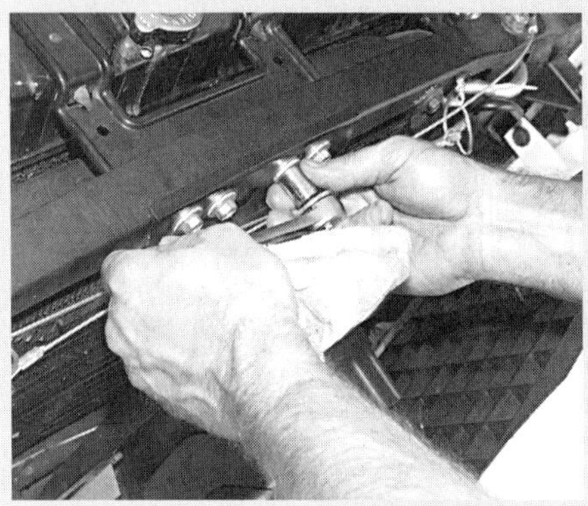

5. Next, remove the radiator's upper retaining bracket using a 1/2-inch socket.

6. The radiator is free once the top bracket is removed.

7. Before you can remove the radiator, disconnect and remove the fan. In this application, it's a clutch fan. Use a 1/2-inch box-end wrench to unbolt the fan from the water-pump hub.

8. Lift the fan and shroud out as an assembly to clear both pump and radiator.

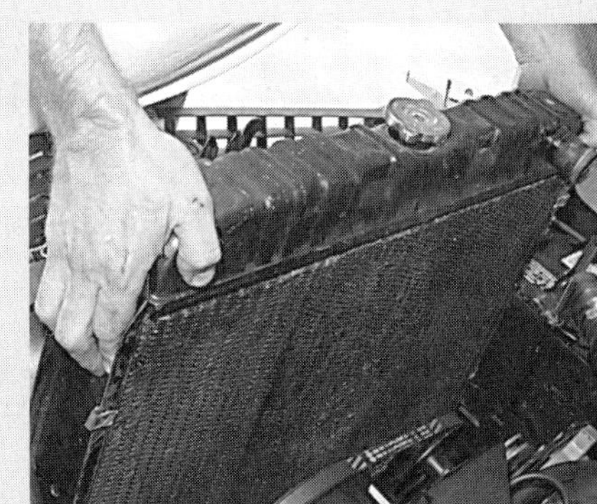

9. Lift the radiator out as shown, taking care not to damage the paint.

10. This is a good time to fit-check the Griffin radiator. Make sure it will clear the fan, shroud, and anything else that might get in the way. Aftermarket electric fans in particular can sometimes create fit problems.

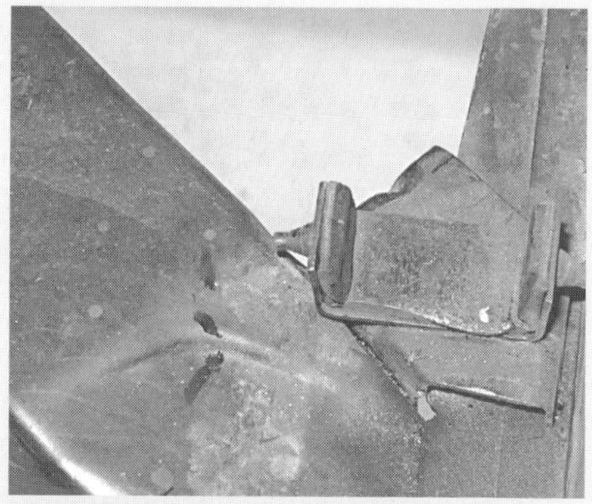

11. While you're at it, check the rubber radiator mounts for deterioration. Now is a good time to replace them, both top and bottom. Rubber mounts are available from Mustangs Plus.

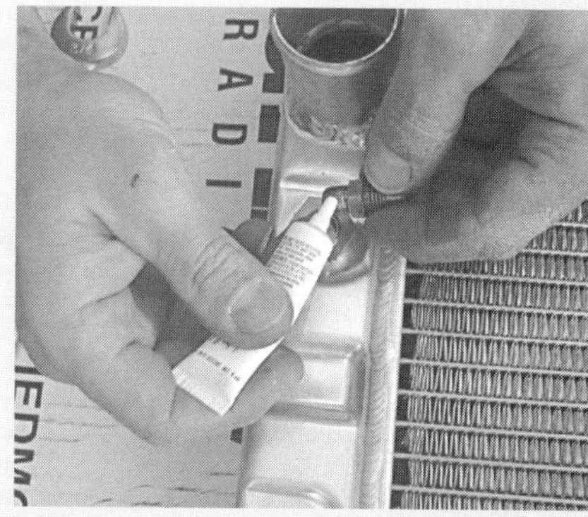

12. Automatic-transmission–cooler fittings must be transferred to the Griffin radiator. Use a high-temp RTV silicone sealer on the threads and snug them.

13. Slip the Griffin radiator into place, making sure it's seated in the rubber mounts.

14. Next, slide the fan and shroud into place. Make sure to torque the fan-clutch bolts to specs in the interest of safety. Use a thread locker like Loc-Tite on the threads.

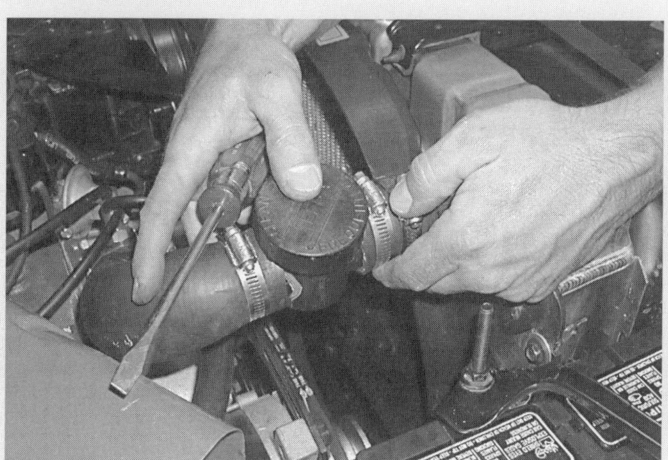

15. A coolant filter in the top return hose is advisable. Coolant filters, if properly maintained, will keep contaminants, such as iron particles, loose silicone, and other impurities, out of your new radiator. Once installed, watch for filter contamination and keep it clean.

16. The Griffin radiator's fan-shroud holes were modified to fit our fan shroud. Once this was done, installation was a snap.

Chapter 4

How to Optimize Your Cooling System

Text by Jim Smart
Photos from the Mustang Monthly *archives*

Cooling system maintenance begins with using the right coolant and checking levels often. You also need to flush it periodically.

Solutions for Your Mustang's Overheating Woes

Your Mustang's cooling system has the overwhelming task of removing excessive engine heat created by internal combustion and the friction of moving parts. Both elements generate a lot of heat. Many factors contribute to the heat an engine makes. Let's look at each of them and how they affect operating temperature.

The cooling system consists of a radiator, a cooling fan, a water pump, a thermostat, hoses, and the water jackets cast into the engine's block and cylinder heads. Heat is introduced to the engine's water jackets where coolant flows. To understand how heat is transferred to the coolant, you have to understand how heat works. Whenever we are cooling anything, whether it's a family in a minivan or the fury inside a hot engine, the process removes heat, it doesn't install cold. Heat always moves to a cooler medium, not the other way around. When the furnace quits in your home in January and the house cools down and becomes chilly, the heat is escaping to the cooler outside air. In July, when it's 95 degrees outside, your home air conditioning is removing heat, not installing cold, even though cold air is blowing from the vents. Your Mustang's cooling system does the same job as air conditioning.

Here is a typical Mustang radiator with an aftermarket electric cooling fan. Hot coolant from the engine flows into the radiator at the upper neck (A). As the hot coolant displaces the cooler coolant inside the radiator's tubes (B), it gives up its heat energy through the tubes to the air flowing through the radiator. When the thermostat opens again and releases hot coolant to the radiator, the coolant in the tubes enters the lower radiator hose (C) and the water pump to the engine to take on the engine's heat. Mustangs fitted with automatic transmissions have transmission coolers (D) built into the radiator. Because an automatic transmission runs hotter than your engine's coolant, transmission fluid gives up its heat to the engine's coolant in the radiator.

Cooling System Function

Understanding cooling system function begins with engine heat. As we light the gasoline/air mix in each of the combustion chambers, heat energy created from that light-off to propel your Mustang down the road also generates energy that can damage engine components. Aluminum pistons can melt at 1,300 degrees F. Steel and iron components begin to break down from heat at 1,500–2,000

The thermostat is located inside its housing or neck, which is connected to the upper radiator hose. Shown here is a technician tightening the bypass hose, which allows some coolant to bypass the thermostat.

This is the larger Mustang radiator common in '67–'70. It is a vertical-flow design with either two, three, or four rows of tubes. It's a big improvement over conventional Mustang radiators because it offers more surface area to transfer heat.

degrees F. Moving parts, despite the engine-oil barrier, create friction as they whirl around and reciprocate inside the engine. From that friction comes heat.

The water-and-antifreeze-mix coolant circulates inside the engine block and cylinder heads to transfer heat from those areas to the radiator. Think of coolant as a heat expressway. Heat energy catches a ride in the coolant, which enables it to escape to the atmosphere through the radiator. The thermostat, located at the upper radiator hose, acts as a heat traffic cop. It keeps hot coolant inside the engine until it reaches a predetermined temperature, ranging from 160 to 195 degrees F, depending on the thermostat. We want our engine to retain some heat energy in order to perform properly. An engine that runs too cool doesn't run efficiently. Engines that run too hot don't perform well either, so engine coolant temperature needs to remain in a specific range for optimum performance.

The humble automotive cooling-system thermostat has had the same basic design since the beginning of internal combustion engines. If you look at a thermostat closely, you will see a small, round chamber filled with wax. There is also a spring and a poppet valve. The spring holds the poppet valve closed. As the wax warms and expands, there is eventually enough pressure from the expansion to push the poppet valve open against the spring pressure. Hot coolant roars out of the engine through the upper radiator hose to the radiator. Coolant from the radiator flows into the engine, displacing the hot coolant. The wax cools and contracts (shrinks), allowing the spring pressure to close the poppet valve.

Thermostat opening temperature depends on the engine type and model year. Classic Mustangs need a 180-degree thermostat, although sometimes a 160-degree thermostat is installed in an effort to run cooler. Late-model, fuel-injected Mustangs need a hotter thermostat—192–195 degrees F—to run efficiently because their computers aren't happy unless coolant temperature is in that range.

When the thermostat opens, hot coolant flows into the radiator, displaced by the incoming coolant from the radiator, which is at a lower temperature. The water pump, located upstream of the lower radiator hose, moves the coolant from the radiator to the engine and back to the radiator again. Water pumps need to circulate a lot of coolant to do their job effectively. With use, water-pump impellers wear down and become less effective. Corrosion also takes a toll on impeller blades and water-pump passages. Think of this corrosion as cooling system cholesterol. Deposits grow on the fins and casting, slowing the flow of coolant. Water pumps also fail whenever the shaft seal and bearing wear out. When the seal fails, the pump leaks coolant. Bearing failure is identified by noise.

Radiators

The radiator's job, as the name implies, is to radiate heat to the atmosphere. Coolant flows

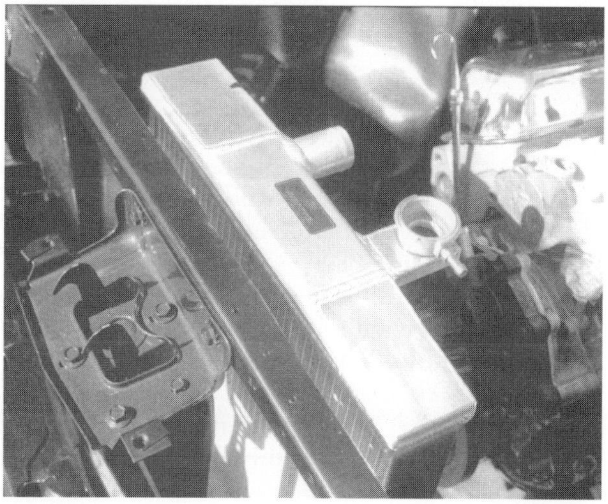

Aftermarket aluminum radiators are popular with racers and cruisers alike. This one, from Be Cool, has a multi-row core, with a lot of cooling surface area. These radiators work well in the tight confines of a classic Mustang engine compartment.

This is the basic four-blade cooling fan that was common with bread-and-butter classic Mustangs with both sixes and V-8s. These noisy, ineffective fans didn't come with shrouds. You'll find them on Mustangs without air conditioning.

Beginning in 1967, Ford began using flex-fans to move a lot of air at idle speed. As fan speed increases, the blades flatten out, consuming less power. However, these are noisy fans. This is a five-blade flex-fan.

Here's the four-blade 289 High-Performance cooling fan, recognizable by its wide blades and riveted design. Although there's a fan shroud here, not all of them are so equipped.

Undoubtedly the best engine-driven fan is the thermostatic clutch fan. This seven-blade fan is effective because it moves lots of air only when needed. When radiator temperature gets high, the thermostatic clutch engages the fan. As the radiator cools, the clutch disengages, allowing the fan to freewheel.

through the radiator's tubes from top to bottom ('65–'70 Mustang) or from side to side ('71–'04). The side-to-side approach is called a "crossflow" radiator, which is a better design than the vertical flow in '65–'70 models.

Airflow through the radiator carries the engine's heat from the coolant to the atmosphere. Between the tubes are fins that increase heat radiation by creating more cooling surface area. Think of those fins as "heat sinks," which help transfer heat. Fins in the radiator do the same thing as fins on an air-cooled lawnmower engine or a computer's motherboard. Fins create more area from which the heat can radiate.

Original-equipment-style radiators, available from U.S. Radiator and Mustangs Plus, are perfect for restorations where correctness is important. In fact, you can go with a four-row-core Desert Cooler radiator and enjoy better cooling while keeping an original-equipment appearance. Griffin and Be Cool aluminum radiators don't look stock, but they cool quite well under the toughest circumstances. These radiators are race-proven. What's more, they are designed and manufactured to fit your Mustang's engine compartment.

Cooling Fans

Air moves through the radiator with vehicle speed. At slower speeds or while stopped at a traffic light, we still need that airflow, which is helped with an engine-driven or electric cooling fan. Classic Mustangs came with engine-driven cooling fans, propelled by the water-pump pulley. In fact, most Mustangs had engine-driven cooling fans until 1994, when all Mustangs went to electric cooling fans. Four-cylinder Mustangs in the '80s and early '90s had electric cooling fans, including the SVO and GT Turbo models.

Mustangs have had a wide variety of engine-driven cooling fans. Most classic Mustangs had a four-blade steel fan, which wasn't very effective. These fans barely got the job done when the cars were new. Mix in cooling-system neglect and you have real trouble in short order. Mustangs with factory air conditioning had more sophisticated cooling fans. Some had six- and seven-blade steel fans, which made a lot of noise and consumed engine power because it took a lot of energy to turn them. The best engine-driven fan is the seven-blade

This is an aftermarket flex-fan from Flex-A-lite. It's considerably quieter than the Ford flex-fan and consumes less power. Note the absence of a fan shroud. This application could use a fan shroud, which increases air velocity through the radiator.

Electric cooling fans don't consume any engine power, and they operate only when needed. The trick is choosing enough electric fan for your application.

version with a thermostatically operated clutch fan. This fan works only when it is needed. When radiator temperature increases, typically when the thermostat opens, the thermostatic clutch fan engages, pulling cool air through the radiator. As the radiator cools, the fan clutch disengages, causing the fan to freewheel until it is needed again. If you must have an engine-driven fan, this is the best way to go.

Another type of fan is the flex-fan, first used by Fords in 1967. At idle, the fan blades move lots of air through the radiator. As engine speed increases, the fan blades flatten out, causing less drag and consuming less energy. The downside is noise, especially with the stock Ford flex-fans. They are noisy and consume a lot of power during acceleration. Aftermarket flex-fans, like those from Flex-A-Lite, are much quieter and consume less energy.

Electric cooling fans, like the thermostatic clutch fan, operate only when needed. A thermostatic switch and relay close at a given temperature, causing the fan to turn on to pull air through the radiator. The radiator cools and the switch opens, turning off the fan until it's needed again. Electric cooling fans also operate when we turn on the air conditioning, which pulls air across the condenser for cabin cooling.

Cooling System Woes & Fixes

Why do our Mustangs overheat? Overheating is simple physics. When an engine overheats, it has overwhelmed a cooling system that cannot get rid of the heat quickly enough. In other words, we're making heat faster than the cooling system can get rid of it. Here are some reasons engines overheat.

The water pump is the heart of your Mustang's cooling system. As water pumps are used, they wear out at the bearings and impeller. They also become corroded, hindering coolant flow. They need to be replaced periodically. Always opt for a high-flow water pump for improved coolant flow.

- Stuck thermostat (stuck open or closed)
- Clogged radiator tubes
- Corroded water-pump impeller and passages
- Slipping fan belt
- Defective fan clutch
- Missing fan shroud
- Clogged engine water jackets and passages
- Inoperative or ineffective cooling fan
- Collapsed lower radiator hose (at highway speeds only)
- Improper spark or valve timing
- Extremely lean fuel mixture
- Cylinder-head gaskets installed backwards
- Crushed header tube(s)
- Defective muffler baffles

Let's look at these issues. Sometimes thermostats don't stick closed, they stick open, and with the same result: overheating. When thermostats stick closed, it's easy to understand why overheating occurs. But when they stick open, why does the engine overheat? Because coolant flow has to be controlled. We have to hold it inside the engine

Overheating problems can be caused by outside factors, such as a dirty air-conditioning condenser. The A/C condenser creates its own problems by being a source of heat in front of the radiator. On a hot day, it's a real problem for maintaining coolant temperature.

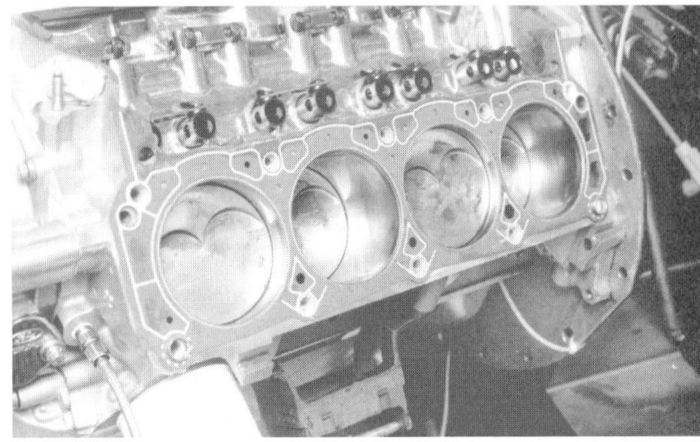

"Front" indicates the direction to install cylinder head gaskets. Always locate the coolant passages at the rear of the block in order to have coolant flow throughout the entire engine. If you install head gaskets backwards, your engine will overheat.

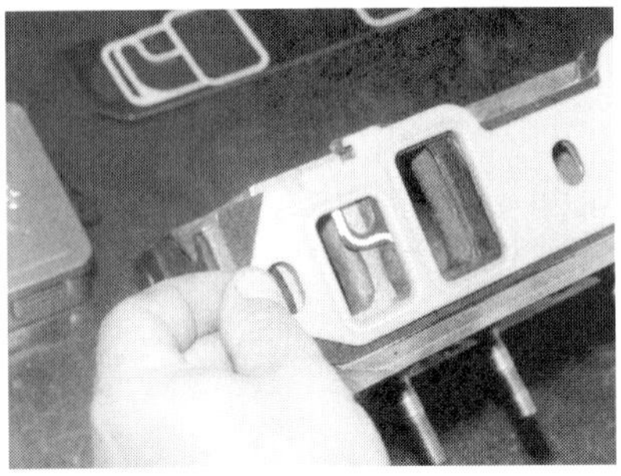

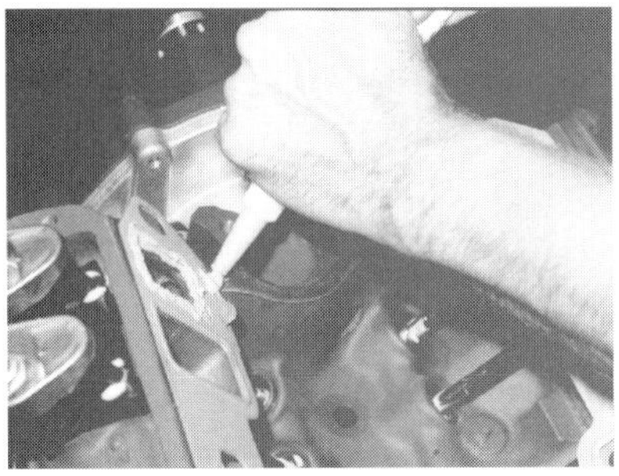

Coolant flow inside the engine can be hindered several ways. When you're installing gaskets, keep gasket sealer use to a minimum because it can block passages. Remember to cut the gasket coolant passage when it's set up for a dog-leg passage, which happens when 351W heads are used on 289/302 blocks.

long enough for heat to be transferred to the coolant. And we have to hold it inside the radiator long enough to cool it. If there is constant coolant flow, the coolant is never in one place long enough to do its job. If a thermostat is stuck open or if it's removed entirely, the coolant stays hot because it never has a chance to cool.

Clogged radiator tubes tend to elude us at times. And we don't want to spend the money on a new radiator, so we keep looking for used ones. Even when radiator cooling tubes look clear, they may be clogged near the bottom where we can't see. They also become lined with scale, which hinders their ability to radiate heat. It's simple: Radiators must be replaced periodically. They do have a lifespan, no matter how often we flush and service them with fresh coolant. The same is true of water pumps; they need to be replaced periodically.

Fan-belt slippage hinders the water pump's ability to circulate coolant, and it also slows the fan speed. If the fan belt slips a lot, you can expect overheating. If you have a faulty fan clutch, the fan will not rotate at the speed where it is designed to function. Check the fan clutch, even if it isn't that old. If it wobbles or is easy to turn, replace the fan clutch.

The fan shroud, which increases air velocity through the radiator, is another issue. Only four-blade steel fans came from the factory without fan shrouds. Anything with more than four blades needs a fan shroud.

Coolant—Engine water jackets and passages become clogged with neglect. Most people neglect their cooling systems, especially given the "maintenance free" mentally we have today. According to John Da Luz of JMC Motorsports in San Diego, your Mustang's cooling system should be flushed and serviced with fresh coolant every two years. The anti-corrosion properties in antifreeze fade away to nothing in one to two years. With no protection, your engine's cooling system rusts and corrodes. Aluminum parts become powdered. Iron and steel parts become rusted. Iron and aluminum do not live happily together, which is why service every two years is vital. Coolant carries small aluminum particles through the iron, and it also carries iron particles to the aluminum. Corrosion begins the minute your engine is fired and coolant begins to circulate.

Coolant can be straight water, which is not recommended. Or it can be a mix of water, ethylene glycol (antifreeze), and some additives designed to prevent corrosion.

Cooling-system corrosion comes from two basic sources: electrolysis and dissimilar metals. "Electrolysis," you say? Corrosion is born from the flow of electrons and rusty body panels. Any time you have dissimilar metals such as aluminum and iron, you have a formula for corrosion. The coolant acts as an electrolyte, just like the acid inside your battery. In fact, your coolant becomes more acidic with time and use, only making the problem worse. That's why it must be changed every one to two years without fail.

There are three basic types of antifreeze. Ethylene glycol (EG) is the one we're most familiar with and is packaged under many brand names. It has a sweet smell and taste, making it attractive to animals. However, it will kill them if consumed. Propylene antifreeze isn't as toxic and yields a horrible taste, making it less appealing to children and animals.

Antifreeze has silicates, nitrate, phosphates, and borates mixed in, which prevent the development of corrosion, but only if replaced on a regular basis. The generic ethylene glycol antifreezes are either green or red in color.

Dexcool, a long-life type of antifreeze, was first introduced in 1995. Texaco also makes this antifreeze, known as Texaco Extended-Life Antifreeze. Dexcool and Texaco Extended-Life are ethylene glycol-based with a five-year life expectancy. Dexcool and Texaco Extended-Life are easily identified by their orange color. These should never be mixed with conventional two-year EG coolants. Don't get Dexcool or Texaco Extended-Life confused with red or green conventional antifreezes. If you mix them, you reduce the life expectancy to two years. You also risk damaging the cooling system.

Evans NPG Non-Aqueous Propylene Glycol is a relatively new type of coolant available from Evans Cooling Systems. Evans NPG works by itself, without water. It isn't cheap, but will completely stop cooling-system corrosion once properly installed. The nice thing about Evans NPG is its ability to transfer heat even better than the water/antifreeze mix we're familiar with. Evans NPG is a toxic coolant, which means it mandates the same caution as ethylene glycol. You must completely purge your cooling system of all coolant before you can install Evans NPG. This means a thorough flush and complete dry-out of the system before installation.

We've learned that most reproduction lower radiator hoses don't have a support spring inside. This spring is designed to prevent hose collapse when the engine is revving. When engines rev, they create a greater demand on the radiator, which will make an unsupported lower radiator hose collapse at highway speeds. Whenever you order a reproduction or aftermarket radiator hose, make sure it has the spring inside. If it doesn't, visit your local auto parts store and buy a spring.

Problems outside of the cooling system will also cause overheating. Improper spark and valve timing will cause overheating. During an engine build, we can sometimes make a mistake with valve timing. Ignition timing can come back to bite us if we crank in too much of it. This can happen with an improperly tuned vacuum and/or centrifugal advance. Cylinder-head gaskets are easily installed backwards if you don't pay close attention. Coolant passages go at the rear of the block, not the front. Blocked water jacket passages anywhere in the block and heads will cause overheating.

Restrictions in the exhaust system, such as a smashed muffler or header tube, will not only cause overheating but also engine damage. The exhaust system scavenges not only hot gasses, but extremes of engine heat. A crushed header tube will cause one cylinder to run considerably hotter. This can cause exhaust-valve damage and failure. Damage can also be done to the piston.

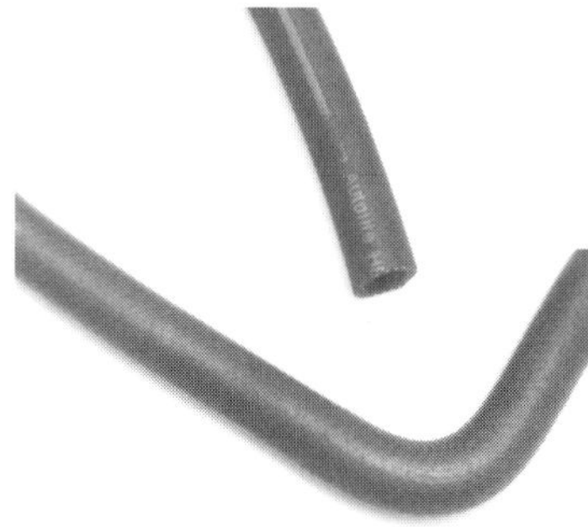

Hoses and belts have been the undoing of many an engine. Not all reproduction and aftermarket radiator hoses are created equal. Poor-quality hoses tend to collapse or burst. Lower radiator hoses must have an anti-collapse spring. When buying belts, always aim for the best quality, not necessarily the best price. Use molded hoses where original equipment called for it.

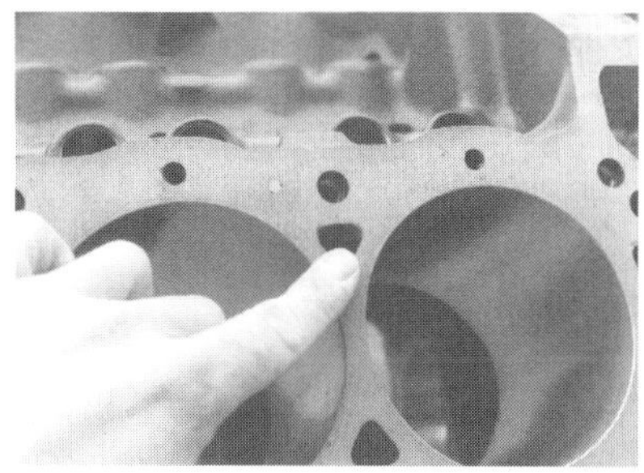

When you're rebuilding an engine or pulling cylinder heads, always check coolant passages. If they're blocked with rust or gasket sealer, overheating can become a problem.

Sources

Be-Cool
www.becool.com

Evans Cooling Systems
www.evanscooling.com

Griffin Thermal Products
www.griffinrad.com

Mustangs Plus
www.mustangsplus.com

U.S. Radiator
www.usradiator.com

Chapter 5

Detailing a 289 Engine

Text and Photos by Mark Houlahan

Our 289 is ready for its close-up, Mr. DeMille. This beauty queen propelled our Project '66 to local shows for the 40th Anniversary Celebration of the Mustang's intro in 2004.

Whether at a local burger-joint cruise or a Mustang Club of America National, one of the first areas everyone checks out is your Mustang's engine compartment. If your Mustang is a daily driver that's serviced by the local auto parts chain without new old stock (N.O.S.) parts, the non-Autolite distributor cap, yellow plug wires, and flex fan can be forgiven as long as things are clean, painted, and presentable. But when it comes to being correct and standing tall next to your competitors on the show field in concours classes, it's going to take some serious elbow grease and a paycheck or two to get that 289 spiffed up and ready for the judges.

Our 289 was not in such great form. The engine builder had painted it, but that's about it. We're not complaining, mind you, because that isn't the engine builder's job. So we had our 289 on an engine stand ready to be dressed; and dressed it would be, with all the right parts to make an MCA judge nod in approval. We once again picked the brains of the folks at National Parts Depot, who helped us determine what we needed for our project. NPD supplied just about everything we needed to get our 289 in top shape.

When it comes to concours detailing, take your time, research your project, and don't be afraid to ask questions and make a decision based on multiple "right" answers. We've seen all sorts of assembly line guffaws, so if your engine had a sticker in a certain place, feel free to put it back where it was originally. And if you're performing a V-8 conversion project like we did, be ready to locate pulleys, brackets, and such if your engine didn't come with them.

Sources

Alamo Classic Mustang
www.alamoclassicmustang.com

Classic Auto Air
www.classicautoair.com

Extrude-Hone
www.extrudehone.com

MCCI/Jet-Hot
www.jet-hot.com

Metro Mustang
www.metromustang.com

National Parts Depot
www.npdlink.com

Pony Carburetors
www.ponycarburetors.com

1. With our 289 fresh from the dyno, there were a few things we had to correct before detailing. On the dyno, we used an electric fuel pump and removed the thermostat, so it was installed first. Then we installed the remainder of the engine's fuel system: fuel pump, fuel-pump-to-carburetor steel line, and the rubber fuel hose between the steel line and our Pony carburetor. Correct crimp clamps and installing them with the right clamp pliers are the perfect detailing touch.

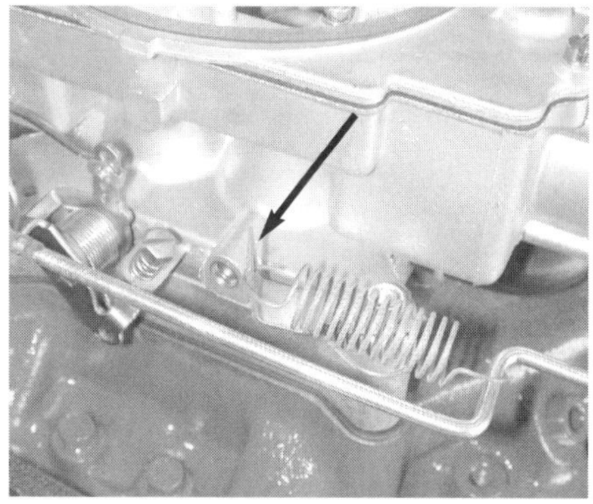

2. On the dyno, the carburetor's linkage was connected to a mechanical lever, so we had to install the correct throttle lever, return spring, and return-spring bracket as well. There's a difference between the 2V and 4V return-spring brackets, so be sure to get the right one. Notice the angle of the spring bracket (arrow), pointing forward at the 11-o'clock position.

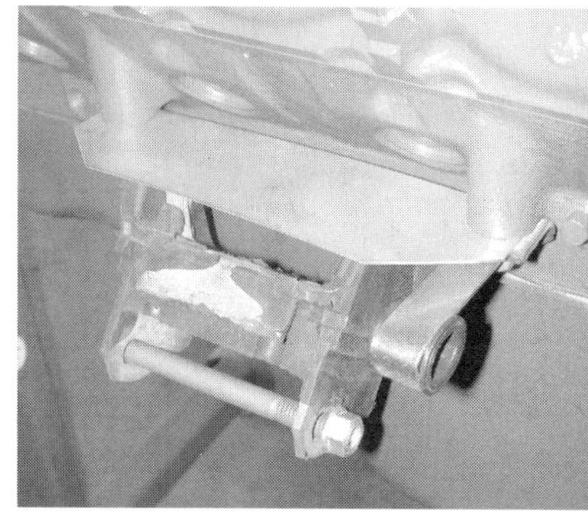

3. Fresh motor mounts are a given when installing a new or rebuilt engine, but getting details right is important. The heat shields were installed by Ford to reduce heat-related wear to the rubber part of the mount. Shown here on the passenger side, Ford used this metal bracket to support and guide the starter cable to the starter. AMK concours mounting bolts secure the mount to the block and are ready to secure the mount to the chassis.

4. Mounting our reproduction yellow-top coil was simple with a new coil bracket, coil detail sticker, and engine calibration sticker, all from NPD.

5. For the front engine dress, we called upon Classic Auto Air for the correct brackets, water pump, crankshaft pulleys, and so on to mount a concours A/C system. We sent some used brackets to CAA as cores to get these nicely painted and inspected ones in return. The included hardware works fine but because we're perfectionists, the brackets were installed with AMK hardware from NPD.

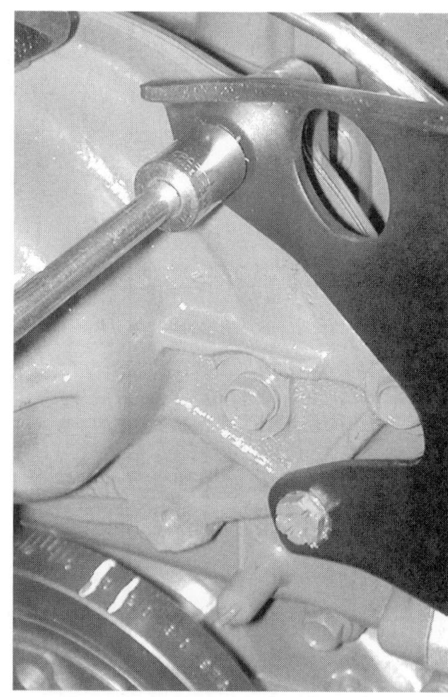

6. The correct power steering-pump mounting bracket is hard to find. We scored ours from our favorite used-parts vendor, Metro Mustang. After a thorough bead-blasting and a coat of paint, it was ready for installation on our 289. She was starting to look the part.

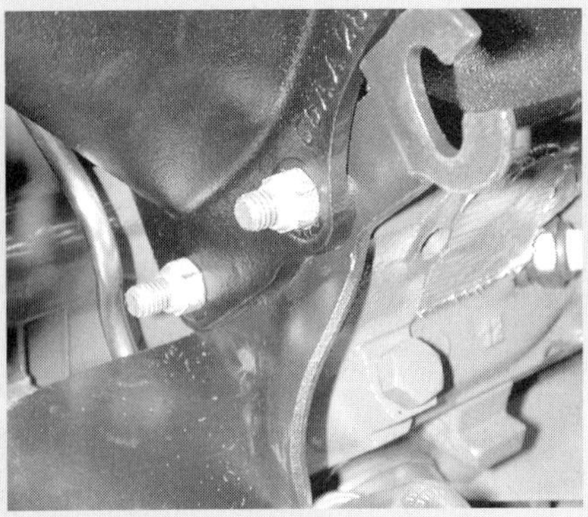

7. We've seen mounting bolts and studs for the power steering pump and A/C-compressor mounting-bracket attachment. We thought it was safe to assume bolts were used for power steering only, and the studs with lock nuts were used when factory A/C was ordered. Either way, AMK has the right hardware kits with the proper bracket spacers.

8. For our power steering pump needs, we picked up a core pump from Metro Mustang and a reproduction power steering reservoir (for A/C-equipped Mustangs) from Classic Auto Air. We sent it all to Alamo Classic Mustang for a rebuild and assembly.

9. Classic Auto Air's rebuilt compressors come with new seals, valves, brushes, and clutch assembly installed and ready to go. The hardware to mount the A/C compressor to the mounting bracket was a bit of a chore to install, but was manageable with the right wrenches.

10. The dogleg bracket, which was loosely installed earlier, was positioned against the compressor, and the retaining hardware installed and tightened.

11. The alternator is a standout item that needs the right detailing. Hanging off the side of the V-8 all by its lonesome, it needs a few concours bits to look right. We used an AMK detailed alternator from NPD with the proper case markings and ink stamps. We added the correct zinc fan, single-groove pulley, lock washer, and retaining nut to the alternator. Finally, the alternator was secured in its brackets with AMK hardware, including the proper aluminum spacer in gold anodizing, all from NPD. Hard to see, but on the rear is an alternator pigtail harness from NPD.

12. These ink-stamped reproduction fan belts from NPD are a nice detailing touch. They are brand new Ford Motorcraft service belts stamped with the proper logo and part number for your application. We ordered the three belts, and all of them fit right the first time (sometimes, when mixing pulleys, the original belt length will no longer work, which is out of NPD's control).

13. Our engine's oil filter from the dyno test was still in place, and we won't take it off until after a few hundred miles of break-in. Instead of spending big bucks for the absolutely correct reproduction oil filter (we're going to drive it, after all), we purchased the Rotunda oil-filter decal from NPD. After careful measuring and taping, we made it fit the filter like a second skin so it would simply slide off and on. We should need only one decal a year, taking it off for oil changes and swapping it to the new filter.

14. To determine how to run the power brakes and automatic-transmission steel lines off the intake vacuum fitting, we looked through our shop manuals, Osborn Reproductions assembly manuals, restoration guides, and Bob Mannel's *Mustang & Ford Small-Block V-8, 1962–1969*. Some books showed both installed, like the erector set we built here. But we got the correct word from Mannel's book: For '66 Mustangs with an auto trans and power brakes, the booster got its vacuum from the intake fitting (steel line running toward the valve cover), and the auto trans sourced its vacuum from the check valve on the brake booster. Now we know.

15. For many, the chrome open-element air cleaner, factory on the Hi-Po but available as an over-the-counter option, is the showpiece of a small-block Ford. We wouldn't think of building a hot 289 without it. From NPD, the air-cleaner assembly comes with the decal and air filter with the proper blue base ready to be installed.

16. While a factory A/C car would have a clutch-type fan assembly, we're installing the extra-cooling five-blade fan and regular fan spacer. When we take our project to Classic Auto Air for the rest of the A/C installation, we'll have them upgrade to the seven-blade fan.

17. Because our distributor houses a PerTronix electronic ignition conversion, we wanted to make the extra wire as discreet as possible. We found an old elbow connector in our spare-wire-connector box, stripped it down, then soldered it to the PerTronix lead that goes to the negative side of the coil. A bit of shrink-wrap covered the connection and, at first glance, the connection looks factory. The positive wire from the PerTronix was routed discreetly behind the distributor and tapped into the factory coil feed wire.

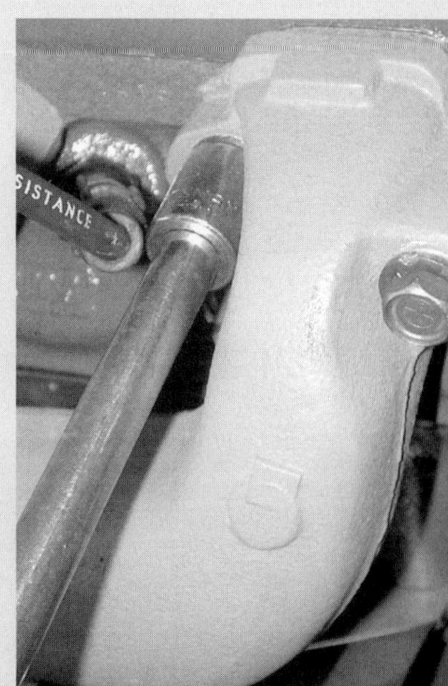

18. On the dyno, we used the dyno headers since they were set up for the thermocouple probes the software uses to "see" exhaust temperatures. For our hidden power scheme on the 289, we shipped our NPD-supplied Hi-Po manifolds to Extrude Hone to be abrasive-flow ported, then dropped them off at Jet-Hot to be coated in their natural gray-iron, thermal-barrier coating.

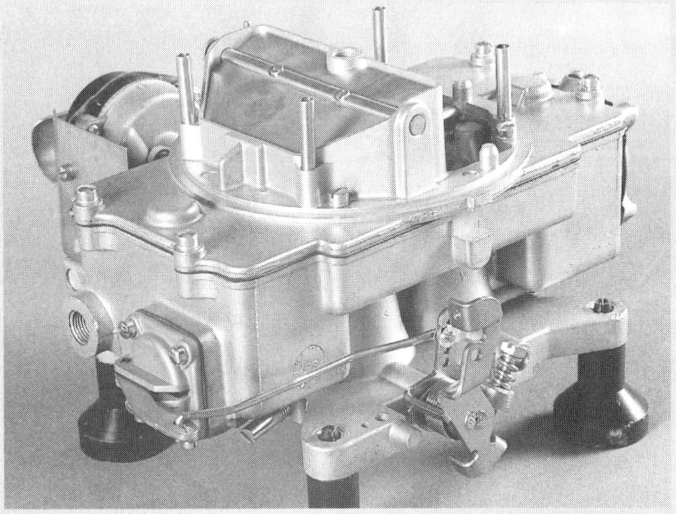

19. The installed manifolds look great, and will for years to come. Some people remove the manifolds for cleaning and refinishing, but that sounded like too much work for us. Plus, the thermal coating reduces underhood temperatures and keeps the heat inside the exhaust for better efficiency. Next time, we'll install the transmission, starter, and other drivetrain components, along with our detailed engine. Then we'll fire up our Mustang project and get it to move under its own power for the first time in four years.

20. Jon Enyeart and his crew at Pony Carburetors can take the grungiest-looking chunk of carburetor aluminum and steel and make it into a show-winning centerpiece for your engine restoration and detailing project. We asked Jon to help us convert to the dealer-installed single 4V Induction Kit (C6AZ-6B068-A). We shipped our tired Autolite 2100 2V to Jon, and he prepared the correct 4100 4V to complete our 4V conversion. While we already had a good, used intake and the proper linkage, return spring, and return spring bracket, Pony Carburetors has everything you'll need to make the switch to 4V induction, including the 4V cast-iron intake. If you want to get even more exotic, Pony Carburetors offers a three 2V setup with the fuel log and linkage.

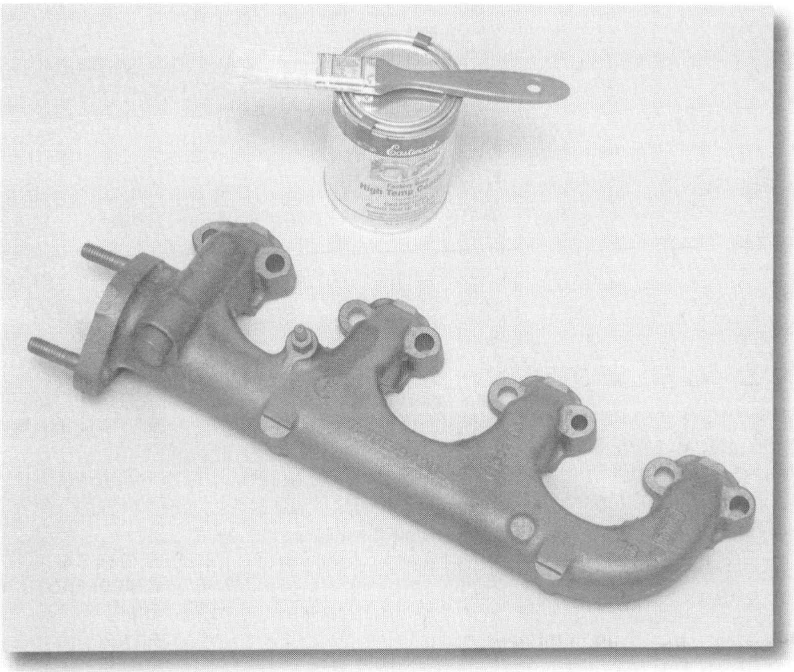

Jet-Hot Coatings and The Eastwood Company both offer products to restore your exhaust manifolds to better-than-new condition.

Chapter 6

How to Restore Exhaust Manifolds

Text and Photos by Jim Smart

Extremely hot surfaces are a challenge for restorers, especially those that are prone to rust, like factory exhaust manifolds. How do we protect these sizzling surfaces from rust? And how do we keep them looking brand new?

First, you need clean surfaces, fresh and uncontaminated for a happy marriage with the coating. Second, think about what you want these surfaces to do. We are going to familiarize you with two options: Eastwood's Factory Gray High Temp Coating and Jet-Hot Coatings, a super high-temp ceramic coating that will never go away.

Eastwood's Factory Gray High Temp Coating is a cast-iron color coating that can withstand up to 1,200 degrees F. Considering most exhaust manifolds and headers run somewhere around 900 degrees, Eastwood's High Temp Coating gets the job done in short order. This specialized coating is also available in silver for those who want to show off their iron manifolds.

Because Eastwood's High Temp Coating has an oily consistency, it smokes like the devil when you first fire the engine. But the smoke won't last long and once it burns off, you'll have a dull gray finish that resembles a brand-new exhaust manifold. The real benefit is, it won't rust. Eventually, you'll have to do this all over again, especially if you drive your Mustang frequently. Weekend show cars last much longer.

Another option for exhaust manifolds and headers is an innovative coating from Jet-Hot Coatings. Jet-Hot ceramic-coats your exhaust manifolds or headers with a bulletproof coating that won't burn off or rust. Jet-Hot is a high-temp coating unmatched in the industry. It's good for up to 1,300 degrees F, but we're convinced it can tolerate even more heat thanks to its space-age technology.

Jet-Hot reduces header and exhaust manifold heat radiation by 50 percent, which means lower underhood temperatures. When you shut down the engine, you can touch the headers in a matter of minutes. What's more, Jet-Hot looks terrific and is available in a variety of colors, including factory gray iron for stock exhaust manifolds.

Sources

Jet-Hot Coatings
www.jet-hot.com

The Eastwood Company
www.eastwoodco.com

Virginia Classic Mustang
www.vamustang.com

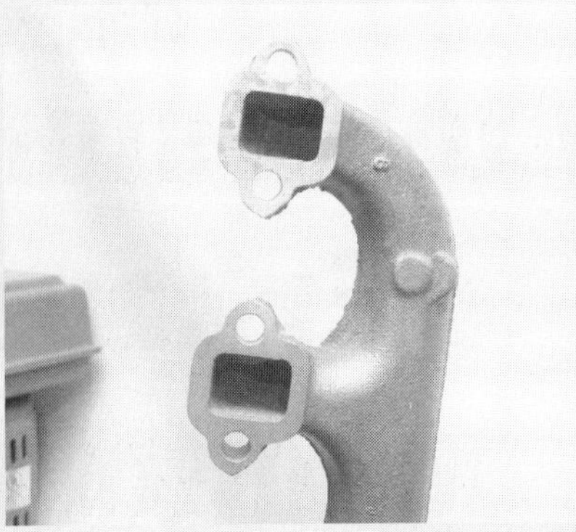

1. Begin with clean surfaces. Our 289 exhaust manifolds were media-blasted, and they look like new. Make sure there are no cracks before media-blasting. They may also be chemically dipped to remove rust.

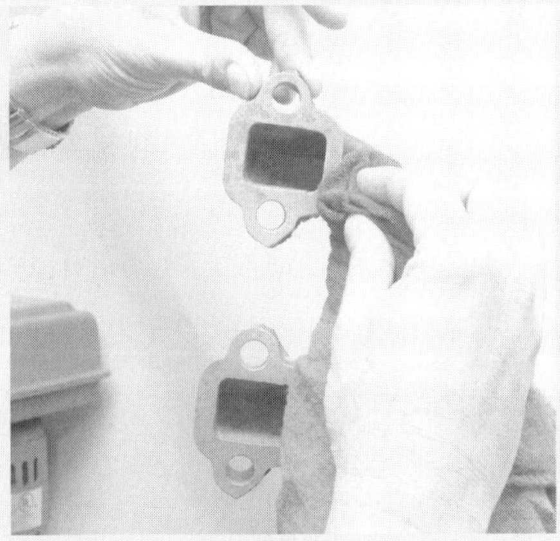

2. Clean all of the skin oil off the manifolds. If you want to avoid this step, never touch clean manifolds without gloves.

3. This is Eastwood's Factory Gray High Temp Coating. Don't spray this stuff on; brush it on with a cheap paint brush. Brush strokes won't be visible.

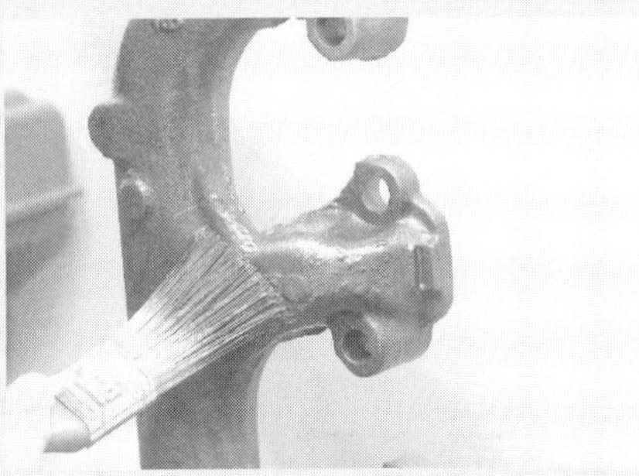

4. Brush Eastwood's High Temp Coating on like this. Work the bristles into the grainy iron surfaces and make sure it penetrates the iron. Apply two coats, and don't miss anything. It isn't necessary to do the inside of the manifold. Don't touch the coating until it's completely dry, which normally takes a day. Until it gets hot, it will be slippery.

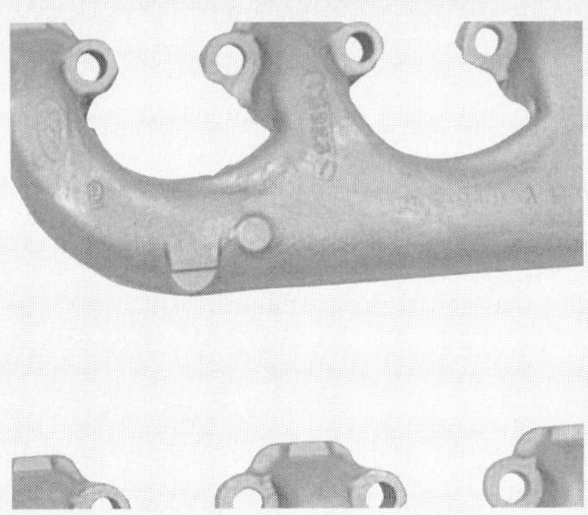

5. Nice, huh? These manifolds stay nice for years and thousands of miles. You can use this coating on just about anything that gets hot, including manual transmissions, steering gears, differentials, and other castings.

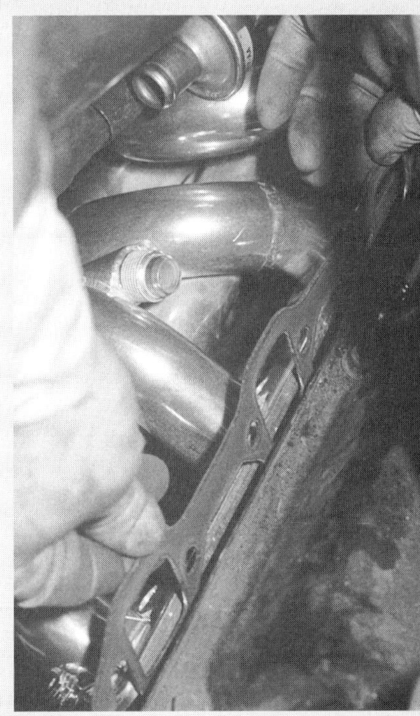

6. This is what Jet-Hot looks like on a set of headers. Imagine what it can do for your headers and manifolds. You can even put it on your Mustang's mufflers, intake manifold, and valve covers.

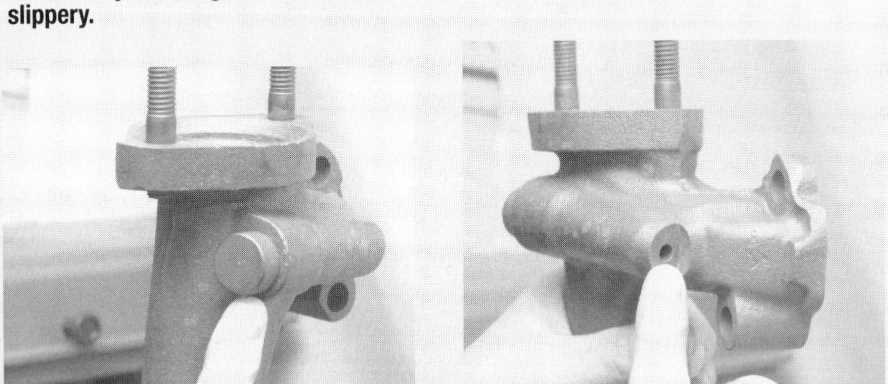

7. While you're at it don't forget to get the choke stove in proper working order while restoring your exhaust manifolds. The choke stove provides heat for the automatic choke's thermostatic coil at the carburetor. The lines typically break off in the exhaust manifold. Using the right size drill bit, drill out the broken line ends. Then call Virginia Classic Mustang and order AMK Products' choke line repair kit, which includes everything you need to get your Mustang's automatic choke working again.

With National Parts Depot's new reproduction exhaust systems, you can get the original factory look and sound for your classic Mustang.

Chapter 7

How to Install a Single Exhaust System

Text and Photos by Jim Smart

Restoring a vintage Mustang has long been a tricky affair because it hasn't always been easy to find the right concours-correct parts and accessories. Unless you're restoring an MCA-judged Thoroughbred using all N.O.S. parts and have the deep pockets it takes to buy them these days, you're looking at reproduction parts to get the job done. For years, aftermarket exhaust systems have been our only choice, even for concours restorations. Exhaust giant Walker, for example, offers nice systems that get your Mustang looking and sounding similar to showroom condition, which works well for weekend cruisers and daily commuters where correctness and originality aren't critical.

However, the Walker systems lack the nuances of the factory-original exhaust systems, like the boloney-slice exhaust tips and proper Y-pipe unions.

National Parts Depot recognized the need for factory-looking exhaust systems for early Mustangs and did something about it with its line of original-style exhaust systems and components. They're affordable, at under $200 for the system we installed. We're going to install one on a '68 Mustang hardtop that competes in MCA-judged competition and see how it looks. Jimmy's Broadway Automotive of Tucson, Arizona, did the job for our cameras.

Sources

Jimmy's Broadway Automotive
www.jimmysbroadway.com

National Parts Depot
www.npdlink.com

This is one of the many original-look exhaust systems from NPD. It's aluminized for corrosion protection, with all the right bends and turns for a perfect fit.

We like the Y-pipe, with its smooth transition from both banks to the intermediate pipe. These pipes are swaged at the manifolds for a leak-proof fit. NPD includes the donut gaskets with the kit.

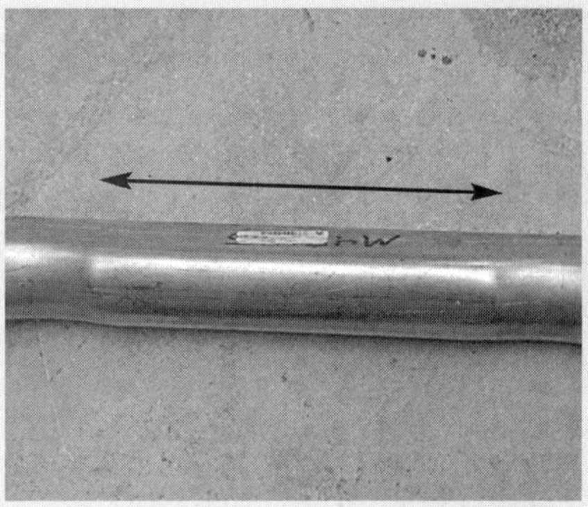

The NPD intermediate pipe has this flat spot, which, when installed, is located along the transmission tunnel for proper factory clearance. This is something most aftermarket exhaust systems lack.

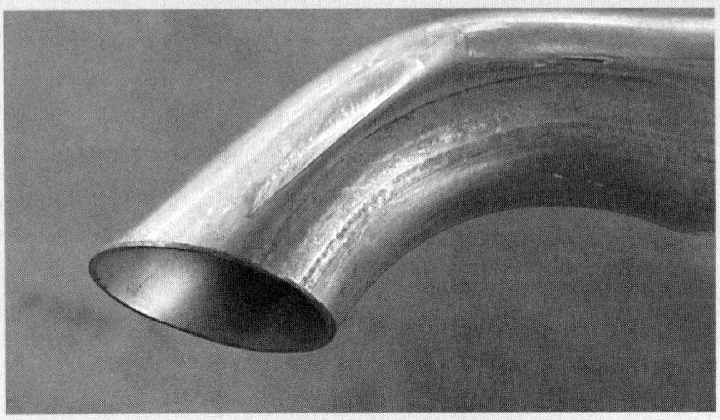

Here's another nuance you rarely find with aftermarket exhaust systems: the correct boloney-slice tailpipe tip. This is like the original factory exhaust tip.

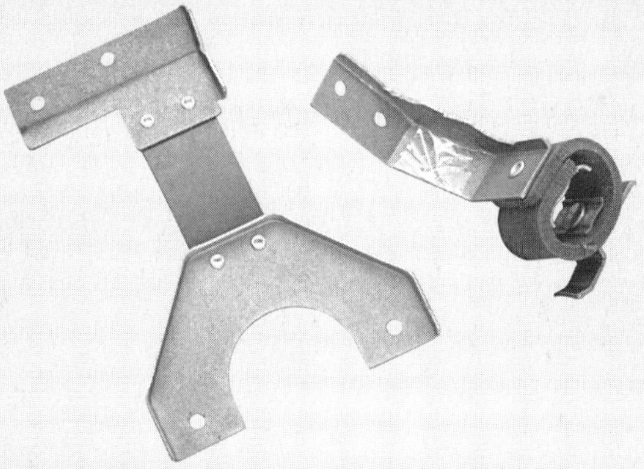

NPD also includes the factory-style hangers, which install using the factory attachment points.

This is the Walker single-exhaust system for '68 Mustangs with the 289-2V V-8. It's a nice system for the driver, but not correct for the concours-restored Mustang show piece. Note the tailpipe tip: no boloney...

The Walker system on our project car was fitted with Ford N.O.S. clamps, which we decided to reuse.

How To Install

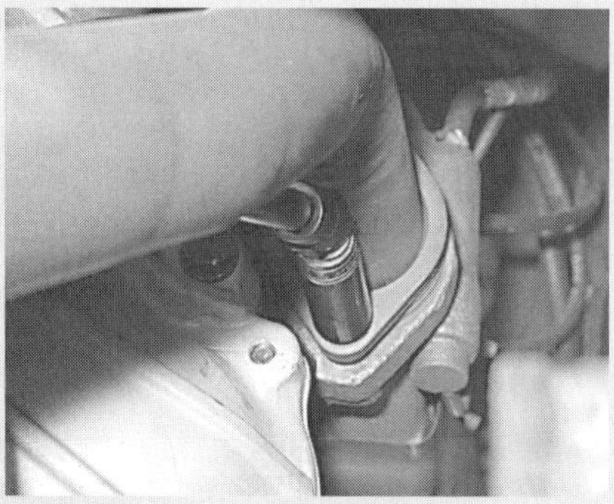

1. The Walker system is disconnected at the exhaust manifolds first using a 5/8-inch deep-well socket. WD-40 on the threads helps.

2. All clamps are loosened (left) and disconnected (right) as shown to free up the system.

3. We had to cut the tailpipe at the muffler to remove the system. Originally, exhaust systems were installed at the factory before the body was dropped onto the rear axle. This is why the factory systems were installed as one long assembly.

4. With clamps and hangers removed, the old system is lowered out of the car.

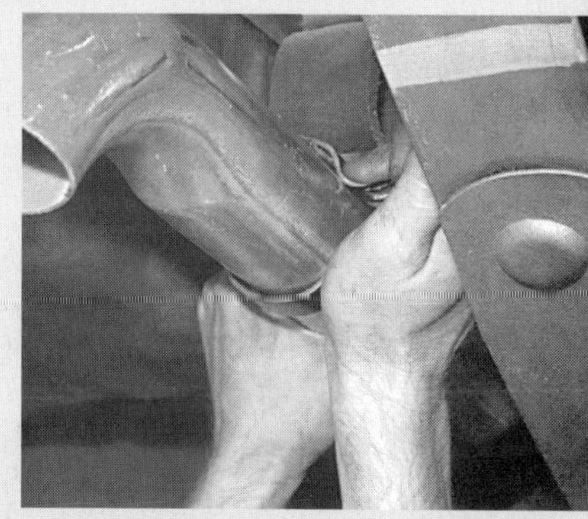

5. The new NPD tailpipe is installed first and loosely secured in the hanger.

6. Next, donut gaskets (left) are fitted at the Y-pipe tips, then the Y-pipe is fitted to the manifolds (right). Secure the pipe, but do not tighten the clamps at the manifolds.

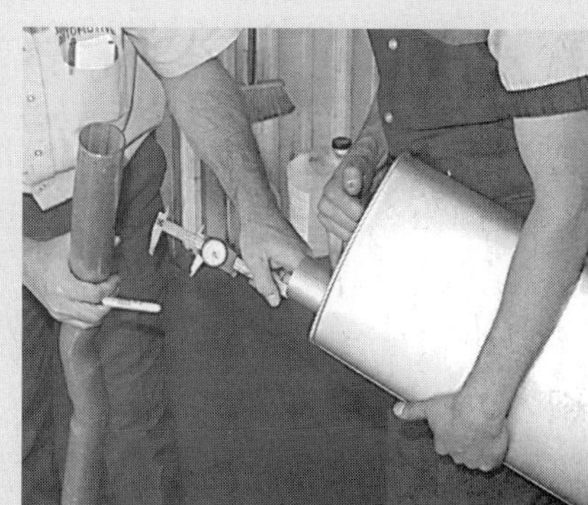

7. Because the NPD original-look exhaust system is designed for both the '68 and '69 Mustang, the intermediate exhaust pipe is a pinch too long for the '68. We're cutting this pipe for proper fit—an easy task for any exhaust shop.

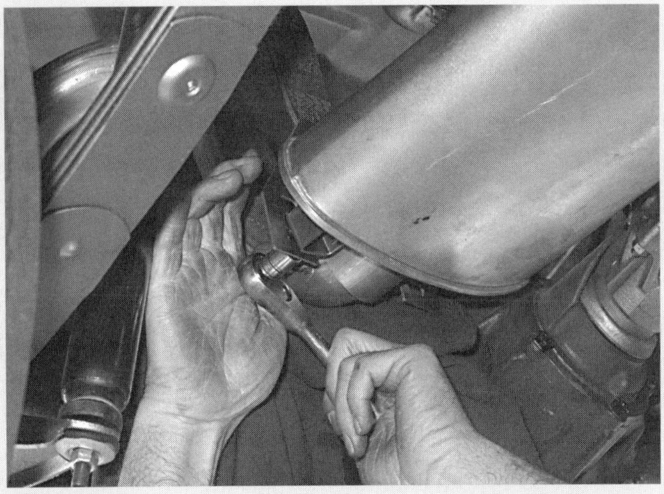

8. The muffler is attached to the rear exhaust hanger and tailpipe as shown. We suggest welding a bead around the seam at the tailpipe and muffler, just like the factory did in 1968.

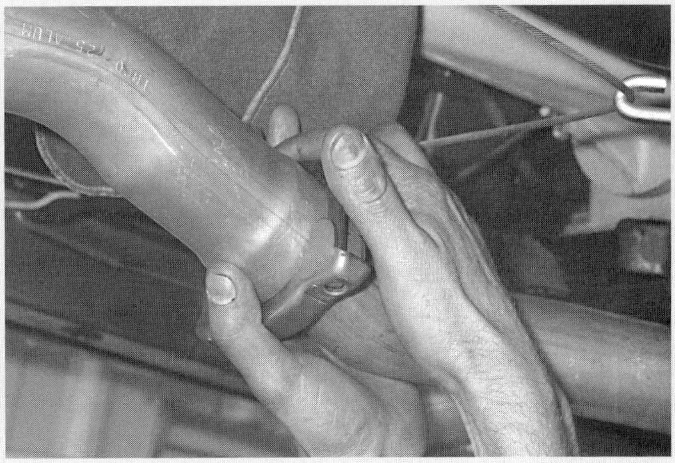

9. The intermediate clamp is installed where the Y-pipe meets the intermediate pipe. Your objective here is to get the two pipes as close to the floorpan as possible. They should be flush with the bottom of the car when viewed from the side, not hanging low.

10. With the Y-pipe and intermediate pipe properly supported near the floorpan, secure the nuts at the exhaust manifolds.

11. Properly installed, our NPD original-look exhaust is good to go. Check the side profile to ascertain proper fit underneath before tightening anything completely. It's a good idea to weld attractive beads where the pipes join to prevent leakage.

Chapter 8

Rebuild & Tune Autolite/ Motorcraft Distributors

Text and Photos by Jim Smart

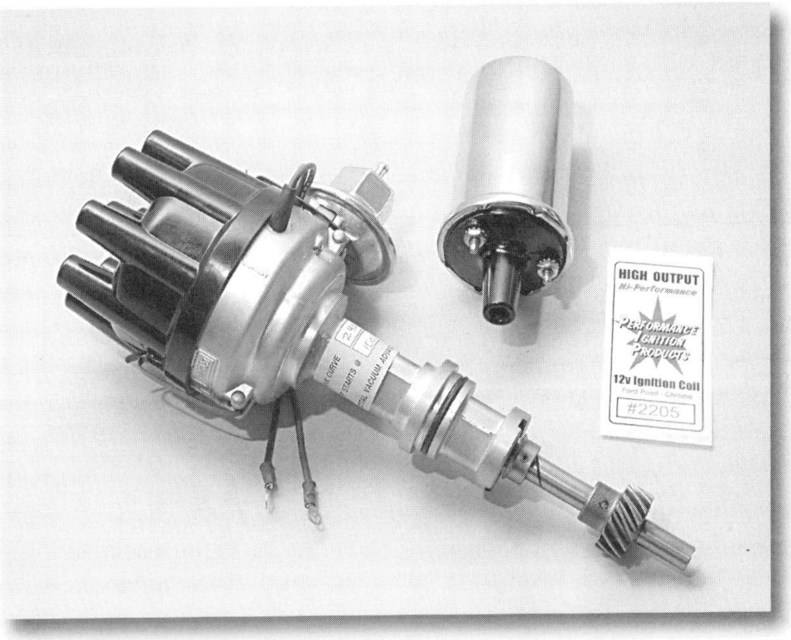

This is the complete Ignitioneering package: a distributor and a PerTronix Flamethrower ignition coil. With the distributor is a tune spec card, which states how the distributor was curved and tuned. Ignitioneering keeps a permanent record for future use.

In an age of electronic fuel injection and seamless engine management systems, it's challenging to remember the antiquated ignition and fuel systems used on our classic '65-'73 Mustangs. In 1965, Mustangs were equipped with point-triggered Autolite distributors and carburetors. In the '70s, they were upgraded with the Motorcraft Duraspark electronic ignition, which was an important first step toward better reliability and fuel economy.

Vintage Mustang owners embrace their cars like few other automotive cultures. But we have our criticisms about them too, including rough idle, sluggish performance off idle, misfires, and inadequate torque.

So how do we find improvement in the older technology? We start with the classic Mustang's point-triggered distributor. The old Autolite distributor is a simple design and easy to service. However, its shortcomings leave it in the dust behind other distributors of the period. GM's Delco distributors were outstanding for their day, and were even easier to tune than the Autolite.

In the early '90s, PerTronix pioneered one of the best modifications for classic Mustangs: the pint-size but powerful Ignitor electronic ignition module retrofit for Autolite distributors. In about 30 minutes, you can install the Ignitor or Ignitor II with astonishing results for your classic Mustang. Expect a smoother idle, crisper throttle response, and unmatched reliability.

We visited Ignitioneering in Fullerton, California, for a close-up look at the Autolite distributor and what can be done to make it a fiercely reliable performer. The company took our '65-vintage Autolite single-point distributor and made it better than new.

Sources

Ignitioneering
www.ignitioneering.com

PerTronix
www.pertronix.com

Virginia Classic Mustang
www.vamustang.com

1. This is our '65-vintage Autolite distributor, pulled from a blown-up 289 that's getting a Jim Grubbs Motorsports rebuild. We're going to rebuild this distributor and fit it with the Ignitor II from PerTronix.

2. After a shaft clean-up for easy disassembly, the drive gear and retainer tension pins are removed.

3. Those old Autolite ignition points and condenser are removed and tossed. Old ignition wiring is removed next. We won't need it later.

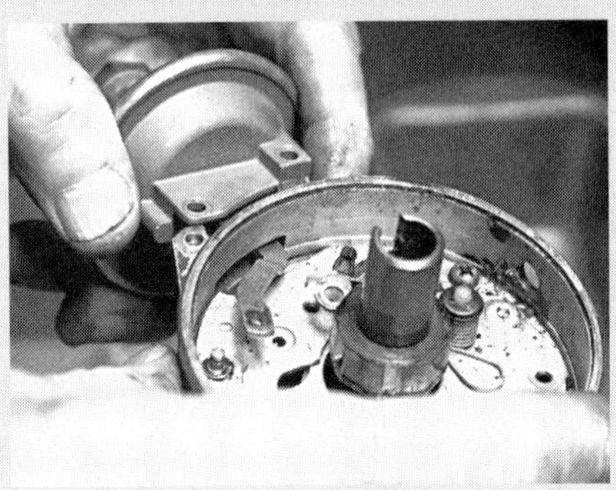

4. The original Autolite vacuum advance is removed next. If you want the correct factory look, opt for a reproduction vacuum advance from Virginia Classic Mustang. The factory vacuum advance is adjusted by adding or subtracting shims. Aftermarket vacuum advances are adjusted with an Allen wrench at the vacuum port.

5. The breaker plate, which appears to be a replacement, is removed next. Don't lose the ground wire.

6. Remove the centrifugal advance (mechanical advance) springs, and make a note of their locations.

7. The rotor head and cam are removed next by taking off the oil wick and C-clip.

8. Centrifugal advance weights are removed by taking out the C-clips and sliding the weights off the shaft.

9. The rotor head's thrust washer is removed next.

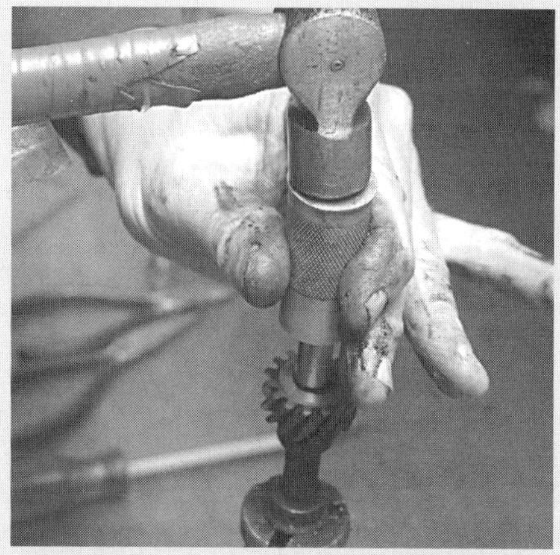

10. With the housing properly supported, the distributor shaft is driven out. Be careful here—too much thrust shatters the housing.

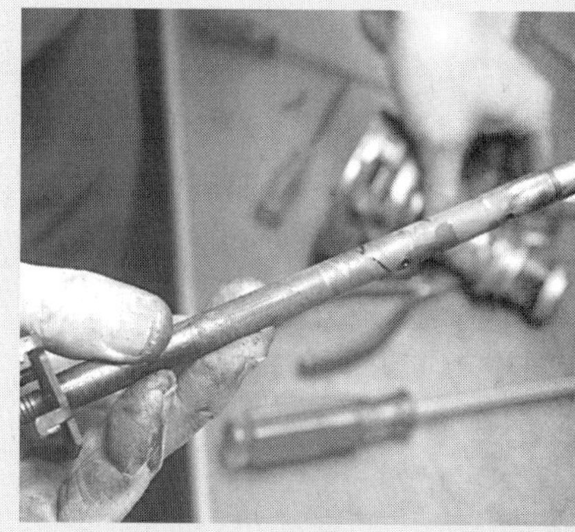

11. Our distributor shaft shows signs of scoring due to poor lubrication. These lines are designed to carry engine oil up to the bushings. But this doesn't always happen like it should. We're going to use oil-impregnated bushings in our rebuild, which means longevity in the future.

12. Autolite and Motorcraft point-triggered distributors have two bushings (although some have only the top bushing) to support the shaft. These shaft bushings wear quickly, which makes the shaft wobble, causing point-bounce and irregular dwell angles. These bushings have to be replaced.

13. Distributor-shaft bushings are driven out with a tap. You can also gently drive them out with a punch.

14. At Ignitioneering, the housing is cleaned in an environmentally friendly ultrasonic cleaner like you use for jewelry. Look at the result.

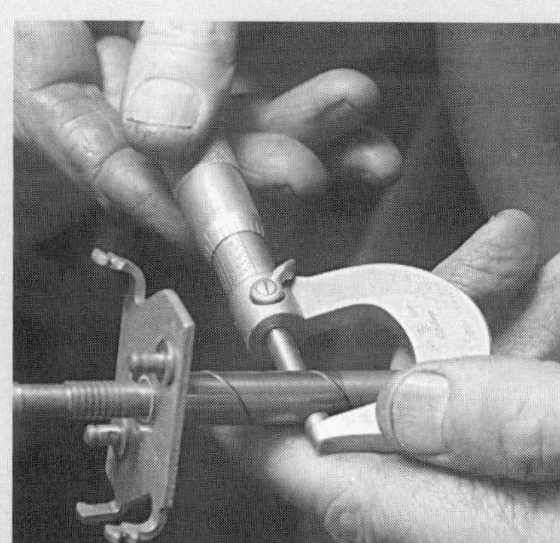

15. The distributor shaft is miked with a micrometer and checked for wear. Ignitioneering replaces the shaft if it's excessively worn. Bushings will be custom-sized for this shaft.

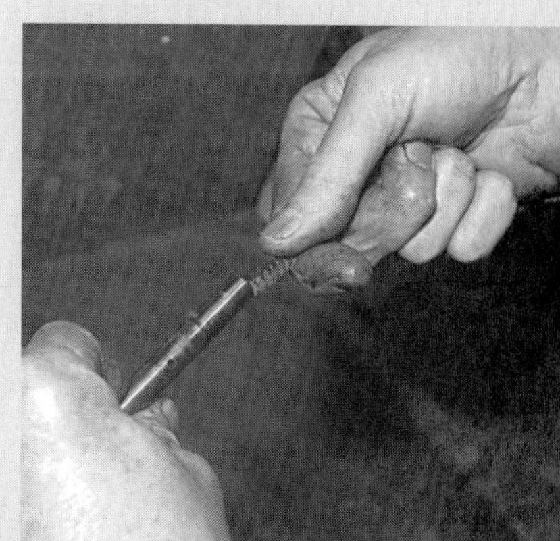

16. The shaft is cleaned up thoroughly with brake cleaner. A trip to the media-blasting cabinet makes the distributor housing like new again.

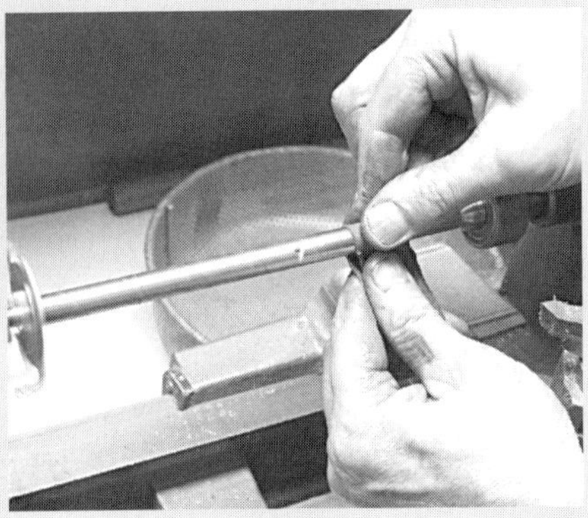

17. The distributor shaft needs two things: a good dressing up on the lathe with 400-grit paper and the tapering of tension pinholes with a drill as shown. Tapering makes tension-pin installation easier.

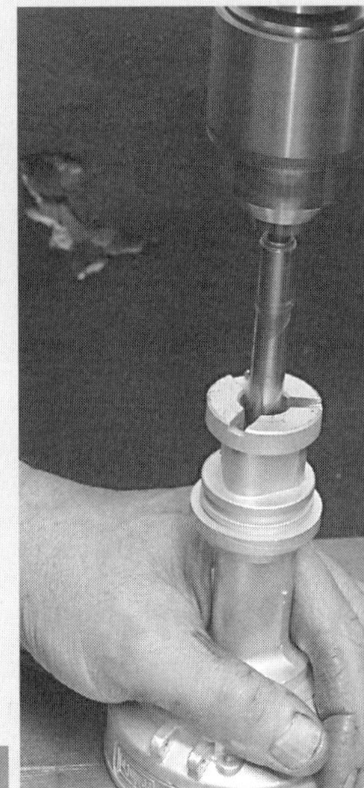

18. Because we're installing custom bushings in this Autolite distributor, the housing is reamed to the proper outside diameter to fit the bushings. Ignitioneering custom cuts new bushings for our Autolite distributor. They'll be wider bushings designed to support more shaft area.

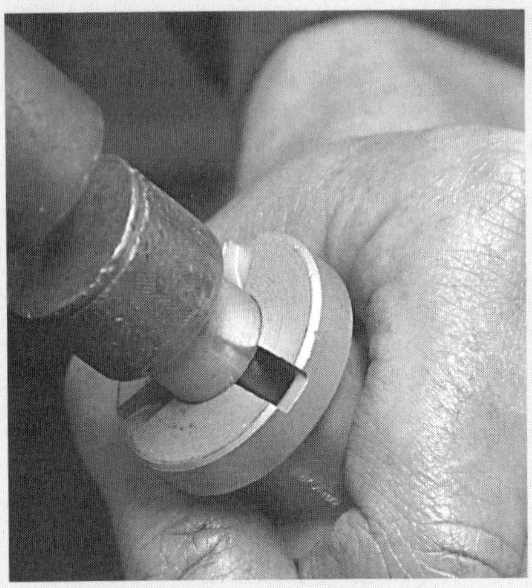

19. New bushings are carefully driven into the housing.

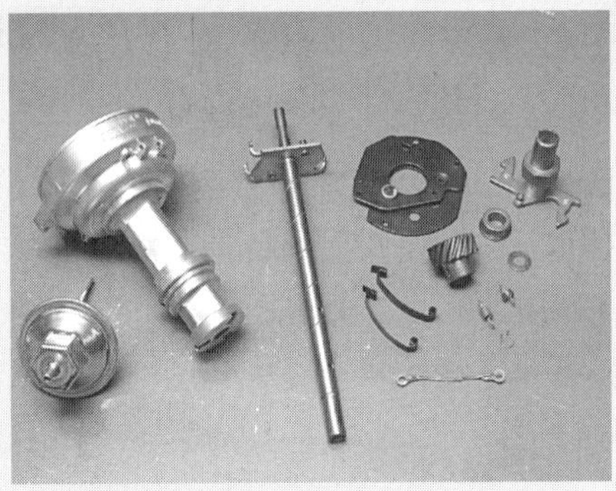

20. These are all of the elements that make up a quality distributor rebuild from Ignitioneering. It's the nicest quality we've seen.

21. The shaft is lubricated with engine assembly lube. Because the bushings are oil-impregnated, lubrication is integral and constant.

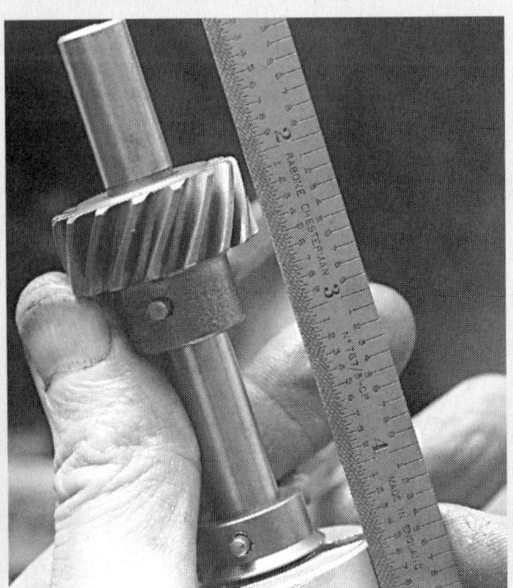

22. Shaft endplay and retainer location are determined here.

23. Ignitioneering uses larger tension pins at the drive gear and retainer for durability. Each is drilled out as shown, with larger pins installed.

24. The rotor-head thrust washer is installed next.

25. Centrifugal-advance weights slide into place along with wavy washers and C-clips. The wavy washers underneath each weight allow the flyweights to float freely. Note the lubrication used here.

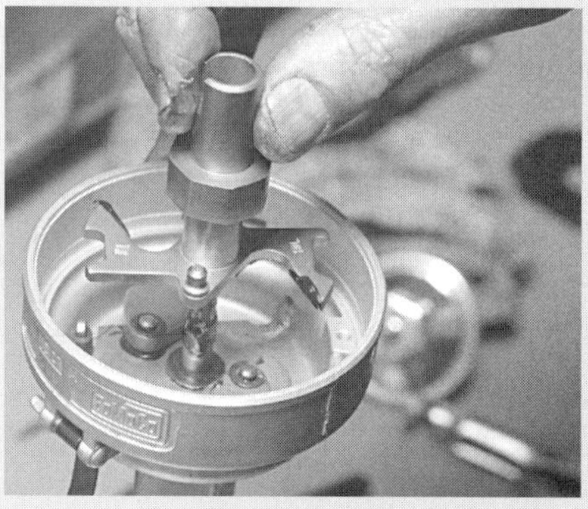

26. The rotor head is installed as shown along with the C-clip retainer inside. Don't forget to install the oil wick inside and give it lots of 30-weight engine oil. The rotor head is slightly modified to allow more advance travel. Ideally, we'll have 36 degrees of total advance at 3500 rpm.

27. Mechanical advance flyweight return springs are next. The springs have various tensions, depending on the rate of advance we're seeking. Stiffer springs delay the advance until a higher rpm. Softer springs allow the advance to come on sooner at a lower rpm range.

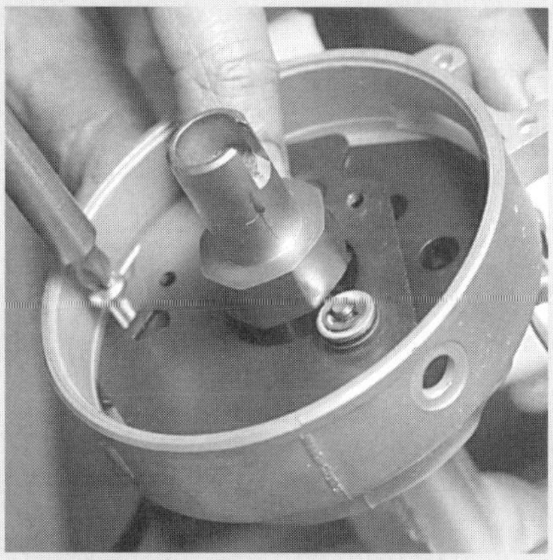

28. A new breaker plate is installed. This high-quality plate functions smoothly and accurately.

29. A new, fully adjustable vacuum advance is installed. Dual-advance units are available for '68 and up applications.

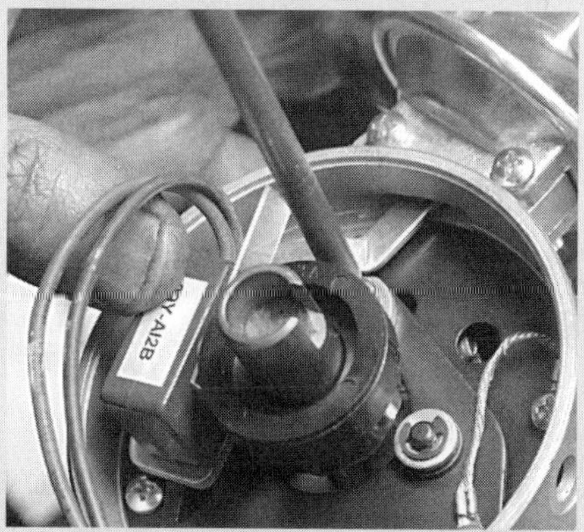

30. We're going with the new PerTronix Ignitor II ignition module. It's more advanced than the Ignitor (which is still available and more affordable) because it changes the dwell as engine revs increase. Call it a "smart" Ignitor.

31. The Ignitor II's red and black leads feed through the same opening as the original ignition wire. We suggest putting this lead inside heat-shrink tubing, which makes it look more original and keeps it protected. Don't forget to use the original ground lead inside the distributor. Without it, the Ignitor II won't work.

32. Using a good, old-fashioned distributor machine, Ignitioneering dials in the mechanical advance by swapping springs and watching where total advance comes in. We dropped this distributor into the 289 engine, set the timing, and ran the engine with excellent results.

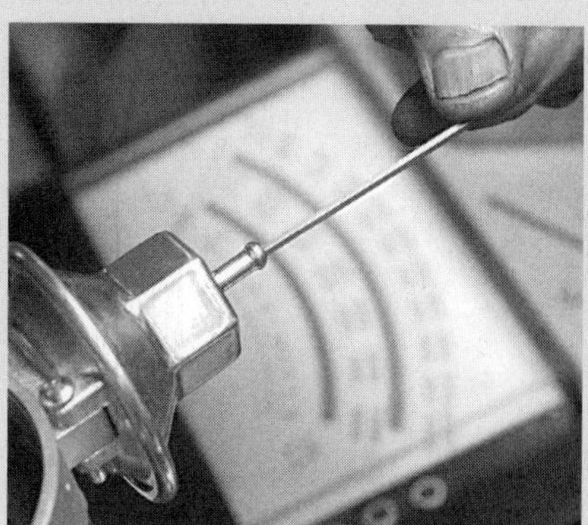

33. The new vacuum advance unit from Ignitioneering is fully adjustable. Turn the Allen wrench clockwise to slow the rate of advance. Go counterclockwise to speed up the rate. The vacuum advance gives us power right off idle.

Chapter 9

How to Replace a Sloppy Shifter

Text and Photos by Jim Smart

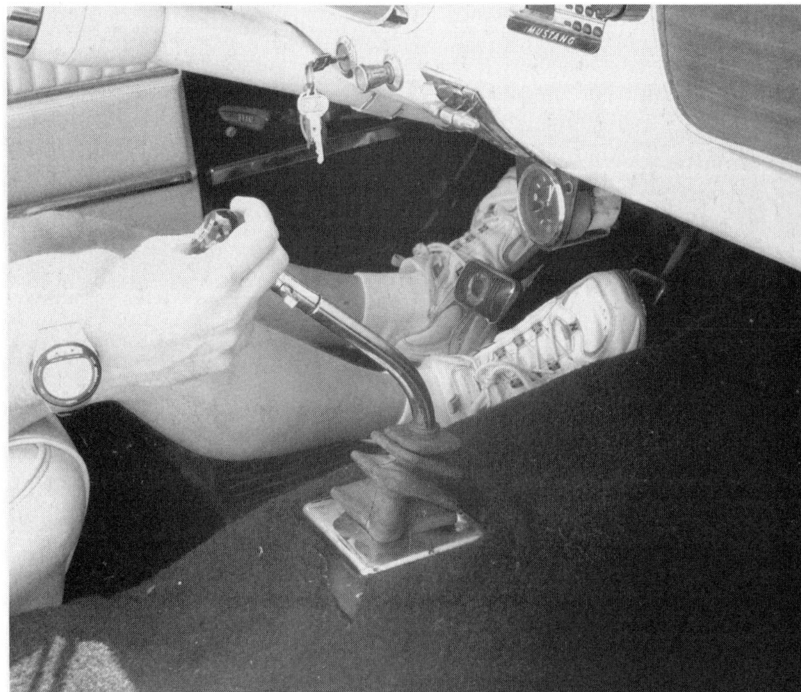

Tired of playing with that broomstick in a barrel? Here's how to fix your sloppy shifter problem.

Isn't it remarkable that a sporty car like the Mustang came with such a crummy shifter? We're talking about a vintage Mustang manual transmission, which had the sloppiest of factory shifters. Find a gear, if you can.

Thankfully, all of the parts to overhaul your '65–'73 Mustang shifter are available from Mustang mail-order companies like Mustangs Etc. If your shifter is sloppy, making it hard to find the gears, there are parts and steps to fix the problem in less than 30 minutes. If your four-speed's Reverse lockout is stuck from years of muck and dirt, like the one in Jeff Fischbach's '65 convertible, it can be cleaned and lubricated at the same time.

Source
Mustangs Etc.
www.mustangsetc.com

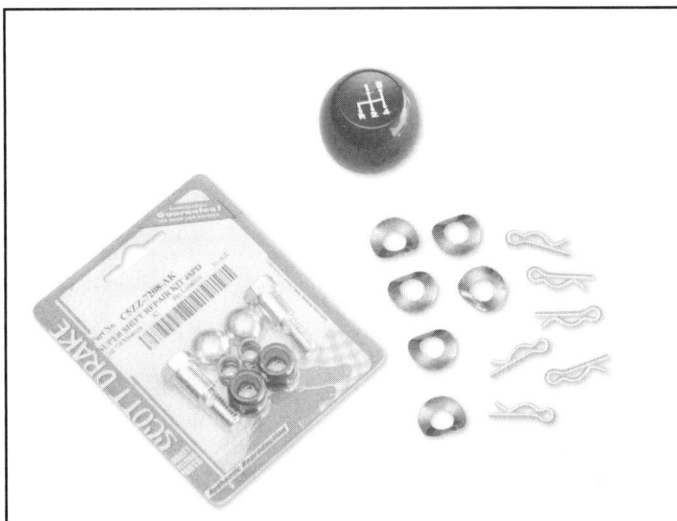

1. Here's everything you need to tighten up a sloppy factory Mustang shifter: the Scott Drake Super Shifter Repair Kit and wavy washers for the linkage underneath.

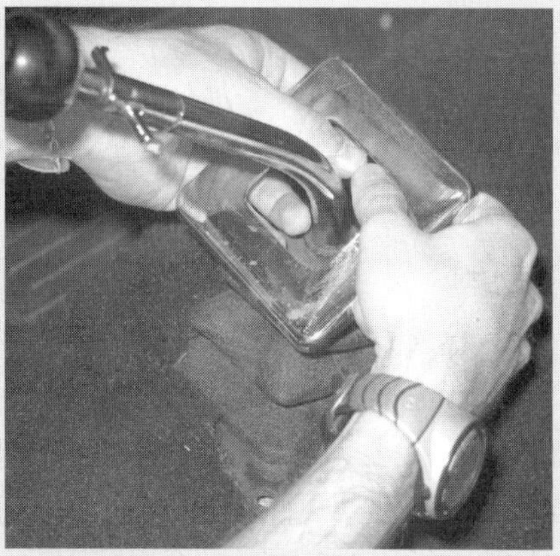

2. First, remove the trim work around the shifter.

3. Four bolts retain the shifter boot beneath the carpet. You'll need a 3/8-inch socket to remove them.

4. We're going to replace the shifter ball, which means this forty-year-old special comes off.

5. Two bolts hold the shifter handle in place. Use a 9/16-inch socket to remove them.

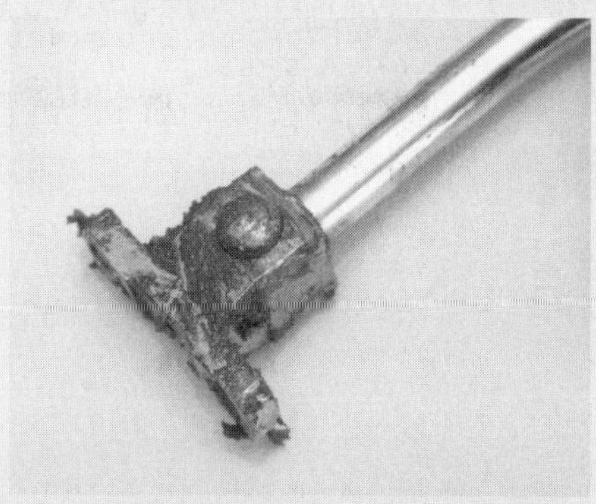

6. A closer look shows us what's wrong with this shifter. It seems worse than it really is. We need to clean this guy up, install new parts, and lubricate for smoother performance.

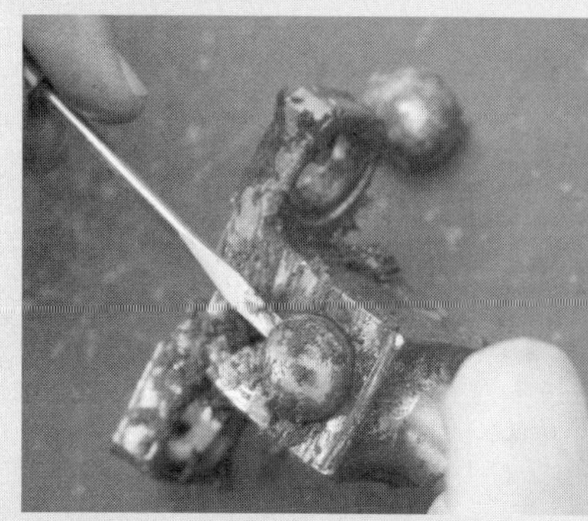

7. Jeff removes the shifter bearing cups, which have worn out with use. There are two cups, backed by compression springs. When the cup is removed, the compression spring pops out. This spring holds the cup against the shifter walls for accurate shifting.

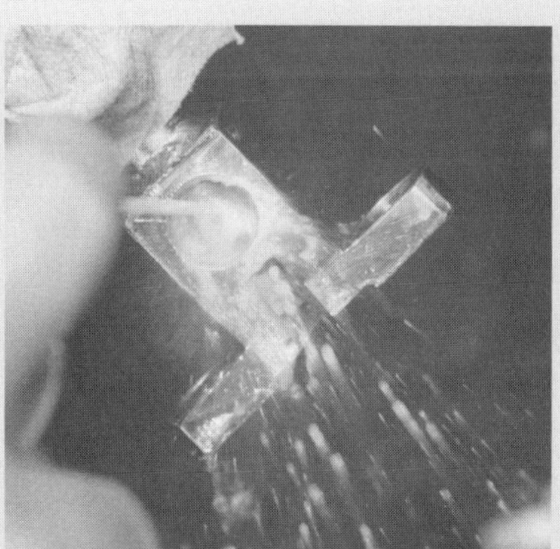

8. Brake cleaner is a terrific grease cutter. This, coupled with a wire brush, gets the job done nicely.

9. Use a spinning wire wheel for final cleanup. It removes any stubborn dirt and grease buildup that's become hard as a rock through the years.

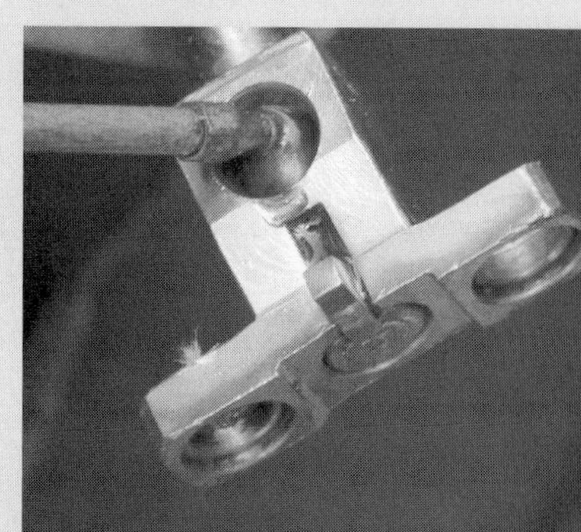

10. Automatic transmission fluid is an excellent lubricant for stubborn components because it has a detergent. It cleans and lubricates, loosening up a stuck Reverse lockout.

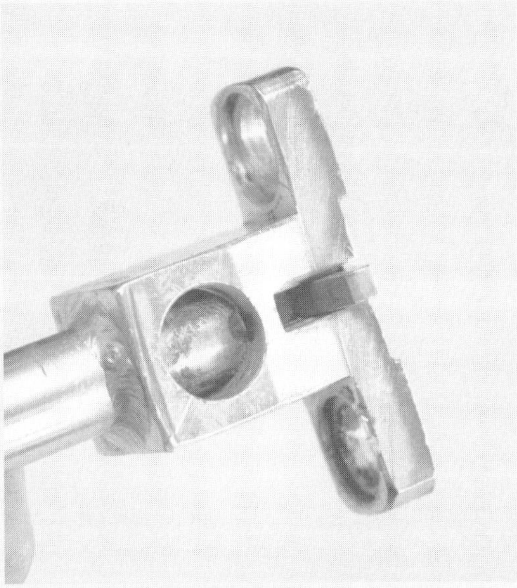

11. In Reverse lockout, the lockout tab should be in this position (four-speed only).

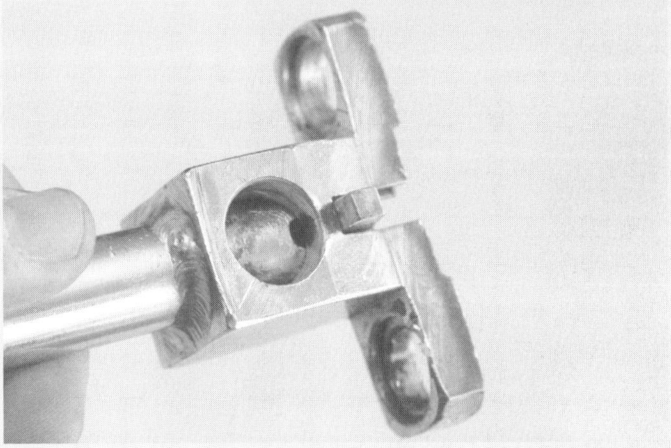

12. When you squeeze the Reverse lockout, this tab moves upward, which clears the way for the shifter to move into Reverse.

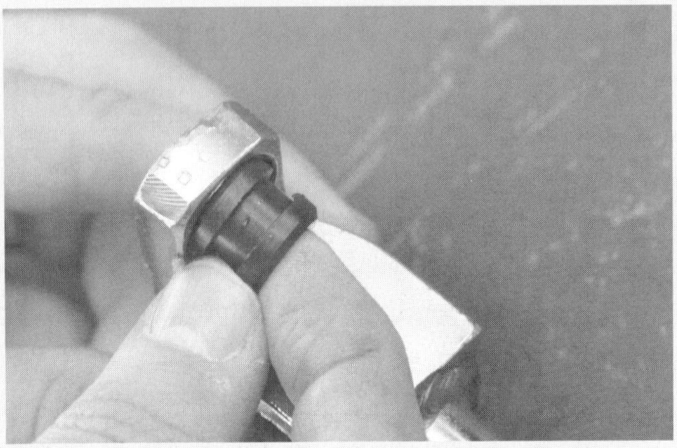

13. The Scott Drake Super Shifter Kit includes these rubber bushings that fit around the retainer bolts. The bushings fit into the handle.

14. White grease is an excellent lubricant and retainer. Pack each of the pockets with white grease.

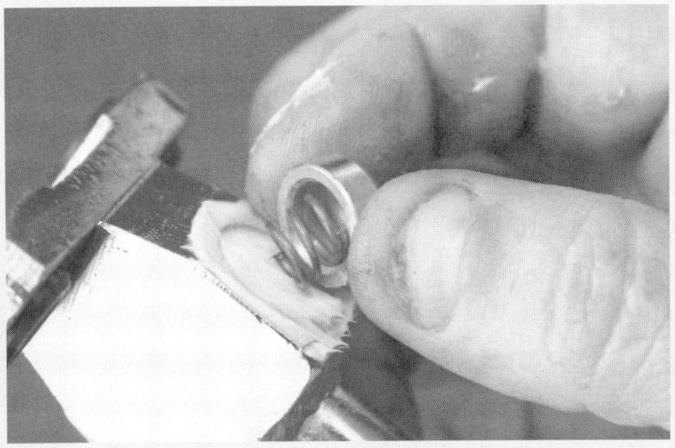

15. The compression spring and bearing cup fit into the pocket like this. There is one set on each side of the shifter handle. This spring and cup keep the shifter handle from rattling.

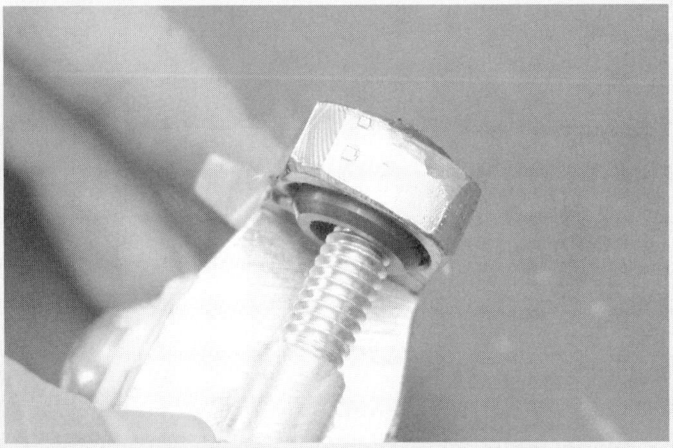

16. Bolts provided in the kit slip into the rubber bushings. You need two lock washers for these bolts, which aren't provided in the kit. Now the shifter handle can be wedged back into place. Sometimes, you need to wiggle the handle into place, making sure the cups and springs don't fall out. Tighten the bolts, then reinstall the boot, trim, and shifter knob.

Currie Enterprises restores and refreshes an original Mustang 9-inch rearend and incorporates a modern limited-slip differential at the same time.

Chapter 10

How to Build a 9-Inch Rearend

Text and Photos by Miles Cook

If you read our magazine regularly, you likely remember seeing our dark green '69 fastback. Our beloved old Mustang has developed into a Mustang Monthly project car of sorts. After having owned it for a little over a year, it's quickly turned into more than just a platform for how-to stories and component testing.

While we've had numerous '65–'66 Mustangs over the years (a half dozen at last count), this is the first of the truly timeless '69–'70 models we've owned, and it's turned into our all-time favorite vintage Mustang. Since it's our greatest love in nearly 20 years of early Mustang ownership (mainly due to the '69–'70 SportsRoof's spectacular styling) we've been on a quest to turn it into a viable daily driver as well as a long-distance road-trip car.

With various engine upgrades already done to the 351W and others planned, we're turning to the car's factory original 9-inch rearend. With a T5 five-speed already in place, we think it's a good idea to go through the rear axle. We want to get this critical component into tiptop shape before thinking about future long-distance journeys.

Our goal is a simple one: a basic but thorough freshening up of the axle and installation of a limited-slip differential while we're in there. Since this rearend is an original piece and is in rebuildable shape, we'll reuse the housing, third member, axles, and existing 3.00:1 gears. To complete the refurbishment, we'll check the housing for straightness and install fresh bearings and seals.

We'll show you how a top-notch rearend shop such as Currie Enterprises can take a tired 9-inch and get it back into like-new condition on a reasonable budget. We'll add Currie's excellent torque-sensing differential (TSD) in place of the non-limited-slip unit. Here are highlights of the procedure.

1. This group of axle housings just came from Currie's welding shop. Our original '69 Mustang housing was among the group on the far right. It was taken to the welding area in case any repair work was needed; everything was OK.

2. Checking the housing for straightness is next. An empty gearcase is installed to replicate an assembled rearend. Next, a perfectly straight steel rod is inserted into the housing.

3. These steel flanges fit over the rod, and exactly duplicate the bearing-race diameter. They will seat in the axle-tube ends only when the housing is perfectly straight. Though our housing is in generally good condition, it isn't totally straight, so next it visits a hydraulic press for straightening.

4. Up on the hydraulic press, carefully applied force is used to make any necessary changes to get the housing spot on and straight as an arrow. Years of experience means t doesn't take Ray Currie and the pros at Currie Enterprises long to produce a perfectly straight and true axle housing.

5. Next, the freshly straightened housing is placed into a rotisserie blasting cabinet for a much-needed steel-shot cleaning.

6. The results from the steel-shot cleaning are impressive. All rust and scale are blasted into oblivion to produce an axle housing that looks as good as new.

7. The preparation process isn't over yet. Here, the housing is cleaned in this pressure steam cabinet. While it cools off, we'll head to the tool bench to look at the center section being rebuilt.

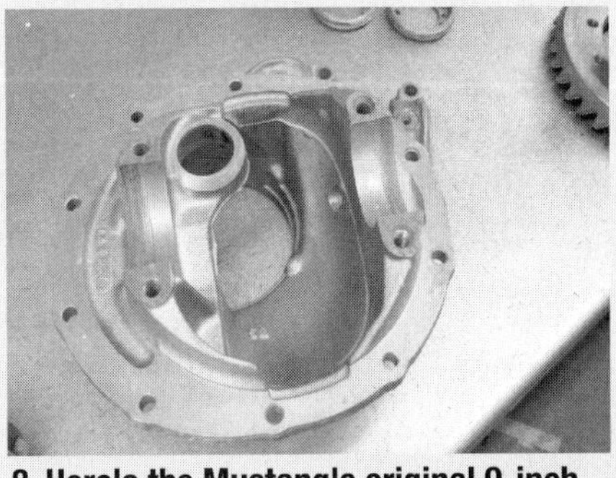

8. Here's the Mustang's original 9-inch gear case. Of course, it's a standard Ford gray-iron unit from 1969. All cleaned up after being taken down to the bare case, it's in perfect shape and ready to be the basis for the complete freshened-up rearend. While this gray iron casting is slightly more brittle that the Ford nodular-iron cases, it will be fine for our relatively mild street-driven car. While aftermarket cases are an option, they start at around $200 and go up to over $1,000.

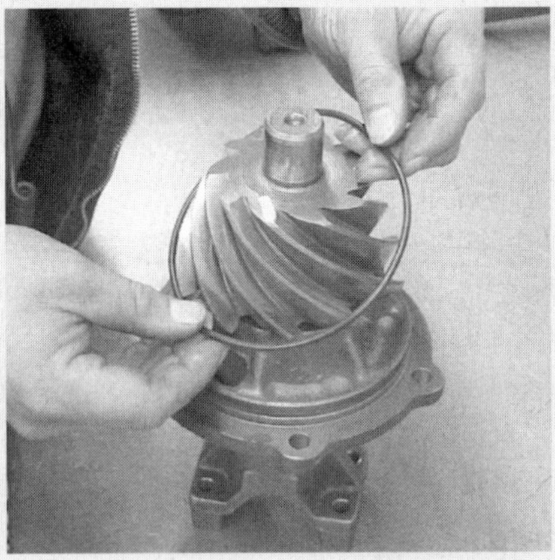

9. The 3.00:1 pinion gear with its bearings is installed into the pinion support. Here, the support gets a new O-ring before going into the gearcase.

10. The gearcase is placed in a special support on the bench and the pinion opening in the case is swabbed with grease. Here, the pinion assembly is lowered into position in the gearcase.

11. The open differential and old ring gear are being separated. We plan to reuse the car's original 3.00:1 gears, but we'll team them up with a new Currie TSD differential in place of the old and worn-out open diff.

12. For the majority of vintage Mustang owners, the Currie TSD is a great way to go. The case is of heavier construction than a stock Ford unit and it has internal clutches. It's a nice upgrade for Mustangs equipped with standard non-limited-slip (open) differentials.

13. The mating of the car's original 3.00:1 gears to the Currie TSD begins with an application of thread locker to the ringbolts.

14. After the ring gear is set in place on the TSD, each bolt is seated in place with an air-impact gun. Once everything is set in place, a torque wrench is used to tighten the bolts in sequence.

15. Next, on the hydraulic press, new bearings are pressed into position on the TSD differential unit.

16. Assembly of the center section continues with the installation of the TSD and ring gear. Here, the bearing races are installed over the differential bearings and are held in place as the assembly is lowered into the saddle on the carrier.

17. After checking to see that the bearing races have seated correctly, these special spanner nuts can be installed. These nuts are threaded on their outside surfaces and the gearcase and main caps are threaded to accept them.

18. When the main caps are installed, as is being done here, the spanner nuts are surrounded in a threaded bore and can be moved in and out to provide adjustment of the differential position within the center section.

19. A torque wrench is used to tighten the main cap bolts to spec. At this point, the spanner nuts have already been checked to ensure they turn and operate smoothly in their bores.

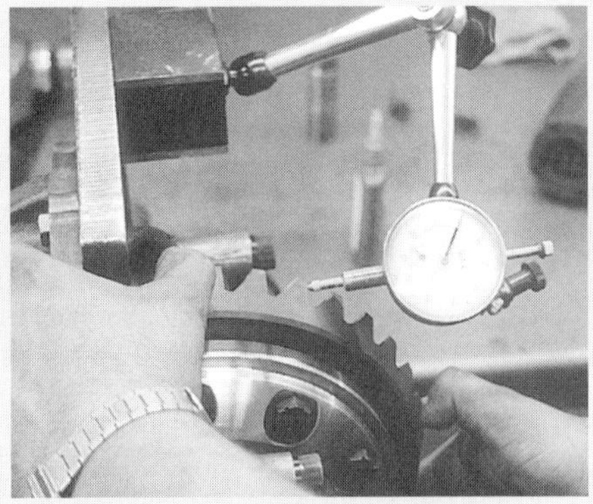

20. After assembly, a dial indicator is used to measure and set backlash in the gears. Backlash is checked and found to be in spec at 0.010 inch. No further adjustment is required.

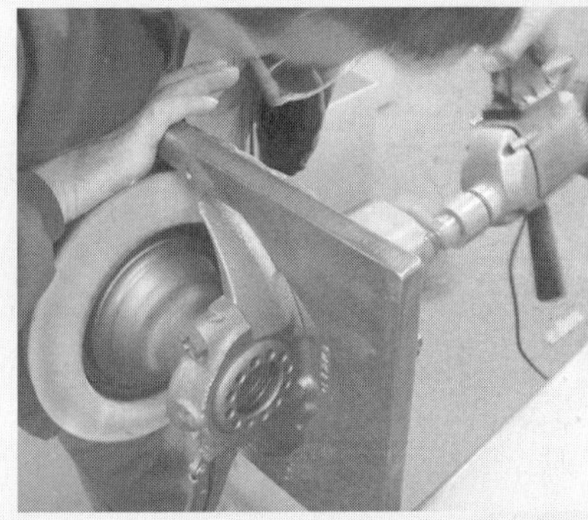

21. Next, the ring gear is coated with a special paint. When the gears are spun with a drill motor, the gear engagement pattern emerges.

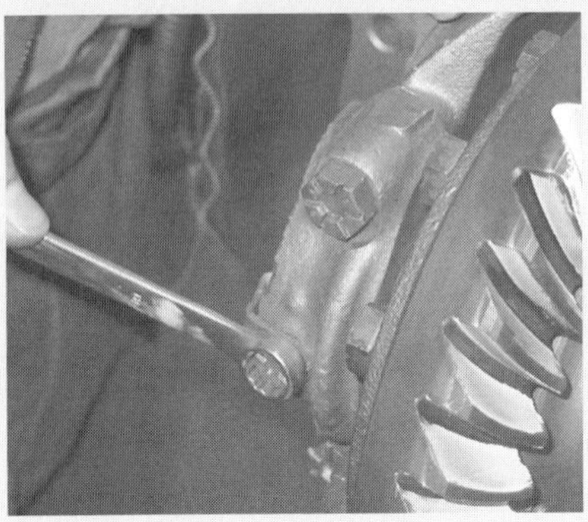

22. With the desired engagement pattern showing in the paint, the spanner nuts are locked into position, preserving the correct differential adjustment. The center section is now ready to go.

23. In the axle assembly area, the inside of the axle tubes are cleaned with a wire brush and a drill motor. This is done as a final precaution against any residual shot or other material. The axle housing is then blown out with compressed air.

24. All new and rebuilt Currie axles get this magnet. It's installed into the bottom of the housing to capture any ferrous material circulating in the gear oil.

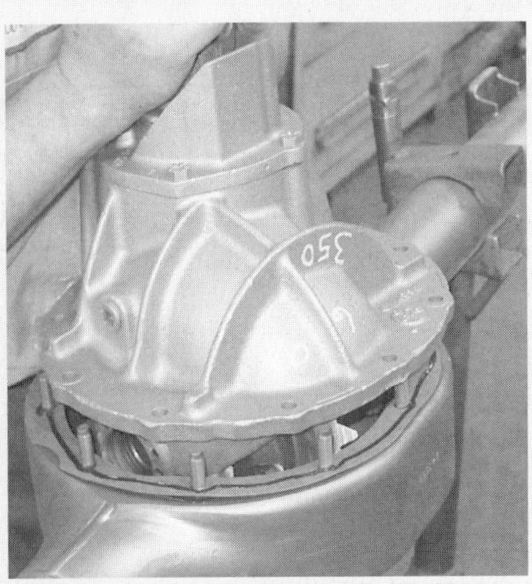

25. After a new gasket is installed on the housing, sealer is used and the center section is lowered into place.

26. New bearing races are installed in the axle-tubing ends and the unit is ready to accept the original 28-spline axles. This Currie-built 9-inch axle will be ideal for our '69 Mustang, and we feel good knowing we have an assembly that has been checked from top to bottom by experts who know the famous Ford 9-inch rearend by heart.

Go fast but look stock with a 347 stroker dressed out in factory trim.

Chapter 11

Swapping a 289 for a 347

Text and Photos by Donald Farr

The "kaboom," followed by the plume of smoke from under the hood, was the first indication the afternoon was quickly sliding downhill. After parking the overheating '66 GT hardtop in a convenient shopping-center parking lot, the second indication of a bad day came after allowing the 289 to cool down. While adding water to the radiator in an attempt to at least limp the car home, the trickling sound of a coolant leak at the back of the engine indicated a blown head gasket or freeze plug, neither of which was easily repairable in a parking lot.

Installed as a Fred Jones Remanufacturing long-block to replace a non-original and very tired 302, the little 289 had served the Mustang well for over two decades. Obviously, repairing the engine was an option, but the urge to replace the old, damaged engine with something a bit more muscular proved way too strong. So out came the 289, and in went a 347 Extreme stroker from Smeding Performance.

Replacing a 289 or a 302 with a stroker small-block is an outstanding way to gain performance in an early Mustang. While the inside dimensions are larger (58 more cubic inches when replacing a 289 with a 347), thanks to a 0.040-inch overbore and a longer stroke crankshaft, the exterior size is identical. For that reason, a 347 replaces a vintage 289 or 302 perfectly. Best of all, you can paint (black for '65 or Ford Corporate Blue for '66 and later) and dress the engine

The Smeding Performance 347 Extreme stroker arrived safely at Classic Creations of Central Florida in its crate shipping container. Request a delivery truck with a hydraulic lift if you or your shop doesn't have access to a pallet truck or hoist.

The 347 Extreme is equipped with Edelbrock aluminum heads and roller-rocker arms, which require tall valve covers or spacers for clearance.

Classic Creations extracted the 20-year-old Fred Jones Remanufacturing 289 from the '66 GT's engine compartment.

Placed side by side, you can see the similarity in exterior size. With the 289 out, we learned that a blown-out freeze plug had caused our problem. Others were leaking as well.

The Smeding Performance 347 Extreme comes equipped with an aftermarket balancer. Fortunately, it's marked on two sides, so by adding a vintage-style pointer to the driver-side timing-chain cover, the engine can be easily timed.

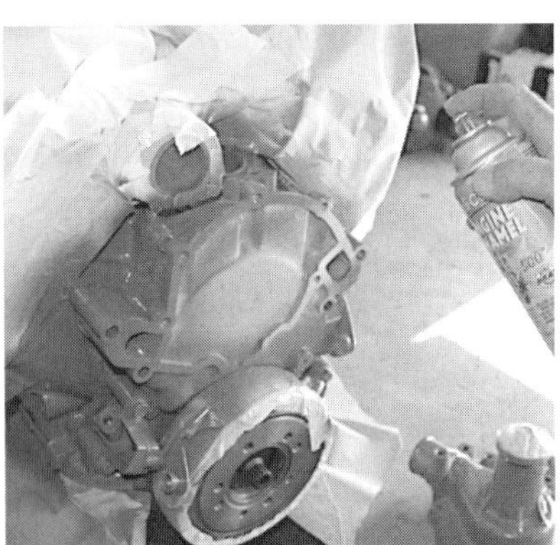

For a factory look, Merv Rego painted the 347 Extreme in Ford Corporate Blue. We could have gone for the all-out original appearance by painting the aluminum heads and intake manifold as well, but they were just too pretty to cover up.

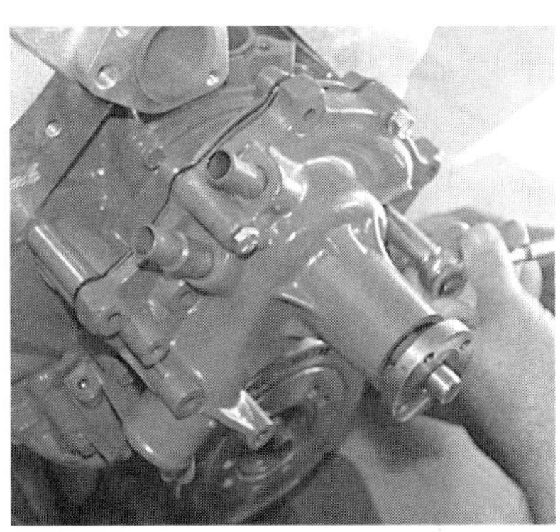

The new water pump, obtained from Virginia Classic Mustang along with all the other detailing items, was also painted Ford Corporate Blue before its installation on the 347. Components from the 289, such as the pulleys, were also transferred to the 347.

for a stock appearance or, as in our case, give it a vintage performance look with Cobra valve covers and an oval air cleaner.

For its 347 Extreme, Smeding Performance starts with a Ford Racing Performance Parts 5.0 block and fits it with a longer stroke (3.400-inch as opposed to the 289's 2.870-inch), nodular-iron crankshaft to obtain the 347 cubic inches. From there, Smeding utilizes 4130 forged I-beam connecting rods with 3⁄8-inch ARP Waveloc bolts and lightweight forged pistons for a 9.7:1 compression. The hydraulic roller camshaft sports 0.493/0.510-inch lifts with 212/222 duration at .050 inch. Edelbrock aluminum heads with roller-rocker arms are topped by Edelbrock's new Air-Gap aluminum intake. Basically, just add a carburetor, distributor, and accessories. Smeding says its 347 Extreme is rated at 380 hp, which will be a lot more fun than a stock A-code 289 rated at 225 hp.

Ben Smeding says he builds his strokers for torque. "Everybody looks at horsepower, not torque," Ben told us. "Two different engines could have the same amount of horsepower at 5000 rpm, but at low rpm, one could have substantially more torque, which is what gets the vehicle moving." He also says he uses brand-new blocks—a Ford Racing 5.0 in the case of the 347 Extreme—to avoid the issues that can plague used blocks, like rust in the water jackets and overbored cylinders. All Smeding crate engines are blueprinted and spin-balanced within one gram to ensure smooth operation and long engine life, and they come with a one-year warranty.

We wanted a stealthy, factory-look for our 347, so we placed an order with Virginia Classic Mustang for the engine accessories. While we were able to reuse some of the components from the old engine, like the starter and front-engine brackets, we decided to go with new radiator hoses and clamps, thermostat housing, water pump with AMK's new water-pump bolts, fuel pump, spark-plug wires and separators, battery cables, engine ground strap, and gaskets. While swapping components from the 289 to the 347, we discovered that the old dipstick tube wouldn't come out of the block, so Virginia Mustang sent us a new one.

We also requested a pair of black-wrinkle Cobra valve covers with a matching oval air cleaner for a single four-barrel, but those plans were dashed when we discovered the standard-size covers wouldn't clear the roller-rocker arms. While there was the option of using valve-cover spacers, we elected to reuse the "tall" versions of the Cobra aluminum valve covers from the 289, and painted the oval air cleaner to match.

Classic Creations' Merv Rego realized something was amiss with the 289's distributor when he attempted to extract it from the block for transfer to the 347. Turns out, the oil-pump shaft had somehow twisted itself inside the distributor shaft, damaging the hex opening beyond repair. We tried to locate a replacement distributor at local parts stores, but the remanufactured housings we found were from the '80s, with an access opening that didn't match our PerTronix wiring. A call to Mike Ulrey netted us a '60s small-block distributor from his stock of vintage distributors. Merv equipped it with the PerTronix Ignitor, and we were back in business.

Be aware that the 347 uses the late-model 5.0 firing order: 1-3-7-2-6-5-4-8; not the 289's 1-5-4-2-6-3-7-8. Ask us how we know.

With our Performance Automatic-prepped C4 transmission, we didn't have to worry about adapting the late-model 5.0 block for clutch linkage. Because the late-model Ford Racing 5.0 block will not accommodate the pivot stud for the vintage-style equalizer bar, owners with manual transmissions will need the adaptor bracket offered by many vendors.

With new motor mounts in place, the 347 dropped neatly into the Mustang's engine compartment. Everything from the 289, including the Hooker headers and A/C lines to the Sanden compressor, bolted right up. Painted Ford Corporate Blue and with vintage-style Cobra valve covers and air cleaner, no one will know it's a 347 in place of a 289.

Sources

Classic Creations of Central Florida
www.classiccreationsfl.com

Smeding Performance
www.smedingperformance.com

Virginia Classic Mustang
www.vamustang.com

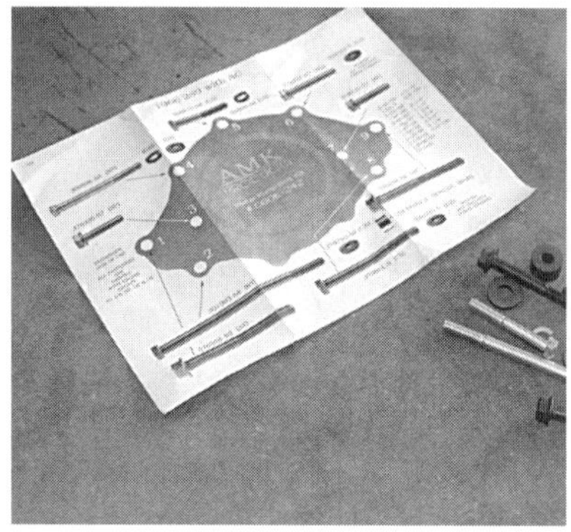

Virginia Mustang recommended a Water Pump Bolt Kit from AMK Products. Available for a variety of applications (A/C, non-A/C, and so on), the new bolts, washers, and spacers are much better (and easier to install) than the old, rusty bolts. AMK even includes a sheet to show where each bolt goes.

Since we're not planning to run at high rpm for any length of time, we went with a reproduction fuel pump from Virginia Mustang.

The 347 Extreme comes with a late-model timing pointer on the passenger side of the engine. With the vintage water pump installed, you can see the interference with the radiator hose inlet.

There was trouble in paradise when the distributor, which we planned to reuse with its PerTronix conversion, proved troublesome to remove from the 289. Turned out the oil-pump shaft had distorted the hex opening in the distributor. We soon learned that the rebuilt distributors at the local auto stores are typically from the '80s, so we located an original '66 distributor.

We originally intended to use Cobra valve covers with a black wrinkle finish, along with a matching oval air cleaner from Virginia Classic Mustang. However, interference with the roller-rocker arms proved to be a problem, so we ended up switching back to the tall Cobra valve covers, available from many Mustang vendors, from the 289.

To prepare the engine compartment for the 347 stroker, the Hooker headers were secured safely out of the way. We also used the opportunity to install new motor mounts. It's a great time to clean up and repaint the engine compartment as well.

With just some minor jostling to line it up with the motor mounts and Performance Automatic C4, the 347 slid perfectly into place. Merv says it was one of the quickest installs he's ever done.

Proving the 347 is a perfect swap for a 289, the headers and other components fell right back into place on the larger-displacement small-block.

Ben Smeding said a 600-cfm Holley is perfect for the 347 Extreme, so we swapped the carb from the 289 onto the Edelbrock Air-Gap intake.

We buttoned up the engine install with factory-style items from Virginia Mustang, like this date-coded upper radiator hose.

Before starting the 347 stroker for the first time, Classic Creations' Merv Rego wisely pre-oiled the fresh engine.

It looks stock, but it's definitely not with 347 cubic inches, Edelbrock aluminum heads and Air-Gap intake, and a performance camshaft. But, with factory look-alike accessories from Virginia Mustang, our 347 could pass for a Cobra 289 from the '60s.

SUSPENSION
&
CHASSIS

Chapter 12

A '65–'66 Power Steering Conversion

Text and Photos by Mark Houlahan

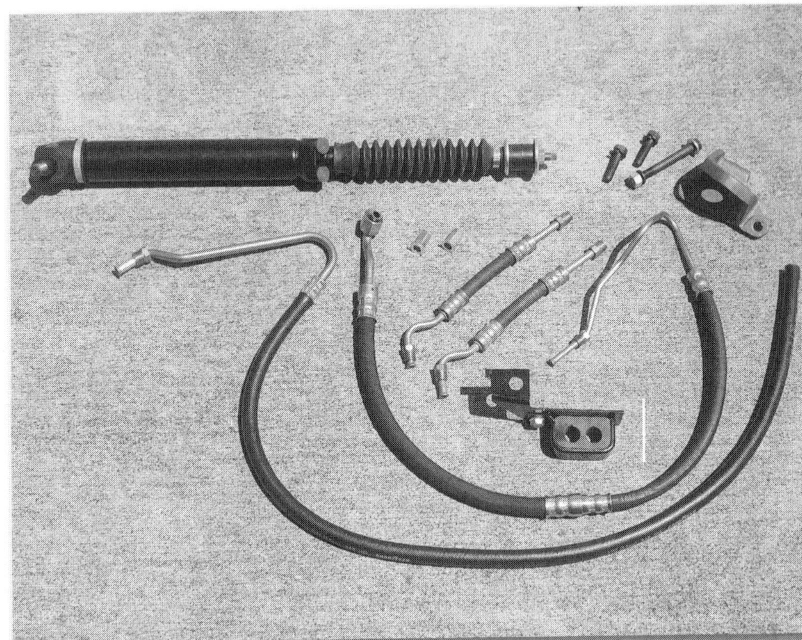

Your money is going to be spent on hydraulic parts, as converting to power steering isn't cheap. Even if you score a complete used setup from eBay for $200, most likely you'll need all new hoses, tie-rod ends, and a rebuild or reseal of the major components. If you want to purchase everything down to the last bolt, already rebuilt and ready to go, plan to write a check for around $1,500.

It would be great to hop in a time machine, go back to 1966, and purchase your Mustang exactly the way you want it. Of course, that's not possible, so owners often take small liberties during a restoration to add options and conveniences that didn't originally come with the car. Unless you have a rare Mustang or you're restoring your car to the letter of the buildsheet, adding power brakes, power steering, a console, Styled Steel wheels, and so on isn't such a bad thing. If the addition is done properly, using all the correct parts, can anyone really complain? We're talking safety and/or driving comforts of the 21st century.

Our Project '66 hardtop is one of over several hundred thousand hardtops sold in the '65–'66 model years. There's nothing particularly rare about it, and with so many hardtops made, taking a few liberties here and there shouldn't be frowned upon. We are, after all, restoring the car and saving yet another Mustang from extinction, whether you like the modifications or not. It was decided right from the beginning the '66 would have power steering, just like the original '66 I owned 14 years ago.

The conversion to power steering is literally a bolt-on project. Approximately 90 percent of what you need to perform the conversion can be purchased new or rebuilt. The few things you might need to purchase used or find from a salvage facility include the power-steering-pump mounting bracket(s) and the hydraulic ram frame-mount bracket. Working with Metro Mustang for hard-to-find used parts, we located the correct power-steering-pump mounting bracket, the frame bracket for the ram, a pump core, and even the pulleys for the engine. We also found a good '66 power-steering center-link core at Metro Mustang, which was restored by Alamo Classic Mustang's Precision Products, using a Stainless Steel Brakes Corporation control valve. We also used one of SSBC's new hydraulic slave cylinders, along with concours hoses from National Parts Depot. Last, we rang the Bramlett brothers at Mustangs Plus for the correct V-8 power-steering tie-rod ends, adjusting sleeves, and idler arm.

Sources

Alamo Classic Mustang
www.alamoclassicmustang.com

Metro Mustang
www.metromustang.com

Mustangs Plus
www.mustangsplus.com

National Parts Depot
www.nationalparts.com

Stainless Steel Brakes Corp.
www.ssbrakes.com

1. While not necessary, we opted for the correct-tagged steering gearbox for a V-8 with power steering because our old gearbox was shot. The steering wheel and column will need to be removed for gearbox R&R. Remove the center cap or horn ring, then remove the retaining nut and steering wheel.

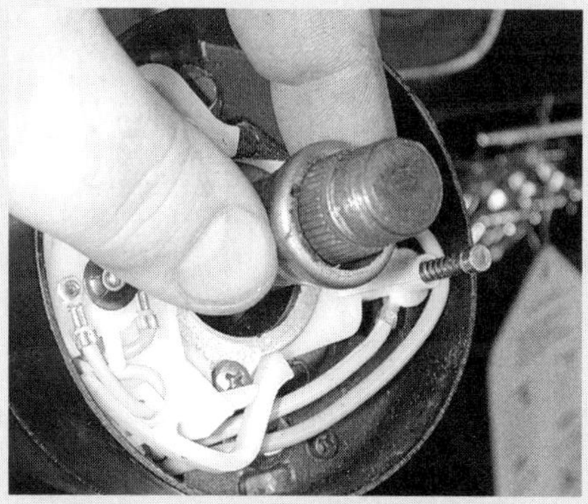

2. Once the wheel is off, remove the upper-column taper bearing. The column tube will not slide up with this in place. Use a small screwdriver to gently pry it free.

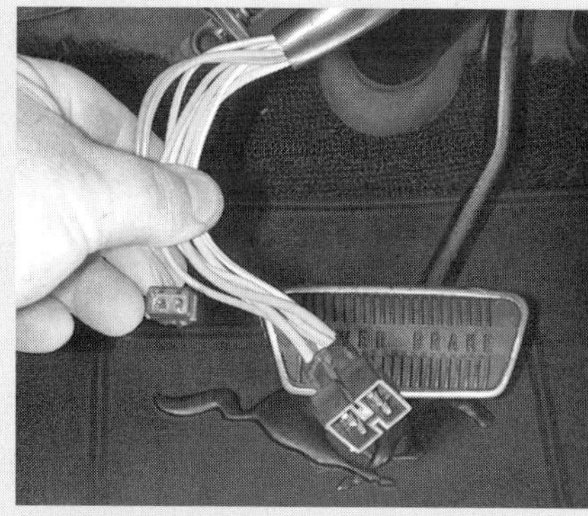

3. At the base of the column, unplug the wiring connectors for the turn signals and horn. Once these wires are free, remove the two nuts that retain the column tube clamp to the dash frame.

4. With the column-tube retaining bracket free, the column tube will slide straight up and off the gearbox's long steering shaft.

5. Since we're getting rid of the steering box and all of the manual steering components at once, we left the major components assembled to each other and removed everything as a unit. The six-cylinder manual-steering idler arm (from the frame) and the passenger-side outer tie-rod end from the spindle are the first to be pulled.

6. On the driver side, simply disconnect the outer tie-rod end from the left spindle.

7. Three bolts attach the steering gearbox to the left front framerail. Once these bolts have been backed out sufficiently, the gearbox will be free, so watch the steering shaft to make sure it doesn't scratch the dashboard.

8. Working carefully, slide the steering gearbox forward to clear the dash, and carefully fold the manual linkage in such a manner to allow removal of the complete system in one piece. If you have a V-8 in place and still plan to swap the gearbox, be prepared to jack up the engine and remove the driver's motor mount for room.

9. The manual linkage is shown here with the new/rebuilt V-8 power-steering linkage we temporarily assembled for comparison. The power-steering linkage mounts in the same locations, but it has provisions for hydraulic slave-cylinder mounting and the power-steering control valve.

10. The rebuilt V-8 power-steering gearbox came to us refinished in a natural finish, so installation required nothing more than lining up the holes and reinstalling the three mounting bolts. At this time, you can reinstall the steering-column tube and steering wheel.

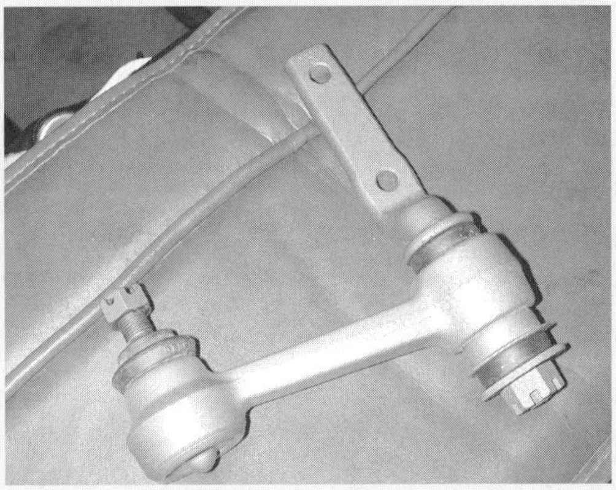

11. The new V-8 idler arm attaches differently to the center link and to the idler-arm mounting bracket. This mounting difference requires a new (or good used) idler-arm mounting bracket. Don't tighten the lower castle nut until the car is aligned; otherwise, the idler arm may pull the steering in one direction or another instead of trailing at an "idle" going down the road.

12. The pitman arm (the curved arm at the base of the steering gearbox, left) is specific to V-8s, so we had to find one at Metro Mustang. Once the pitman arm is slid into place, the retaining nut and lock washer can be secured. The rebuilt power-steering control valve and center-link assembly is then attached to the pitman arm and idler arm (right).

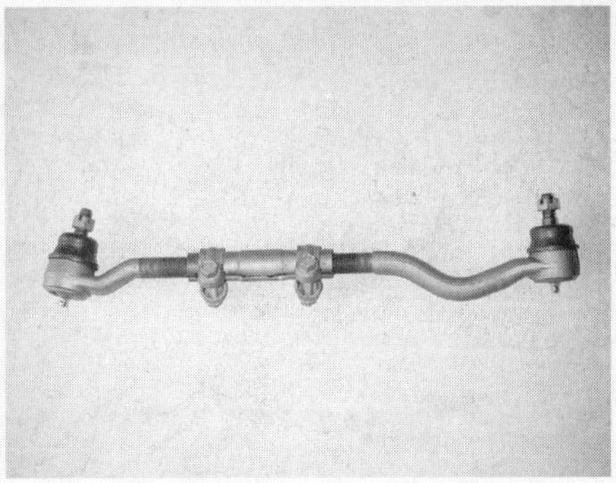

13. The power steering conversion requires different tie-rod ends; since our steering was completely shot, we ordered new inner and outer tie-rod ends and the proper adjusting sleeves for both sides from Mustangs Plus.

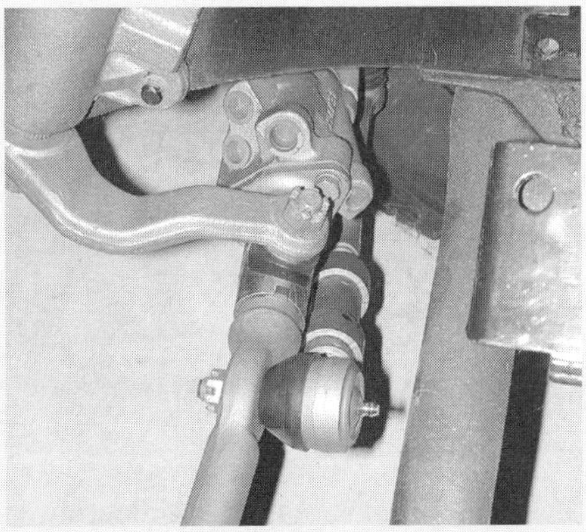

14. The tie-rod assemblies attach to the center link and spindles with the provided castle nuts. Notice that none of our attaching points have cotter pins installed. We'll wait until our project is aligned and all of the castle nuts torqued properly before we secure them with cotter pins for safety. Just don't forget to install your cotter pins.

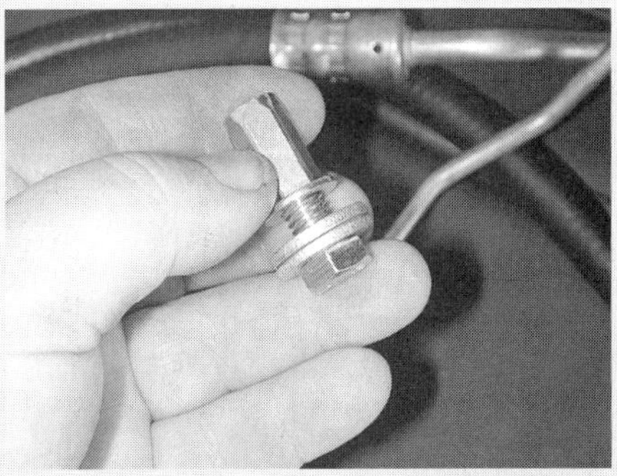

15. To mount the hydraulic slave cylinder, the slave-cylinder mounting bracket will have to be added to the left framerail. These crush-nut inserts (two required) will be installed first. Use a bolt and two washers as a bearing surface to install these into the frame. Ours went right in and collapsed evenly, but we've heard of some that had to be welded to the frame.

16. The cleaned and painted slave-cylinder mounting bracket is installed using correct AMK Products hardware from National Parts Depot. The AMK fasteners save time because the hardware is the correct length, strength, and design.

17. Our new Stainless Steel Brakes slave cylinder is prepared with a boot kit, new mounting bushings, and new pressure hoses. These small hoses often cause problems if they are installed crossed, causing the steering wheel to spin back and forth on its own. Remember: The hoses do not cross, but go straight from top port to top port and nestle together when tightened.

18. The slave cylinder is attached to the center link via a castle nut. This nut is not provided with the new slave cylinder, so you'll have to score one from Metro Mustang too.

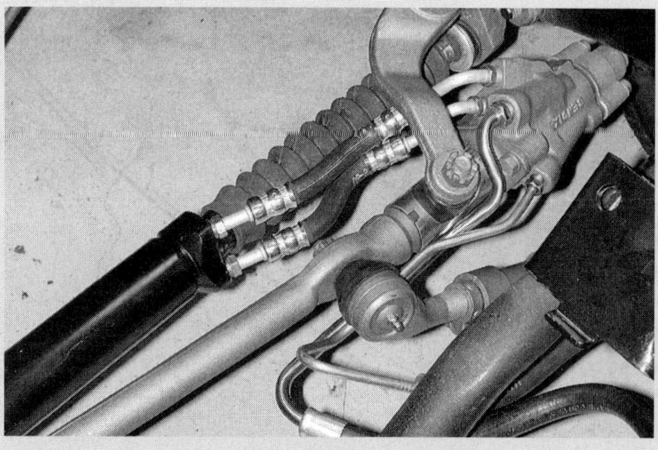

19. Once the small hoses from the slave cylinder are routed and installed into their respective ports on the control valve, all that remains are the main pressure and return hoses between the pump and the valve itself. These hoses route in front of the control valve and under the inner tie-rod end as shown here. Ford used a special clamp to secure the hoses to the valve to prevent rubbing and vibration, but the clamp (as far as we know) isn't available. For daily drivers, a regular worm-drive hose clamp will work.

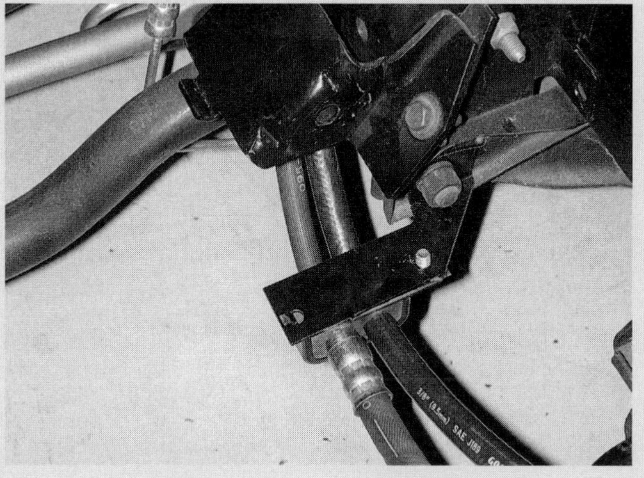

20. The last detailing touch is the reproduction power-steering hose-retaining bracket. This bracket attaches to the lower control-arm pivot bolt as shown here, and supports the hoses as they are routed to the power-steering pump mounted on the engine. The pump and pump bracket are not shown, but are added later.

Chapter 13

How to Replace Power Steering

Text and Photos by Jim Smart

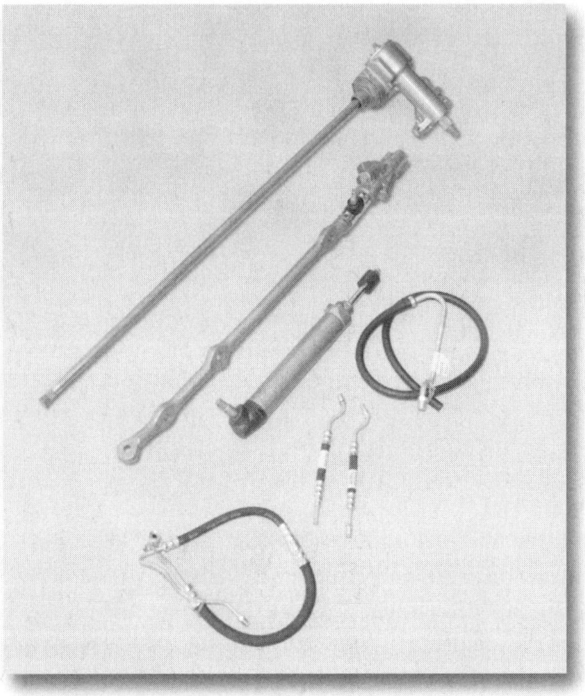

This is the complete power-steering system from Mustangs Plus. It includes a 16:1 steering box, a control valve, a power ram, and pressure and return hoses. The Bendix control valve has been completely rebuilt and tested, and the result is precision, leak-free operation.

Vintage Mustang power-assisted steering doesn't have the best reputation for reliability and easy servicing. The old Bendix power-steering systems, as used on a wide variety of classic automobiles, including Chevrolet's Corvette, usually wear out in a big way with leaks and erratic operation. You can rebuild them, which can be hit-and-miss depending on what's wrong with the system. Or you can order a complete remanufactured Bendix power-steering system from Mustangs Plus.

To do the job right, you should replace every part of the power-steering system. You'll need a control valve, a power ram, two directional pressure hoses, one pressure hose, and one return hose. In some cases, you'll need a new pump. Replace the pump if the fluid appears burned or there is leakage. It's generally a good idea to replace the pump when you replace the entire system. Contaminated fluid in an old pump can ruin new components.

Because power steering is a high-pressure hydraulic system, it runs decidedly hot. And because it experiences extremes of stress under a wide variety of conditions, components wear out, and seals and hoses get hard. When seals become hard, they don't seal properly, which causes fluid leakage and erratic operation. Old hoses can rupture without notice, as many of our readers have experienced.

With heat and pressure also comes hazardous operation. Your Mustang's power-steering pump makes approximately 1,800 pounds per square inch (psi) of fluid pressure, which can cause injury. Be careful when you're working with hydraulic hoses, control valves, and rams.

From 1965 to 1970, Mustangs had power-assist steering, which utilizes hydraulic pressure to help move the steering linkage left and right. A hydraulic ram (also called a power ram) applies pressure on the steering linkage. The ram works in two directions: left and right.

The power-steering pump sends fluid under pressure to a control valve, which is tied to the pitman arm off the steering gear. When we steer, we move the control valve ball, which slides a valve piston back and forth inside the control valve. Steer left and you move the control-valve piston to the right, which directs fluid under pressure to the appropriate side of the ram piston. Pressure on the ram piston moves the steering linkage to the right, which helps us steer left. When we steer right, the opposite happens.

Classic Mustang power steering is little more than manual with hydraulic assist to make steering easier. Aside from a difference in steering gear ratio, power-assisted steering isn't much different from manual.

Mustangs Plus provided everything we needed to renew the dated and worn-out power-steering system on a '65–'66 Mustang. We also opted for a remanufactured 16:1-ratio steering gear while we were in the neighborhood. These new components make steering easy and improve our safety.

Source
Mustangs Plus
www.mustangsplus.com

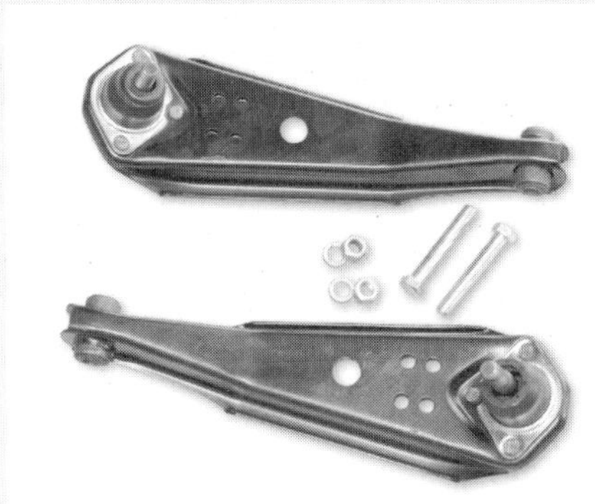

1. While we're at it, we'll replace the lower control arms because the ball-joint dust boots and bushings have rotted badly. These guys get a grease zerk at each ball-joint grease cup.

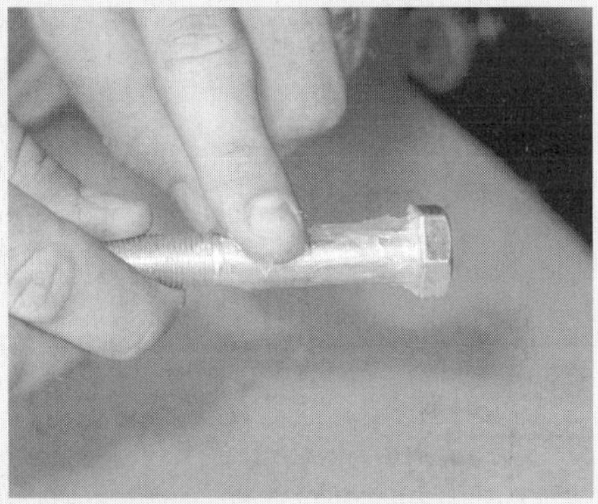

2. New AMK Products hardware is greased thoroughly with chassis lube. This is the lower-control-arm pivot bolt, which comes with a lock washer and nut.

3. New Grab-A-Trak lower control arms are fitted into place and tightened.

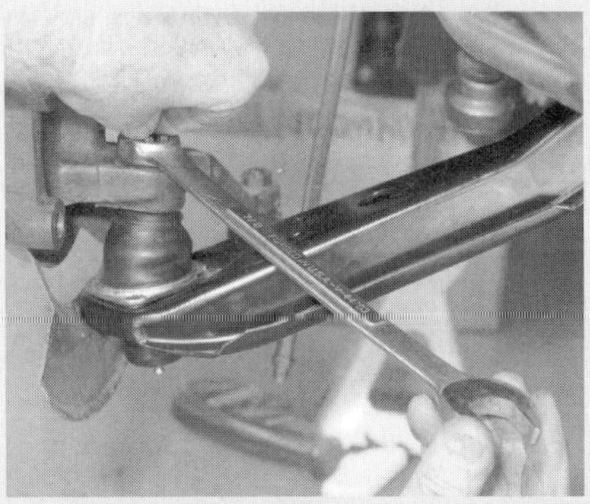

4. The new lower control arm is attached to the steering spindle as shown (left). The nut is completely tightened and the cotter key installed. Your cotter pin should look like this (right). Finally, install the front stabilizer-bar end links.

5. Our remanufactured and adjusted 16:1 steering gear is installed next. This is tricky business, especially if the engine is in the car. For '65–'67 Mustangs, we have to deal with a long steering shaft that has to be aimed through the firewall and below the dashboard. Three bolts through the framerail retain the steering gear. Make sure they're tight.

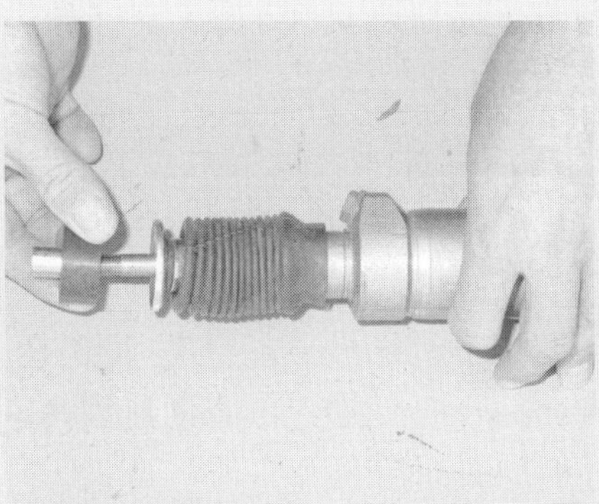

6. Next, prepare the power ram for installation. The ram dust boot goes on first. A lip at one end of the boot is secured at the ram tip. Slide the boot retainer clamp onto the end of the boot, but don't secure it just yet. Next, install the flange and rubber bushing.

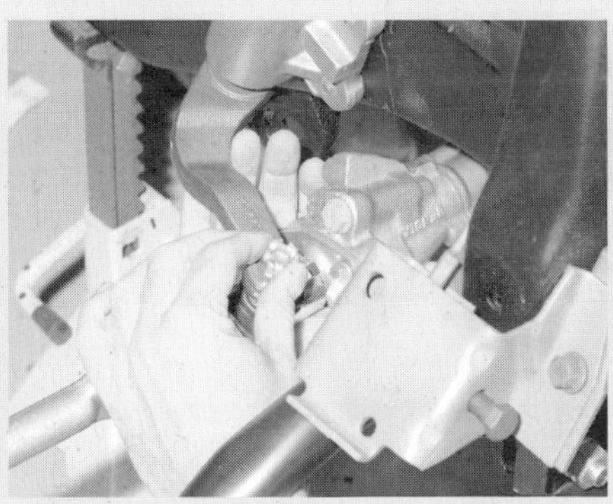

7. For '65–'66 Mustangs, the control valve and center link are one long assembly. Secure the control valve as shown at the pitman arm. Secure the other end at the idler arm. The '67–'70 control valve screws onto a power-steering-specific center link.

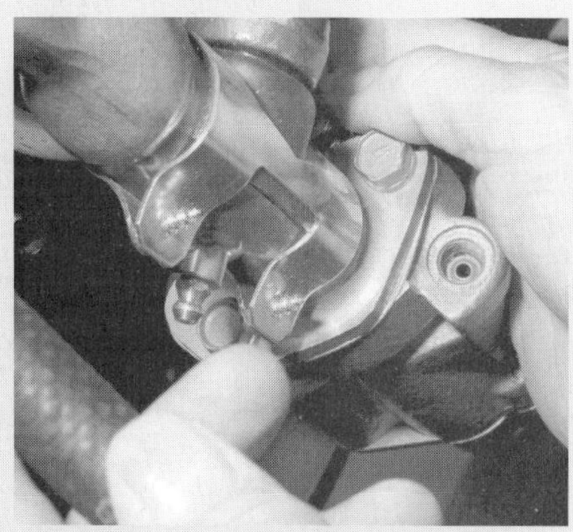

8. The power-steering control valve return line installs like this: The larger fitting on top of the control valve ties into a long low-pressure hose, which vectors fluid back to the pump reservoir.

9. The high-pressure hose from the pump to the control valve threads into this port.

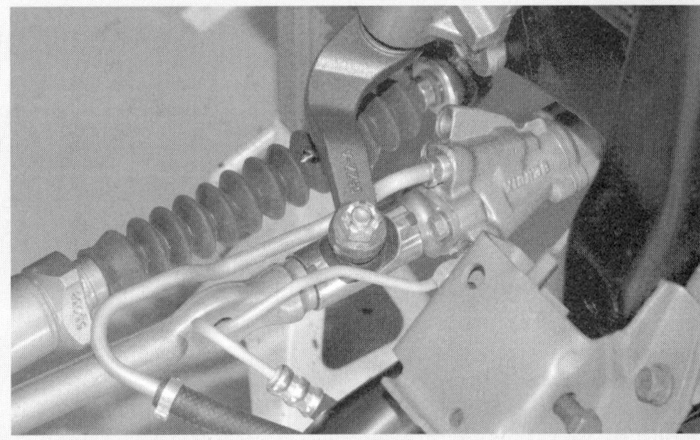

10. When both the return and pressure hoses are installed, they should look like this. Line routing may be different from what you see in the Ford shop manual. This is an improved line path compared to original equipment.

11. The return and pressure hoses from the power-steering pump are routed through this insulated bracket. The restrictor on the pressure side is fitted into the rubber insulator before the lines are secured by the bracket. Make sure you have plenty of flexibility in both lines between the insulator and control valve. You want these lines to move smoothly with the control valve and linkage.

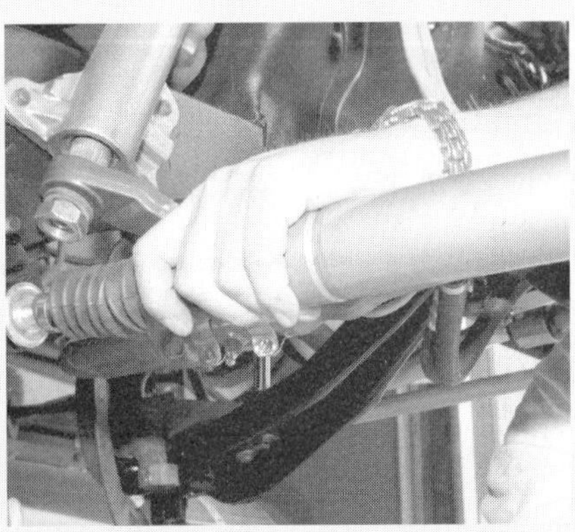

12. Next, the power ram is secured at the left framerail bracket. If you're running long-tube headers, you'll need a Hedman power-ram bracket extension to replace the factory bracket. This moves the ram away from the hot header tubes.

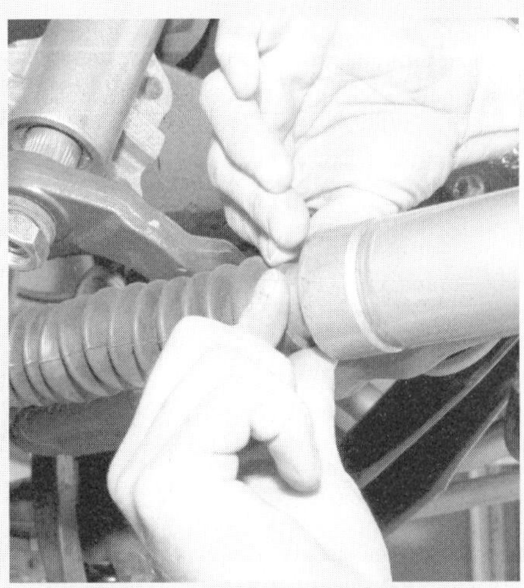

13. Check the ram dust boot for proper security once the ram is installed.

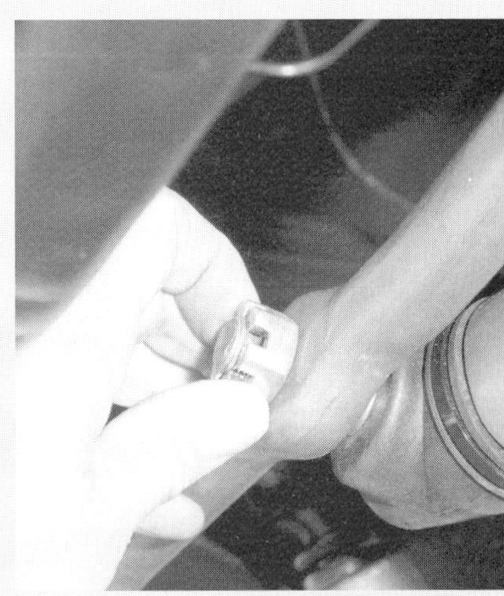

14. The power ram is secured at the center link as shown. Tighten it and secure the castle nut with a cotter pin.

15. Crimp the dust-boot clamp at the bracket end with a pair of diagonal cutter pliers. Do not overtighten. Crimp just enough to secure the clamp.

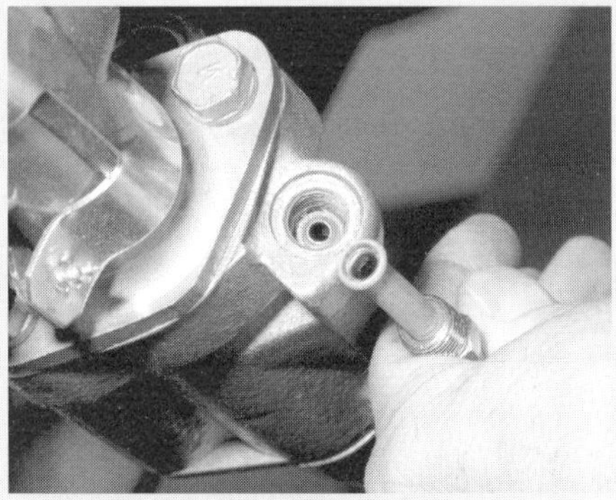

16. Bendix power steering gets a bad rap for leaks, but not necessarily because it's a bad system. Most leaks occur because we don't check the line ends and seats for damage. Small nicks in the line end and seat cause leakage regardless of how tight you get the fitting. Examine the line ends and seats for damage before installation. If you find excessive nicking in either location, do not use the control valve, hoses or ram. Return them for a replacement.

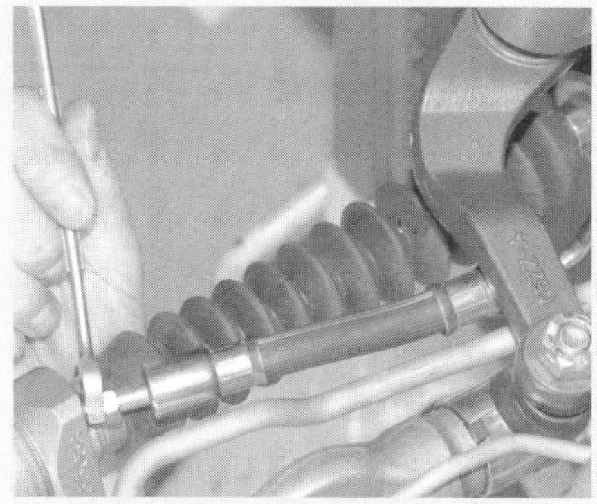

17. Install the top directional pressure hose. The top ram port goes to the top port on the control valve. Do not install these directional pressure hoses backwards. If you do, steering left will cause a violent jolt to the right, possibly resulting in injury and vehicle damage. As you tighten the fittings, make sure the hose does not twist. Also make sure the hose clears everything. The bottom directional pressure is installed last. The same rules apply: proper port to proper port.

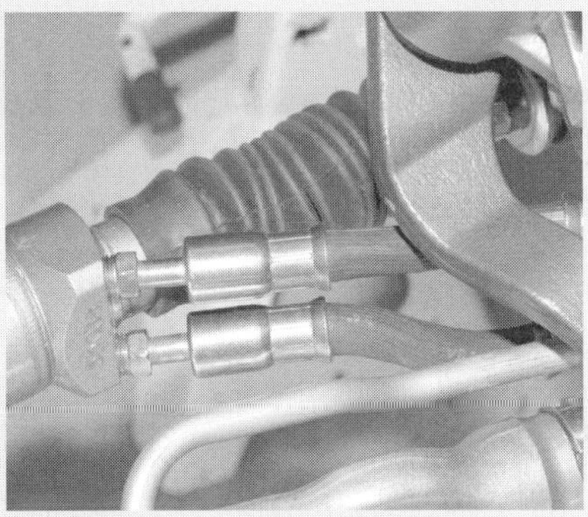

18. Line routing should be nice and clean, with no twists or contact.

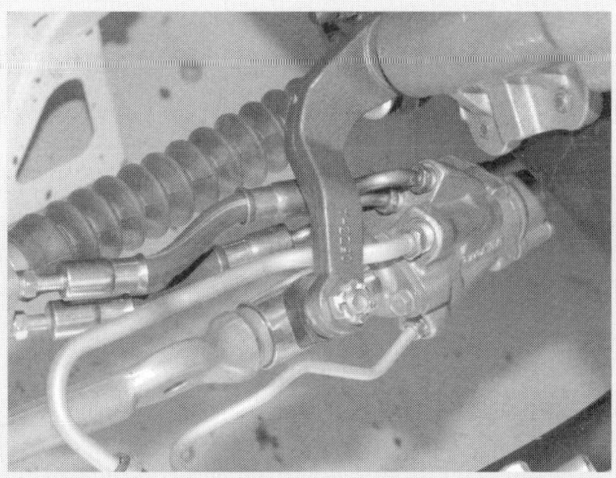

19. Here's the completed control valve/hydraulic line package. Note the clean and uncluttered installation. None of the lines should be touching anything.

20. One thing that makes '65–'66 Mustang power steering different from '67–'70 is the tie-rod ends. On the left side, the outer tie-rod end is offset to clear the control valve. For '67–'70 Mustangs, inner and outer tie-rod ends are the same on both sides. Longer tie-rod ends go inboard. Shorter tie-rod ends go outboard.

21. This is the pressure line from the power-steering pump, designed for the Ford/Thompson pump. The Eaton pump calls for a different fitting and hose.

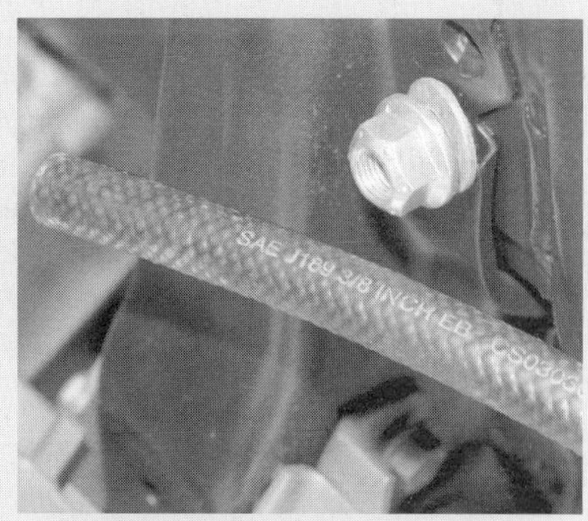

22. The return hose doesn't experience high pressure like the pressure hoses. As a result, it's secured to the reservoir with a worm-gear clamp. This line returns fluid from the control valve and ram to the pump.

Chapter 14

How to Replace the Rag Joint

Text and Photos by Jim Smart

The rag joint is your steering system's most vital link.

There is one, often overlooked component that is considered by many experts to be the most important to your Mustang's steering system. We aren't talking about the steering gear, power-steering control valve, steering ram, tie-rod ends, idler arm, or Pitman arm. We're talking about the rag joint.

The rag joint is the steering-shaft-to-steering-gear coupling, also called a flex joint. You won't find a rag joint on '65–'66 Mustangs, and rarely will you find one on a '67 Mustang unless it has Tilt-Away steering (a rare option) or a collapsible steering column (late in the model year). In '65–'67, the Mustang's steering shaft was a solid shaft that ran from the steering box to the steering wheel.

From a safety standpoint, the '65–'67 Mustang solid-shaft steering column was a bad idea. So, beginning in 1967, a few Mustangs and all Cougars had a collapsible steering column that was connected to the steering gear with a flexible, reinforced rubber joint. Through time, use, and extreme underhood heat, the joint dry-rots and deteriorates. Eventually, it shreds and fails. When it fails, the steering wheel and shaft become disconnected from the steering gear—not a good thing.

Source
Mustangs Etc.
www.mustangsetc.com

During installation, we were introduced to a then-new urethane disc. Although it probably stands up to the elements and heat better than a rubber disc, we aren't completely convinced the urethane disc won't fail more catastrophically. The rubber disc has a fiber reinforcement like that in tire sidewalls, which provides some link to the steering gear should the rubber break down. Urethane doesn't have the fiber reinforcement as a backup should the urethane fail. The choice, of course, is yours. Urethane doesn't isolate road shock and vibration as well as rubber, but it provides a more solid link to the steering gear for better feel.

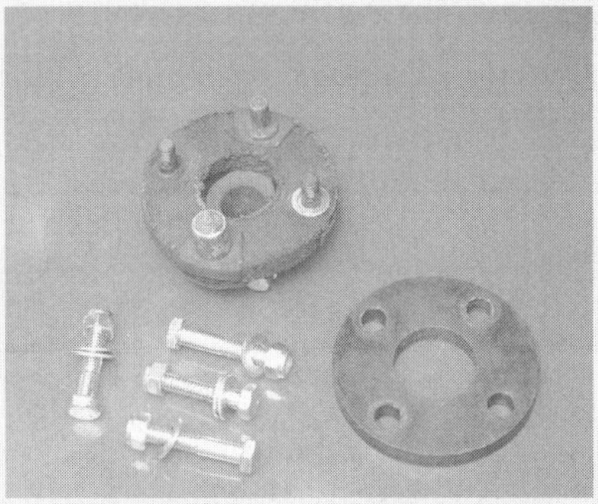

1. Disassembled, the humble rag joint doesn't look like much. Made of little more than tire-sidewall material, it's a rubber disc with four holes. It keeps the connection with the steering gear.

2. The rag joint consists of the steering-gear coupling, rubber disc, tangs, and bolts. The tangs keep the steering shaft connected to the steering gear should the bolts ever work loose or escape entirely. These tangs also ensure the steering shaft is tied to the steering gear properly.

3. Replacing the rag, or rubber disc, calls for the removal of the permanently installed factory tangs. Each of the tangs is mushroomed for a safe, permanent installation. The mushroom heads must be ground off to remove the tangs.

4. Each of the tangs is driven out with a punch as shown. This separates the rag joint.

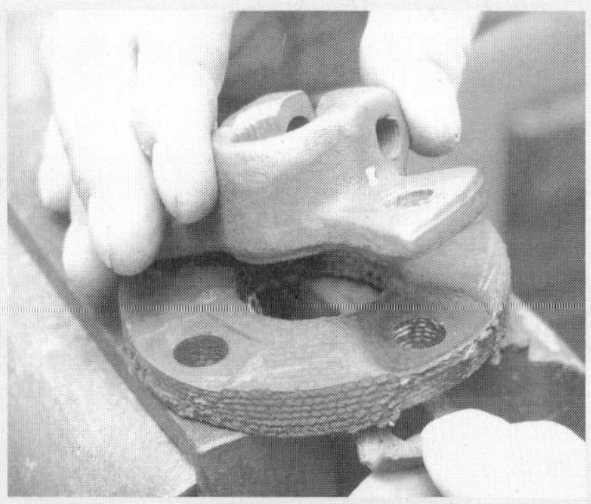

5. With the tangs removed, the rag joint is disassembled. Throw away the old rubber disc.

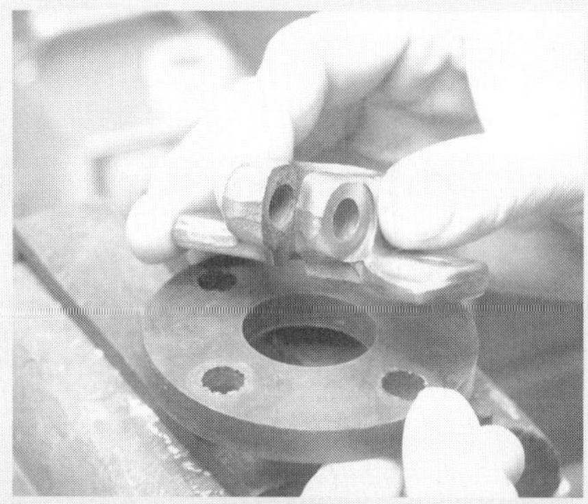

6. The new rubber disc from Mustangs Etc. replaces the one we tossed. The rag-to-steering-gear coupling is positioned as shown.

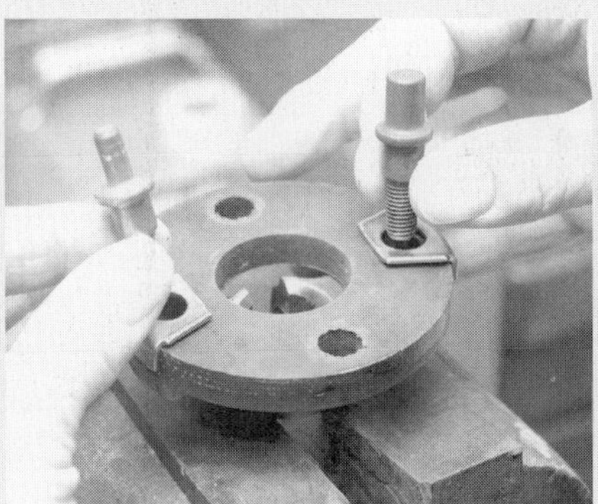

7. Included in the new rag-joint kit are new tangs—one small, one large. The tangs fit into the steering-shaft flange. Here, they fit through the rubber disc and through the rag-to-steering-gear coupling, which provides a solid connection between the steering shaft and steering gear. It's a backup system should the rag joint fail.

8. With the tangs in place, the rag joint should look like this.

9. Self-locking nuts (preferably the compression-type shown here) are used to secure the tangs, rag disc, and coupling.

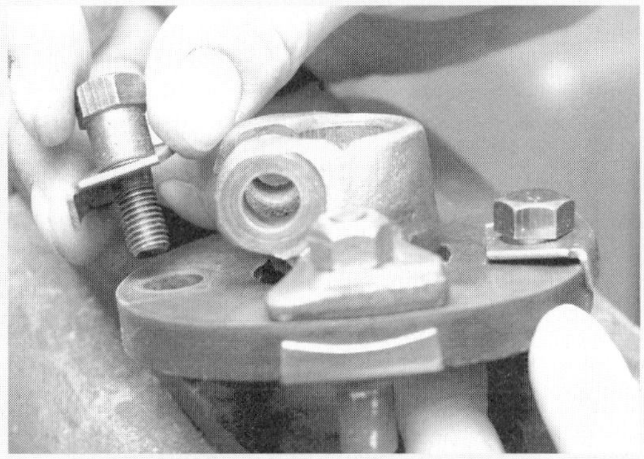

11. These bolts are specialty fasteners originally used to secure the factory rag joint. They have a half-inch head with a large shank/shoulder for security in the rag disc. Use these bolts for your rag-joint installation. Anything less would be courting trouble. These bolts secure the rag joint to the steering-shaft flange. Although most rag-joint kits include fasteners, we suggest using original equipment.

10. The assembled rag joint should look like this.

12. This is our rebuilt rag joint, complete with the correct fasteners, ready to install.

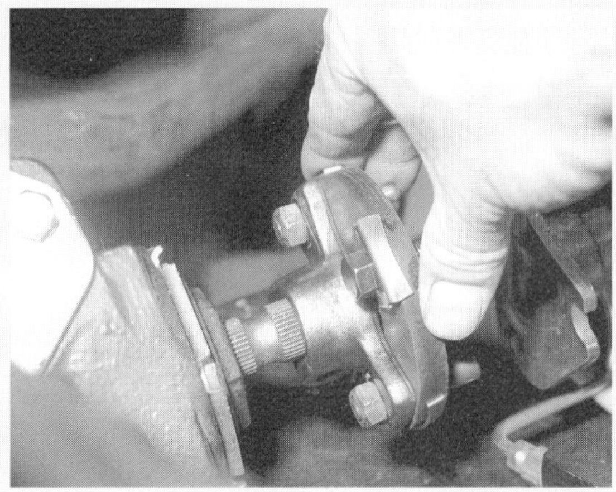

13. The rag joint slips onto the stubby worm-gear steering shaft. Because the coupling slips on only one way, it's virtually foolproof. Make sure the bolt alley lines up with the retaining bolt.

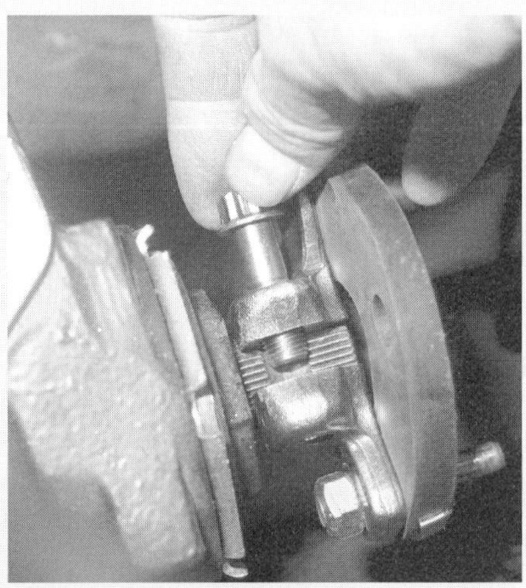

14. Here's the rag joint (left), with the bolt alley lined up on the worm-gear steering shaft. Next, the bolt is installed and tightened to specs (right).

15. The installation is complete. The two tangs are in the steering-shaft flange. Two bolts are secured to both the steering-shaft flange and the rag joint. Mission accomplished.

Chapter 15

Rack-and-Pinion Steering Upgrade

Text and Photos by Miles Cook

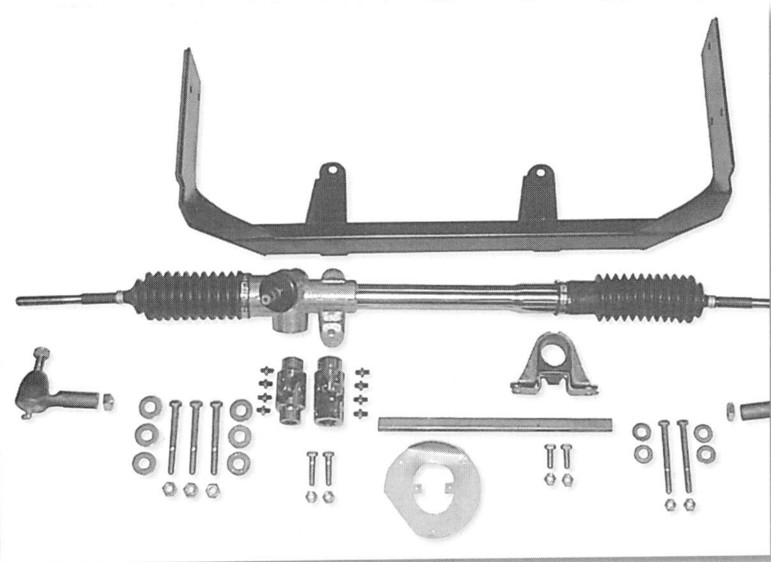

With Flaming River's new bolt-in rack-and-pinion, it's now possible to have modern steering on '65–'70 Mustangs without cutting or welding. The system is a complete bolt-in setup and comes with the cradle (above), rack, steering shaft, shaft U-joints, and mounting hardware. Billet and polished racks are also available as upgrades.

Source
Flaming River Industries
www.flamingriver.com

For those of you who own a vintage Mustang and drive it often, this upgrade might be of interest. Flaming River recently released a bolt-in rack-and-pinion steering system that precludes any cutting, drilling, or welding to the car. Of course, any modern steering rack is going to provide greatly improved steering compared to the early Mustang's outdated worm-gear and tie-rod steering, which is the main reason for going to a rack-and-pinion setup.

If you want to make your car more enjoyable for the street or a twisty back road, this system is ideal and is designed to fit '65–'70 Mustangs. The two main components are a slick mounting cradle and Flaming River's quick-ratio (16:1) manual rack. A tilt steering column is also available, which is a nice option for '65-'66 cars that never came factory equipped with a tilt wheel.

After having the standard kit on the market for a while, Flaming River made a system available for cars with headers. That was followed by another version for big-block cars that also might have headers. However, since it's impossible for Flaming River to offer a system for every engine, header, and transmission combination, first check to see if the kit will work in your car. Custom cradles can be made on a case-by-case basis, such as the one made for our unusual application.

Our trusty '69 fastback was once again our subject car. With the rack-and-pinion, its 351W and four-speed clutch linkage made for something considerably more unusual than the typical '65–'66 car with a 289 and an automatic. Further muddying the waters are the car's JBA headers, which are something akin to a medium-length design. That is, they're not full-length headers, but they're longer than stock manifolds or compact shorty headers, so clearance issues with the system's steering shaft U-joints had to be addressed.

Flaming River made the system work on our car by building the custom cradle for our specific application. However, the following installation highlights, for the most part, pertain to installing the package in any '65–'70 Mustang.

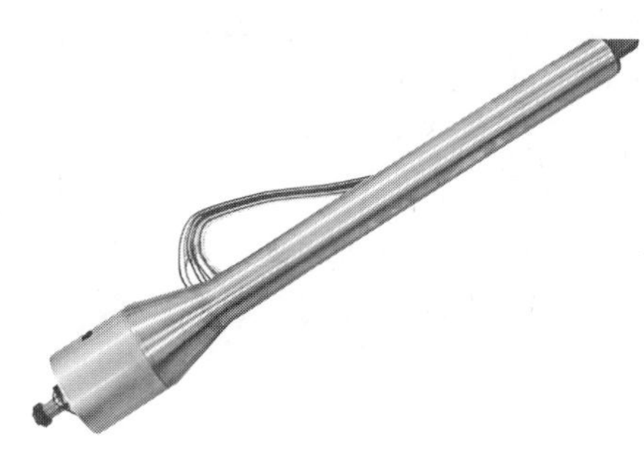

You can adapt the system to work with your car's original column or run an optional tilt column in either a paintable or polished finish. Columns are also available to work with a car's original steering wheel or an aftermarket version, like a Grant or a LeCarra.

1. Shown here are the two cradles available for cars with stock exhaust (bottom) and with headers. Automatic versus manual transmissions need to be considered. For example, if your car has a manual transmission with the stock clutch linkage and clutch cross-shaft, unusual headers, an aftermarket oil pan, and an unusual engine such as a 351 Cleveland or an FE big-block, a custom cradle might be required and can be made by Flaming River. We had one made for our car because of its JBA headers and four-speed clutch linkage, which required offsetting the rack further than the standard header cradle.

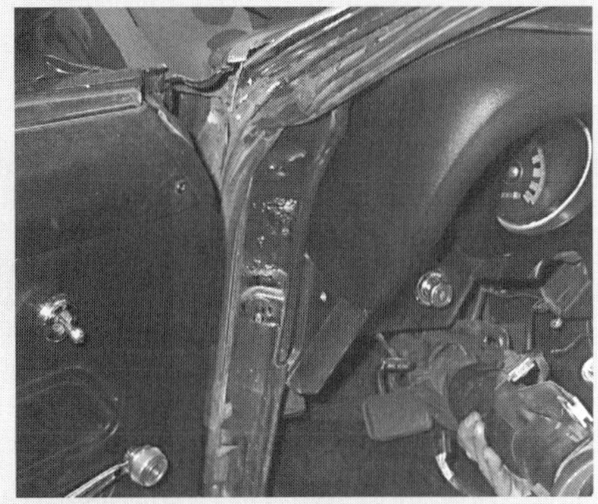

2. The header steering rack at the bottom is offset one inch, compared to the standard rack on top. Each comes with the correct corresponding cradle.

3. First, all the stock steering parts need to come off. We begin this process by disconnecting the steering box from the steering column under the hood in preparation for removal of the box and column.

4. Next remove the column by removing the steering wheel and unbolting the column from the dashboard and floor.

5. The steering column's wiring harness unplugs from the car, at which point the column can be completely removed.

6. Moving underneath the car, remove the outer tie-rod ends from the spindles with a tie-rod-end puller or by carefully tapping the ends out of the spindle with a hammer. We removed these ends without a puller because they were recently replaced. Once separated from the spindle, unscrew the ends from the tie-rod sleeves. The outer ends will be used with the new steering rack. Procure new ones if needed.

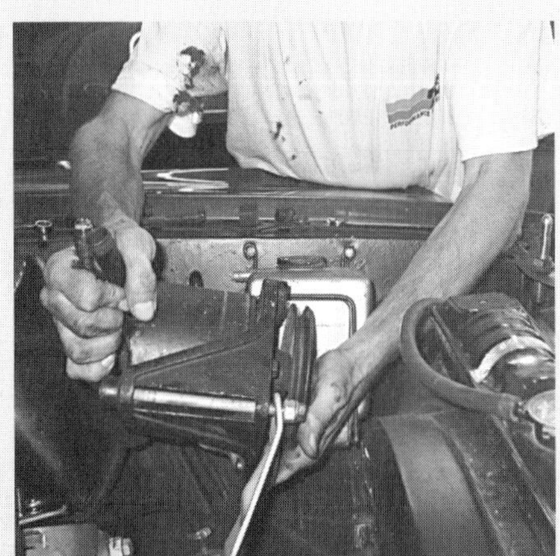

7. If your car has power steering, remove the pump from under the hood.

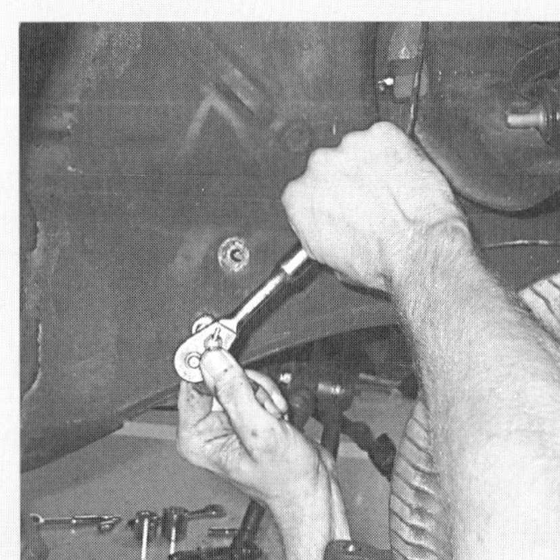

8. On the passenger side, unbolt the idler arm. The bolts are accessed from inside the fender and are easier to reach with the wheel removed.

9. On the driver side, remove the pitman arm from the steering box using a pitman-arm puller. It can be rented from a local auto parts store.

10. With the idler and pitman arms disconnected and the tie-rod ends and power-steering-pump lines removed from the pump, the whole factory steering linkage can be removed.

11. Finally, remove the steering box by removing the three bolts that hold it in place. The bolts are accessible from inside the driver-side front fender.

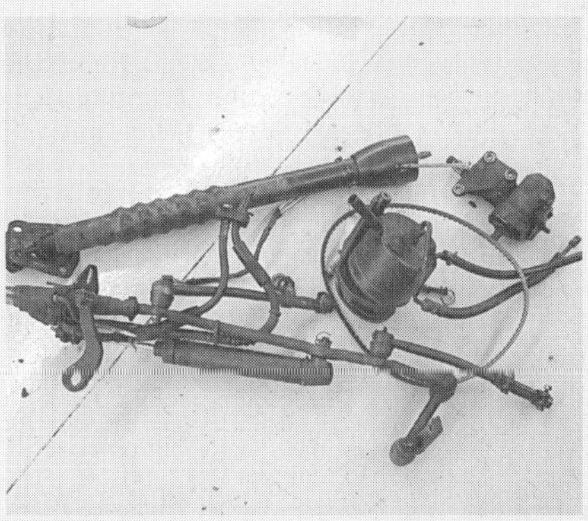

12. Here's the factory steering system as removed from the car: linkage, slave cylinder for the power steering, power-steering pump, steering box, and steering column. These components can be reinstalled to either sell the car without the rack-and-pinion steering or it can be installed on another car. In any case, we recommend keeping the stock

13. With the coast clear of all the stock steering bits, the cradle bolts right into place using the holes in the car's front subframe that originally were for the idler arm on the passenger side (two holes) and the steering box on the driver side (three holes).

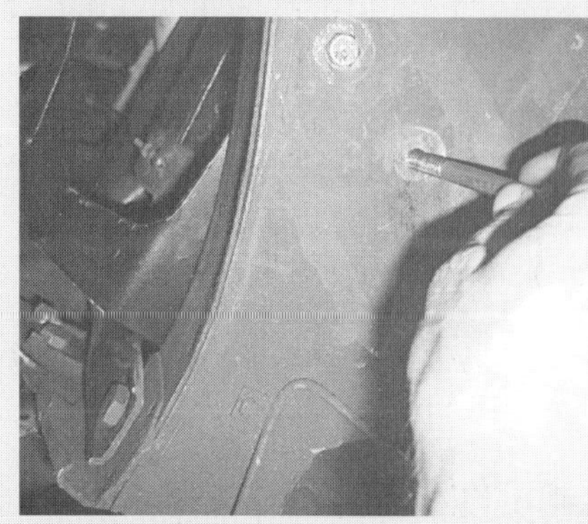

14. With the cradle in place, you might have to use the bolts to help align it with the holes in the subframe. Each car is different. Once the cradle is lined up, install the new bolts, washers, and nuts to secure the cradle in place. Torque the five bolts to 50–65 lb-ft.

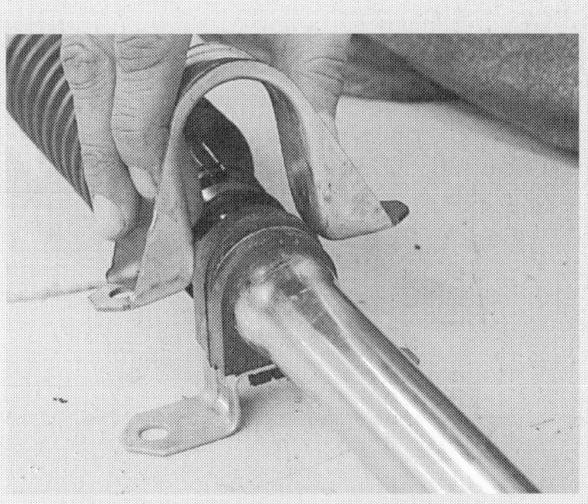

15. With the cradle installed, the steering rack is next. To fit the rack onto the cradle, install the passenger-side rack mounting clamp to the larger section of the rack tube (right side) using the provided hardware.

16. The steering rack then bolts to the cradle with the supplied hardware.

17. At this point, the steering rack is installed. Installation of the tilt column is the next step after connecting the rack to the spindles via the tie-rod ends.

18. The spindles thread directly into the steering rack and reattach to the spindles as shown. Reinstall the cotter pins through the hole in the tie-rod-end threads and through the castle nuts, then bend the tabs over on the cotter pins so the nuts cannot back out.

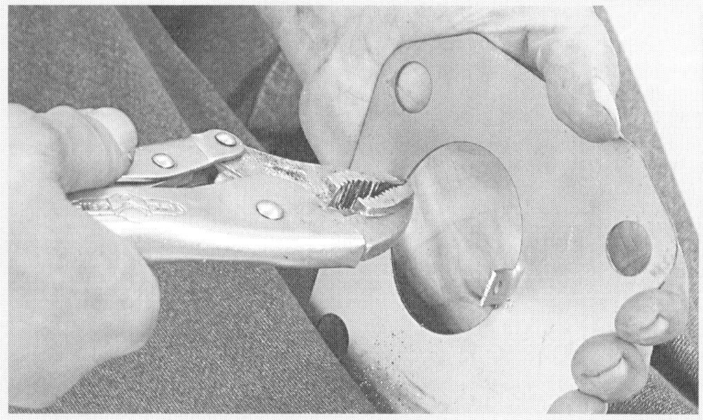

19. Installation of the steering column begins by bending the tabs on the floor mount inward toward the car's interior so the column can fit through it.

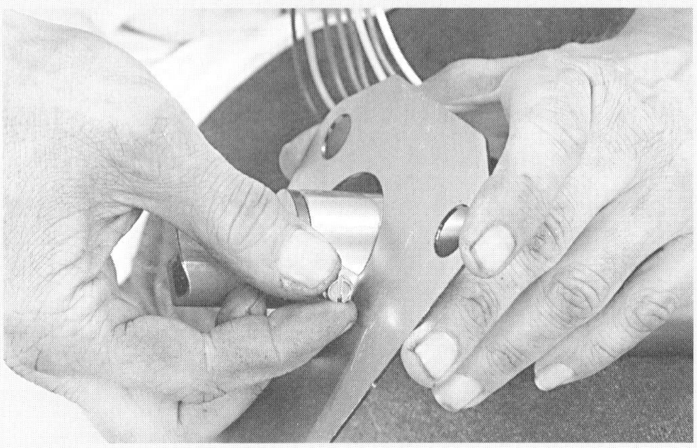

20. With the supplied hardware, secure the floor mount to the column. Mark the location of the tabs on the column and drill a pair of small pilot holes for the screws.

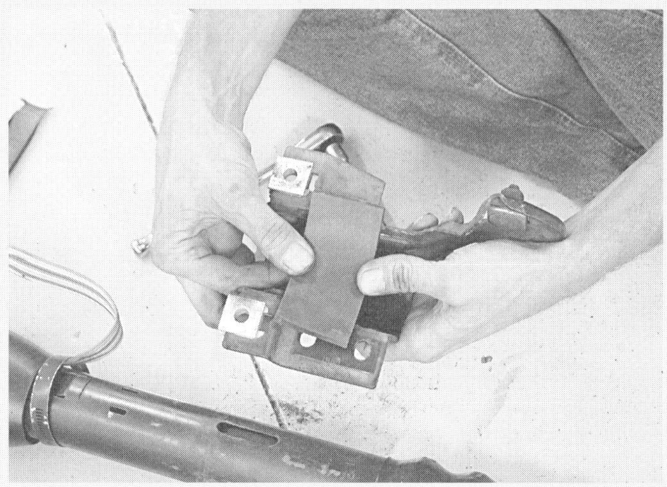

21. Using the dash support bracket from the original column, place the supplied rubber shim between the bracket and column to ensure a tight fit. The new column is installed using the original dash support bracket and hardware.

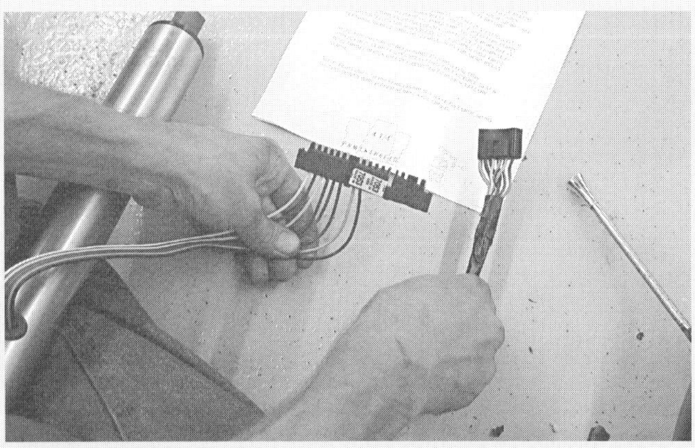

22. Flaming River provides the wiring information needed to interface the new column with the original wiring harness. It's up to you to match the wiring on the column with the wiring in the car. One option is to acquire another factory plug to connect the new column into the car's underdash harness. You could also cut the plug off the existing column, but do this only if you can't find another one.

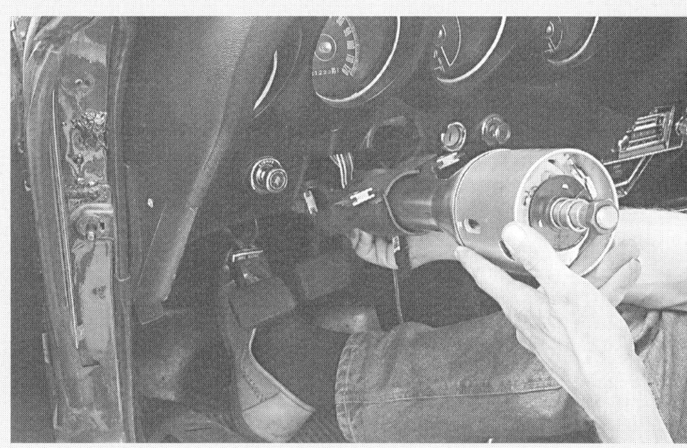

23. With the dash support bracket and floor mount in place, installing the new column is basically a reversal of removing the original one.

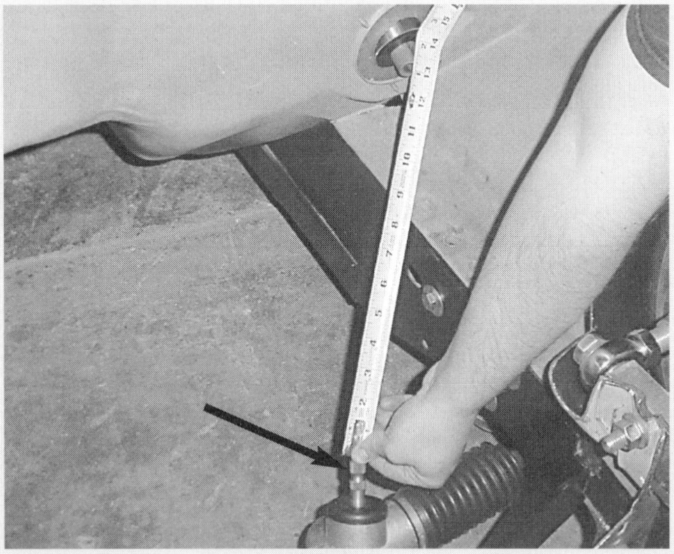

24. With the column and rack installed, the next task is to measure and cut the supplied steering shaft to length. Measure from the end of the steering rack's pinion shaft (arrow) to the end of the steering column shaft, then subtract 3 1/4 inches from that figure, which compensates for the length of the two shaft U-joints. Be sure the shaft doesn't extend through the bottom of the U-joint yokes. For header kits, the steering shaft is cut into two pieces and three U-joints are used. There's also a support bearing for the steering-shaft section that connects to the steering column. Measurements have to be taken for each individual car before the shaft is cut.

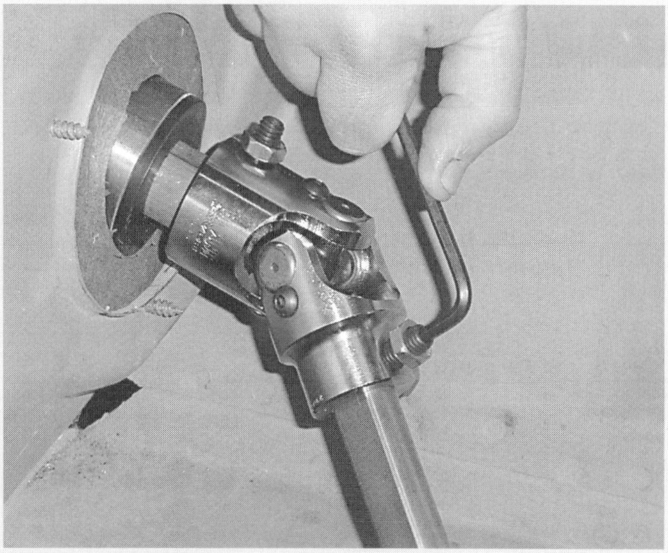

25. For installation of the U-joints, use the set screws in the U-joints to mark the steering shaft. The shaft needs to be dimpled to allow the set screws in the U-joints to be securely tightened.

26. Using a 1/4-inch drill bit, dimple the steering shaft where the set screws made a mark on the shaft. Reassemble the shaft and U-joints and use red Loctite to ensure the set screws don't loosen. Tighten the set screws to 25 lb-ft and tighten the jamb nuts securely.

Chapter 16

How to Install Power Rack-and-Pinion Steering

Text and Photos by Jim Smart

The bolt-on system from Randall's Rack and Pinion makes it easy to upgrade vintage Mustang steering.

Classic Mustang steering was nothing to write home about even when the cars were new. They may have looked like Mustangs, but they drove like Falcons. Mustang manual steering was and still is a chore, especially with a dinky aftermarket steering wheel. Power-assisted steering remains a nightmare because it leaks and is prone to other failures. However, with the bolt-on rack-and-pinion steering from Randall's Rack and Pinion, you can improve your Mustang's steering without being obvious. Best of all, you can install it in one day and be ready for cruising that night.

What we like most about Randall's system is the attention to detail and forethought in the design. These guys thought about every aspect of rack-and-pinion in '65–'70 Mustangs and figured out how to ease installation. We were shocked at how easy it was once we took off all the old stuff.

Tucson's Laurie Slawson is a Mustang Club of America board member and show judge who also grabbed an MCA Gold with her concours '68 hardtop. A few years ago, she decided to step out and build a Mustang restomod. Laurie wanted a big-block fastback with a "rumpity-rump" demeanor. One of her goals was rack-and-pinion steering, but the challenge was installing it in a big-block car. Not all of them fit. Randall's answered the call with its bolt-on system.

For the installation, Laurie enlisted the help of Jimmy's Broadway Automotive in Tucson. Because her 390 had been pulled for a teardown and build-up, it was easier for Jimmy's to complete the conversion while we shot our photos for this how-to.

To install the Randall's rack, remove the existing steering system: steering gear, idler arm, inner and outer tie-rod ends, and so on. With power steering, also remove the control valve and power ram. In fact, it's a good idea to drain the power-steering system before disassembly begins.

Mustangs from '65–'67 have a long, solid steering shaft to be reckoned with along with the steering gear. Raise the car enough to clear the steering shaft. To get there, remove the steering wheel and column, two locknuts, and a collar, disconnect the turn-signal switch, and loosen the column at the firewall. This frees up the steering gear for removal. On '68–'70 models, disconnect the rag joint at the end of the steering shaft at the gear.

Overall steering-column length must be 28 3⁄8 inches. This is from the top of the column to the top of the shaft assembly provided by Randall's. Carefully place your steering column in a vise, and cut it to size with a saw. Don't use a pipe cutter as a shortcut because that will make it impossible to fit the bearing-supported shaft.

Sources

Jimmy's Broadway Automotive
www.jimmysbroadway.com

Randall's Rack and Pinion Ltd.
www.randallsrack.com

1. When our Randall's rack arrived, everything necessary to complete the job was included. This is a simple power rack available from any auto-parts store. When it wears out, you can get one from Randall's or from a local auto-parts store. We like the exceptional engineering exhibited here: A power rack designed to fit virtually any classic Mustang out there.

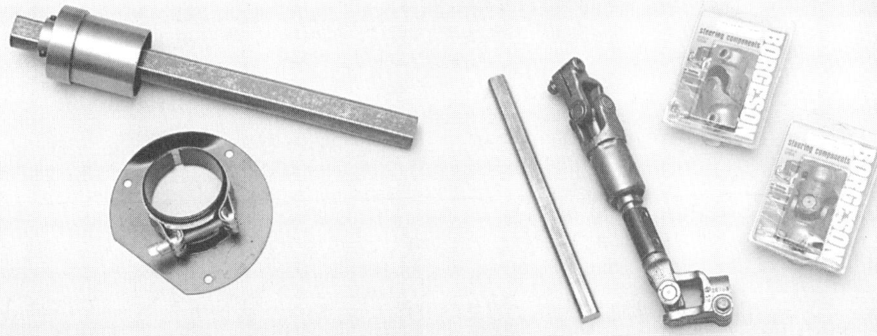

2. The stock Mustang steering column is adapted to the rack with a variety of articulating shafts and universal joints provided by Randall's Rack and Pinion. Count on an easy connection if you have a stock column. Aftermarket columns such as IdidIt or Flaming River require some fabrication work.

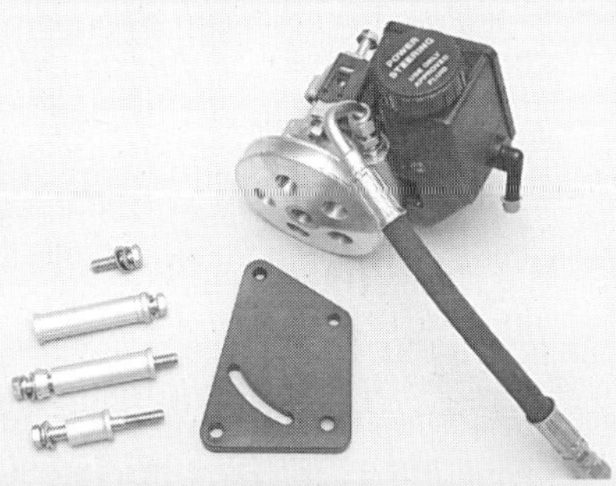

3. Use a factory Eaton or Thompson power-steering pump or opt for one from Randall's Rack and Pinion. Randall's aluminum pump bolts onto your engine's front dress and is easy to adjust.

4. First remove the old steering linkage. Here, technician Derek Torres begins removal of the vintage Bendix power-steering system, beginning at the control valve. The control valve is tied to the steering gear via the Pitman arm.

5. Derek disconnects the rag joint from the steering gear as shown. Once the rag joint is disconnected, remove the three steering-gear retaining bolts using a 1/2-inch drive-socket wrench with an 11/16-inch socket. For the rag joint, use a 9/16-inch box-end wrench.

6. Derek knocks out the tie-rod ends, disconnecting them from each spindle. Don't toss the outer tie-rod ends, as you'll need them. It's a good idea to replace them, especially if they're worn.

7. Two bolts retain the idler arm, which ties the steering linkage to the righthand framerail. Then Derek drops the entire steering linkage.

8. To begin the installation of the Randall's rack, Derek wrestles the rack into place to start the bolts. The Randall's power rack replaces the factory tubular crossmember Ford originally used. It bolts into the crossmember bolt holes with Grade 8 fasteners.

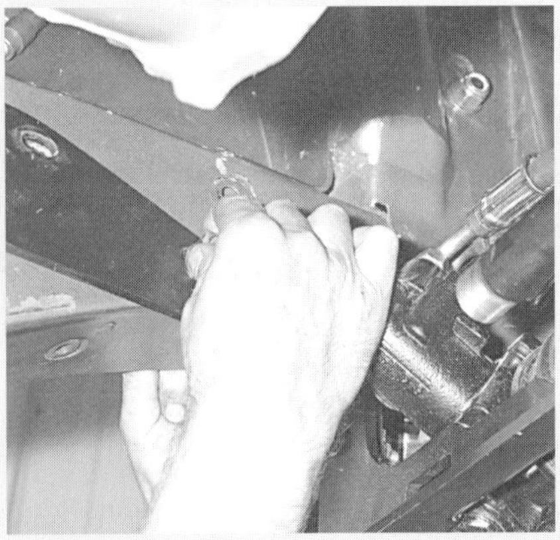

9. Each side bracket ties the rack to the idler-arm and steering-gear locations. Grade 8 bolts provided by Randall's make this a secure installation. Take extra care not to over-torque these bolts.

10. Because the Randall's rack replaces the Mustang's factory crossmember, it bolts where the crossmember originally fastened to provide strength and support.

11. The Mustang's outer tie-rod ends transfer over to the Randall's rack. Don't get them mixed up: left-to-left and right-to-right

12. Steering tie-rods provided by Randall's screw onto the rack as shown. Derek has a special tool for this purpose, however, you can use a large Crescent or open-end metric wrench.

13. Randall's provides a billet-aluminum coupler to tie the rack to the Mustang's outer tie-rod end.

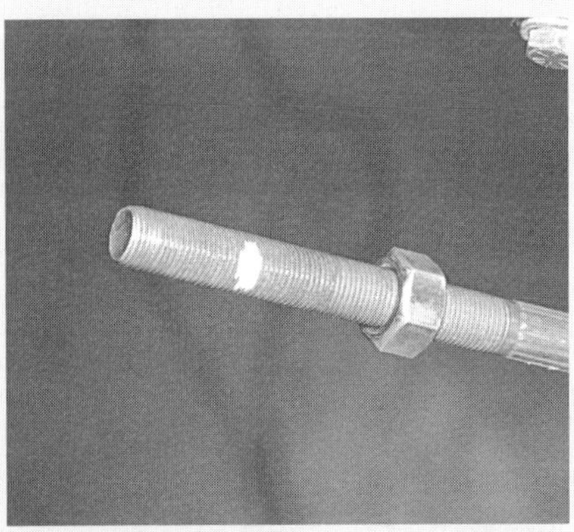

14. In our situation, we had to shorten the inner tie-rods by 1/2 inch each. What is shown here is closer to 1 inch, which was later changed to 1/2 inch. Shortening the inner tie-rod may be necessary for improved toe adjustment. Do not cut until you are certain it needs to be shortened.

15. One rivet on each side secures the inner tie-rod assemblies. Insert it, then depress the pin with a hammer.

16. Dust boots go on each side once the tie-rod installation is complete. These boots keep road dust out of the rack.

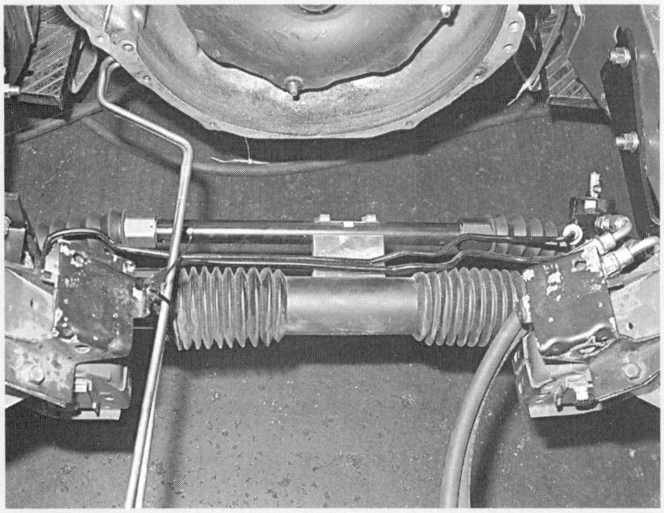

17. Rack installation is complete, shown here from the top side because the engine has been removed for a rebuild. All that's left now is to install the hydraulic hoses for the pressure and return lines to the pump.

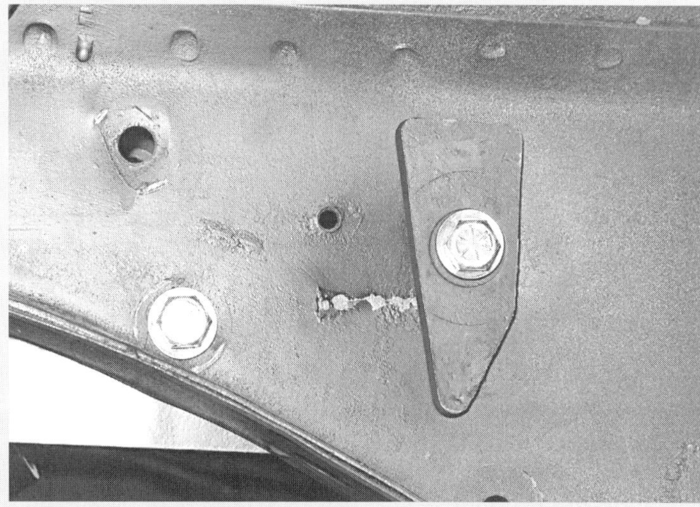

18. When we ran down the rack bolts, we found out the hard way about frame cracking. Laurie's '68 fastback had a defect. The internal support (or bridge) inside the rail, which keeps the rail from collapsing during steering-gear and idler-arm installation, fell down inside the rail. When Derek ran the bolts down, the rail cracked. We installed a temporary gusset until we could fix the rail for Laurie. We're convinced the bridge's spot-welds failed, causing it to fall down inside the rail when we removed the steering gear.

19. A challenge for this installation was Laurie's IdidIt steering column. It was too long by about 2 inches. Laurie will have to send the column back to IdidIt and have it shortened by 2 inches for smooth operation with the Randall's rack. For stock rag-joint applications, cut the coupling off the shaft and tie it into the constant-velocity package Randall's provides.

20. Our greatest challenge was Laurie's long IdidIt column. The angle, as you can see, is too steep. When the column is shortened 2 inches, it will be a perfect fit.

'65–'67 Solid-Shaft Steering Columns

When you're doing a tilt-away or '68-up collapsible steering column, a Randall's conversion is straightforward. For '65–'67 solid-shaft columns, this is a test of fabrication skills. Randall's suggests cutting the steering column as shown to take 3 inches off the tube. The steering shaft has to be cut and machined into a double-D shape to fit to the steering coupling. This mandates a visit to a machine shop. The steering column tube should be 28⅜ inches from the collar at the steering wheel to the tube end. The accompanying images should give you some idea of what's involved. Detailed instructions come with the Randall's Rack and Pinion system.

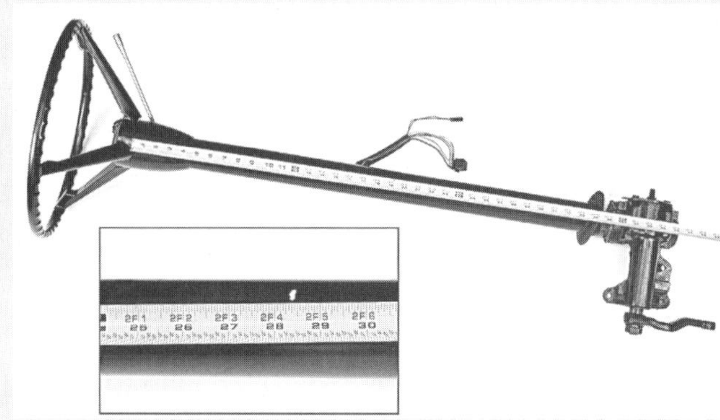

Here's a quick illustration of what's involved with solid-shaft column Mustangs. From the collar to the cut is 28 ⅜ inches. The tube is cut at the 28 ⅜-inch mark, not the shaft. The shaft is cut longer to reach the adaptor and flexible shaft.

This is the column tube cut at the 28 ⅜-inch mark with the specially modified double-D shaft. A machine shop has to machine the shaft to a double-D shape in order to fit the adaptor.

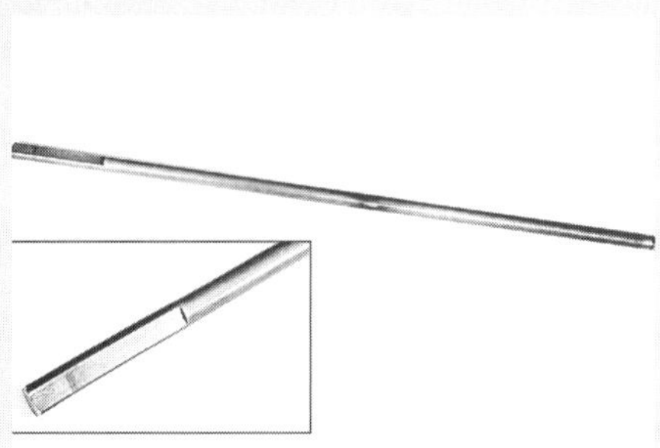

Here's a look at what a '65–'67 Mustang's solid-shaft should look like once machined.

To safely tie the double-D shaft to the flexible coupling, drill a ¼-inch hole for the roll pin as shown here.

Drill a ⅛-inch hole for a sheetmetal screw to secure the adaptor to the steering column.

This is what a modified solid-shaft column should look like upon completion.

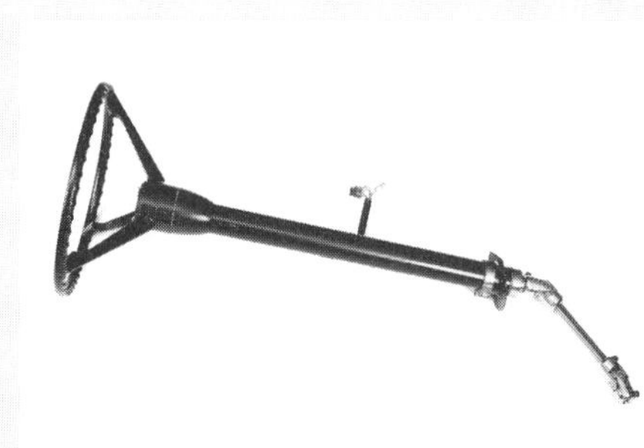

Here's a completed solid-shaft column with the flexible shaft installed.

Proper front-end alignment is key to handling and tire wear

Chapter 17

Front-End Alignment

Text and Photos by Jim Smart

For black-magic mystery, front-end alignment is right up there with air conditioning, starter-motor or alternator rebuilding, fuel injection, carburetor tuning, and automatic-transmission overhauls.

What is it and how do you do it?

It's easy to write off front-end alignment as something best left to professionals. But did you know you can do front-end alignment yourself? And did you know you can converse intelligently with your front-end alignment professional, assuming you can find one that still aligns vintage cars such as '65–'73 Mustangs?

Front-end alignment is nothing more than simple geometry: How all four tires interact with both vehicle and pavement. Alignment is an understanding of how the suspension and steering behave as the car cruises down the road. How tires interact with the pavement depends on how they are adjusted when the vehicle is at rest and when we start driving. A Mustang's front suspension behaves differently when cruising straight than when cornering.

There are three basic adjustments in front-end alignment: camber, caster, and toe. If you possess absolutely no knowledge of front-end alignment, these are terms that won't make a bit of sense, so let's explain them to you.

What Is Camber?

Camber is the angle of the tire/wheel as it relates to the road. Think of the front wheels as your own legs. Positive camber makes you bow-legged. Negative camber makes you walk pigeon-toed, on the insides of your feet. So how does camber affect your Mustang? When you have positive camber, your Mustang won't corner well because each turn runs the front tires onto their sidewalls, which is why you'll never see positive camber in a front-end alignment-specs chart. When a Mustang corners, negative camber keeps more tire contact patch (tread) in contact with the pavement. Negative camber, with the vehicle at rest, assures some negative camber in a turn. Mustangs need 1⁄4-degree negative camber minimum, and 1 1⁄4-degrees maximum if you're going canyon chasing.

Camber can be tricky because it changes as you drive. This is why Ford provided a range of camber to play with. At rest, approximately 1⁄4-degree negative camber is good for tire wear. If handling is more important than tire wear, you can add negative camber at the sacrifice of tire life.

First-generation '65–'66 Mustang camber is adjusted at the upper control arm via shims at the two control-arm attachment points. Add shims to push the arm out toward positive camber and subtract them to allow the arm to move in toward negative camber.

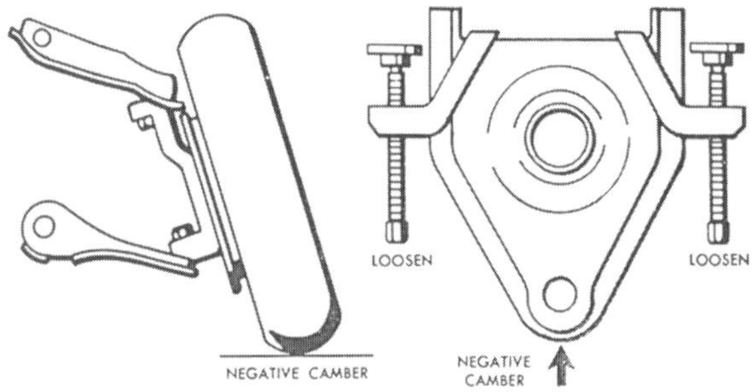

Camber is the angle that both front wheels are tilted at the top and bottom as they roll down the road. This is negative camber, measured in degrees. The top of the tire is pointed in and the bottom is pointed out. Negative camber keeps good tire contact patch on the road in turns. Mustangs do well with 1/4 degree to 1 1/4 degrees. For street use, 1/4 degree of negative camber is best. For road racing, you may want a pinch more negative camber.

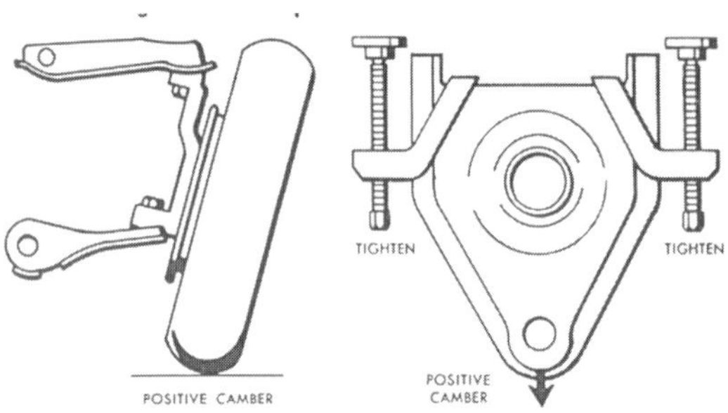

This is positive camber, where the top of the tire is pointed out and the bottom is in. Think of positive camber as bowlegs. Hard to walk that way, isn't it? Positive camber makes a Mustang plow all over the place in turns because it rides on the sidewalls instead of tread. The difference in camber from side to side must never exceed 1/2 degree. Ideally, you want less than 1/4-degree of difference in camber from side to side.

Alignment shops used to use a bubble gauge mounted on the wheel to adjust camber and caster (spindle angle and wheel attitude). One bubble gauge indicated camber while the other showed caster. Technology has done away with the bubble gauge and brought us laser-guided, four-wheel alignment at most shops across the country. Marlo's Frame & Alignment in Chatsworth, California, still gets outstanding results with the old-fashioned bubble method. The best front-end alignment tool is still a seasoned professional.

First-generation '65–'66 Mustang camber and caster adjustment takes place at the upper control arm using shims. The control arm is moved out via shims; arm angle (caster) is controlled via shims as well.

From '67–'73, Ford revised the Mustang's front suspension for easier camber and caster adjustment. Instead of using shims on the upper control arms, Ford went to an adjustable lower control arm. Adjustment is facilitated via an eccentric on each lower control-arm fulcrum. Roll the eccentric to move the arm in or out to change camber.

From '74–'78, the Mustang's front suspension system changed completely to rack-and-pinion steering, with coils and shocks located between the upper and lower control arms. This makes the Mustang II's front suspension more like the '67–'81 Chevrolet Camaro and Pontiac Firebird. The upper control arm on the Mustang II is adjustable, much the same way as its GM counterparts. The upper arm moves in and out to change camber.

In '79, Ford greatly simplified the Mustang's front-end alignment process by eliminating the upper control arm and going with fully adjustable MacPherson struts. To change camber, the strut is moved in and out. The same can be said for '05–'07 Mustangs, which also have MacPherson struts.

What Is Caster?

Caster is spindle angle from top to bottom, or the positioning of the upper and lower ball joints as they relate to one another. If the spindle is tipped forward, it creates negative caster. If tipped aft, it creates positive caster. This controls spindle geometry as the car turns. Think of caster as how the tire/wheel articulates around the axis that connects both ball joints.

Caster is challenging to understand because it's set up based on how the car will be moving at speed. If you could closely study the front tires while they are in motion, you'd notice the tire

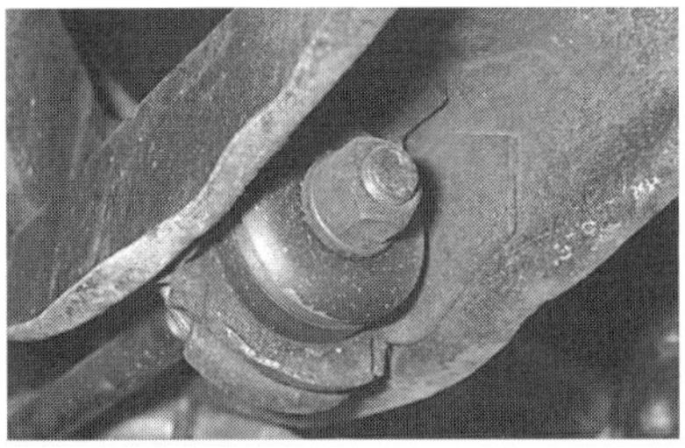

Caster for '67–'73s is adjusted at the strut rod by moving the rod (and lower control arm) back and forth. This eliminates the use of shims on the upper control arm, which is a fixed position for '67–'73.

Camber and caster adjustment changed forever for '67–'73 with an adjustable lower control arm that is adjusted using this eccentric at the fulcrum. Camber is adjusted at the eccentric; caster is adjusted at the strut rod.

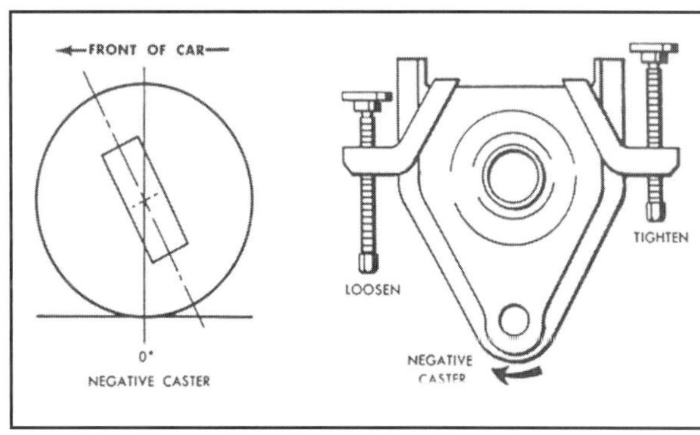

Caster is spindle angle from top to bottom. Negative caster occurs when spindle angle is tipped forward at the top as shown.

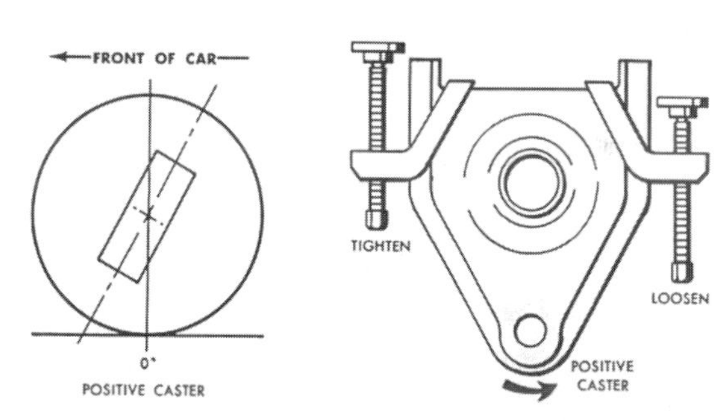

When the spindle is tipped to the rear, this is positive caster. The '65–'66 Mustangs use shims on the upper control arm for caster adjustment. To shift caster to positive or negative, load shims at the front or rear attachment point. For '67–'73s, adjust the strut rod to move the lower control arm fore or aft to control caster. In either case, it's adjusting the spindle angle known as caster.

contact patch is different from when the vehicle is stopped. In motion, the tire contact patch moves rearward, which changes caster. Caster becomes more negative in degrees once we get going. This becomes even more apparent with radial tires, which make the car wander more from side to side. The solution is caster adjustment whenever tire type is changed from bias-ply to radial or vice versa. In fact, it's a good idea to have a front-end alignment any time you rotate or replace tires.

Mustangs like 0 degrees of caster, meaning both ball joints are 12 and 6 o'clock from one another. Ford calls for a minimum of 1 degree of negative caster and a maximum of 1 degree positive with no more than 1/2 degree of variation between the two on V-8 models. In short, caster must be within 1/2 degree from side to side for best results. Six-cylinder models need 0 degrees minimum caster and 2 degrees positive maximum due to the front-end weight difference.

First-generation '65–'66 Mustangs caster adjustment takes place via shims at the upper control arms. Again, shims are added or subtracted to control caster. From '67–'73, the strut rods were adjusted to change caster. The same can be said for '74–'78 Mustang IIs, which also use a strut adjustment to change caster. The Mustang II's strut rods are attached to lower control arms from behind.

For '79 and later, caster adjustment happens via MacPherson strut movement around the top caster/camber plate, making alignment a cinch for late-model fans.

What Is Toe?

Although most of us think of toe as those piggies on the ends of our feet, toe takes on a different meaning with cars. Toe is the last adjustment checked during front-end alignment. As its name implies, toe-in is the angle of the front wheels as they relate to a car's longitudinal axis.

Toe is adjusted at each of the tie-rod end assemblies in the steering linkage. When adjusting toe-in, you are adjusting each front tire's attitude as they relate to each other. Proper toe adjustment begins with the steering wheel at dead center, exactly in the center of lock-to-lock travel. If your steering gear is three turns from lock-to-lock, steer all the way left, then clockwise back one and a half turns to center. Once the wheel is centered, lock it in the centered position during toe-in adjustment.

When measuring toe-in, first determine the distance between the right and left tire sidewalls (or rims) with the wheels straight ahead. Measure front

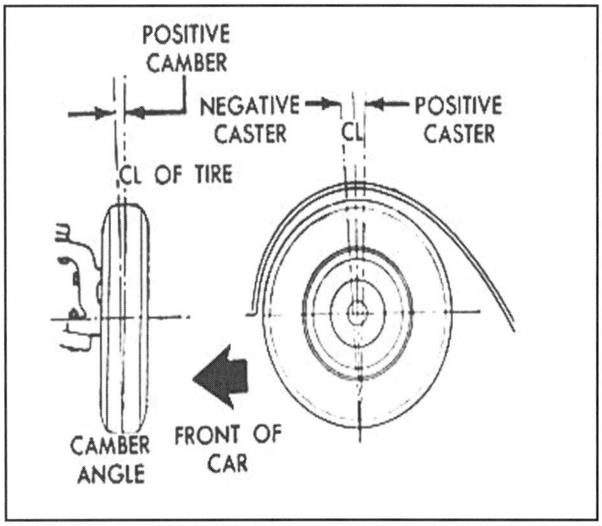

Camber and caster can be summed up in this simple illustration. Camber and caster have to work together as a team for exceptional handling and control. Without proper adjustment, you're just wasting time and tire.

Camber is especially crucial for handling because it determines how much contact patch we have. Contact patch, all of the tread on the pavement, is what enables us to cut apexes without winding up in the weeds.

and rear distances, and do the math. Toe is always set as toe-in, with both front wheels pointed a pinch inboard to allow your Mustang to stay on course and make light work of steering wheel return-to-center. When steering wheel return-to-center is sluggish, it's usually a toe-out problem.

Ford calls for a minimum of 3⁄32-inch toe-in and a maximum of 11⁄32-inch toe-in. We're talking about a finite amount of toe-in, just enough to bring the wheel smoothly back to center with proper tracking. Too much toe-in will return the steering wheel to center aggressively and cause a sawtooth tire-wear pattern. This is why it's so important that toe-in measurement is as accurate as possible. Toe-in also determines where the steering wheel sits when it returns to center. Sometimes, poor adjustment returns the steering wheel to 11 or 1 o'clock instead of noon. It's necessary to make identical, finite tie-rod end adjustments to center the wheel. Don't remove and adjust the steering wheel: Properly adjust the tie-rod ends and toe-in.

Commonsense Alignment

A good rule of thumb is to always take a test drive before doing a front-end alignment. This gives you a good idea of what you're up against. Any front-end alignment shop that doesn't do a test drive beforehand is cutting corners (no pun intended). Not taking a test drive is like assuming food needs salt and pepper before tasting it.

If the car wanders right or left, toe and caster are the biggest suspects. Sluggish steering wheel return is typically due to inadequate toe-in; aggressive steering return is too much toe. Tire scrub in turns is most likely too much positive camber; excessive inboard tire wear can mean too much negative camber.

Another item is component wear. Never approach a front-end alignment until you have checked all the components for wear. Excessively worn ball joints allow camber and caster adjustment to go right out the window the minute you hit the road. Worn tie-rod ends allow toe-in to wander. What's more, worn-out components are dangerous.

Proper wheel-bearing adjustment is also important. Loose or maladjusted wheel bearings will throw off front-end alignment. Make sure the front wheel bearings are properly packed, installed, and adjusted. Spin the brake drum/rotors and torque the adjusting nut to 15–20 ft-lb. Spin it again and feel for tightness. If the rotor is hard to turn, loosen the nut and spin it again. By loosening the nut, you're allowing the bearings to seat properly; torque to 15–20 ft-lb and check again. The rotor should spin smoothly. Slowly back the nut to where you have cotter-pin hole alignment. Torquing the nut to 15–20 ft-lb firmly seats the bearings. Loosening the nut to secure the cotter pin takes the load off the bearings.

Sources

Marlo's Frame & Alignment
www.fly-ford.com

The Eastwood Company
www.theeastwoodco.com

Checking Front-End Wear

Front-end alignment is pointless if you're dealing with worn-out components. Upper ball joints don't live as long as lower ones. In fact, lower control arms generally last twice as long as upper arms. As a rule, you should replace both at the same time.

There are two ways to check ball-joint wear, but it's best to have a front-end professional check ball-joint and tie-rod endplay. They have the right equipment to do the job.

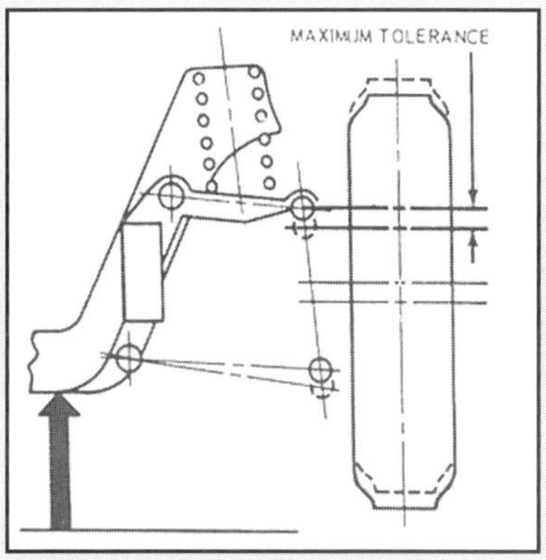

Upper ball-joint axial play is checked by taking all the weight off the tire. Axial play is ball-joint endplay.

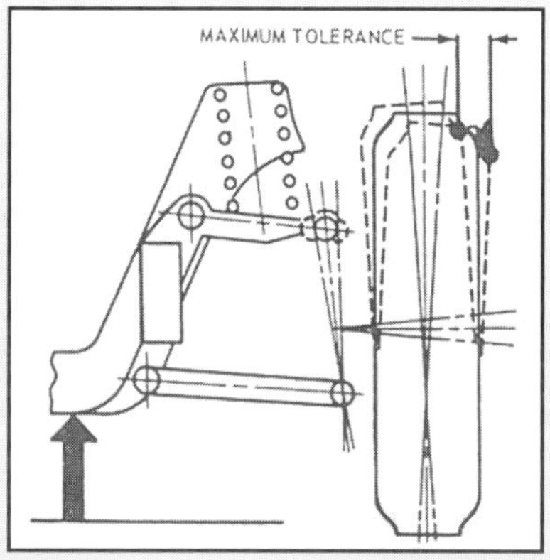

Upper ball-joint radial play (sideplay) is checked by lightly placing weight on the tire without a full load. Again, let a seasoned professional check front-end wear issues.

Toe-in is checked by determining the distance between the tire sidewalls or rims fore and aft. Marlon Mitchell of Marlo's Frame & Alignment, who has been doing front-end alignment for 25 years, checks toe-in fore and aft. He is seeking 3/32-inch to 11/32-inch toe-in.

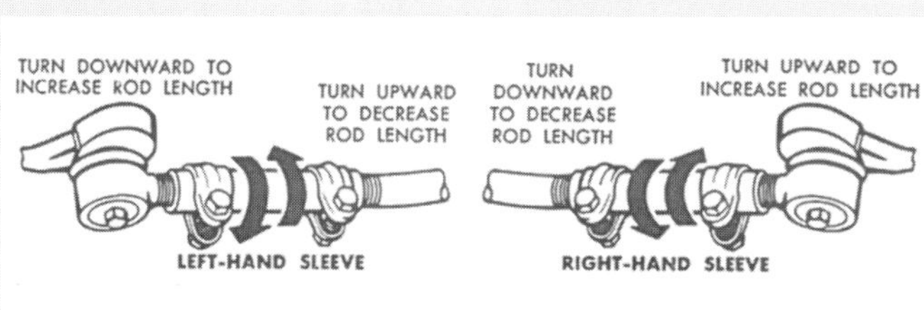

Toe-in is adjusted by twisting the tie-rod ends in or out. Marlon loosens both tie-rod ends and twists them accordingly to get toe-in spot-on. Road testing is a regular part of Marlon's routine. Nothing goes on the rack without a road test, and nothing leaves without one.

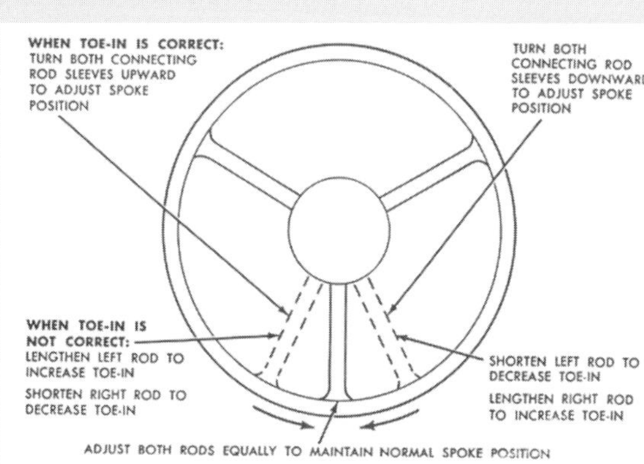

Ford suggests this approach for all toe-in adjustments. Follow this illustration to the letter to achieve proper toe-in in your home garage.

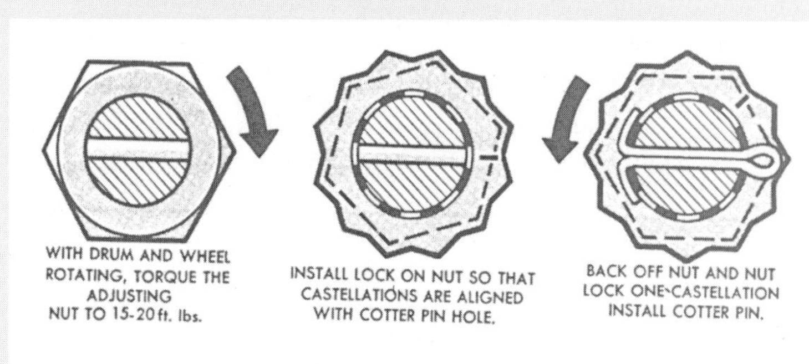

Wheel bearing adjustment has a direct effect on front-end alignment. Loose wheel bearings behave like bad ball joints and tie-rod ends. Alignment doesn't stay put when wheel bearings are loose.

Home Alignment

The Eastwood Company makes it possible to perform a front-end alignment in your home garage. You'll need the Fastrax camber/caster gauge from Pole Position Racing & Rod Products and the wheel alignment gauge from Manco Products. As its name implies, the Fastrax camber/caster gauge enables you to set camber and caster. All you have to do is roll the car back and forth once adjustment is made. Then, check camber and caster again to confirm adjustment.

The Manco wheel alignment gauge is for setting toe-in, which makes light work of front-end alignment. All you have to do is center the steering wheel and get to work underneath.

All front-end alignment work must be done with the full weight of the vehicle on the tires. Never perform a front-end alignment with the wheels off the ground. Again, when you set camber and caster, roll the car back and forth to stabilize adjustment, then check again. A road test is the final confirmation of your performance.

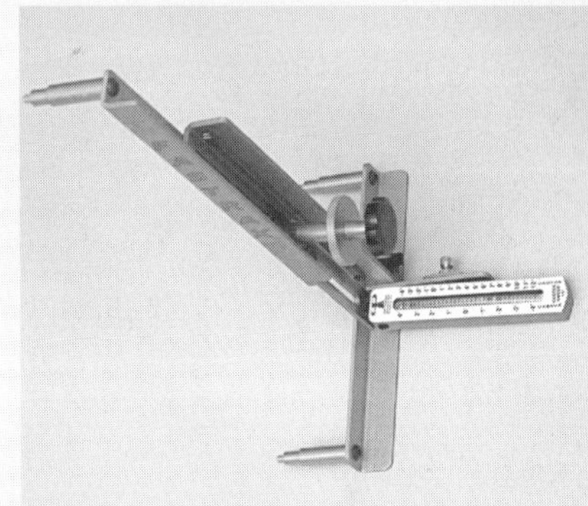

This is the Fastrax camber/caster gauge from The Eastwood Company. Easy to use with detailed instructions, the Fastrax gauge enables you to set camber and caster adjustment on your garage floor. If you're lacking confidence on this one, visit a trusted alignment shop afterward to have your work checked.

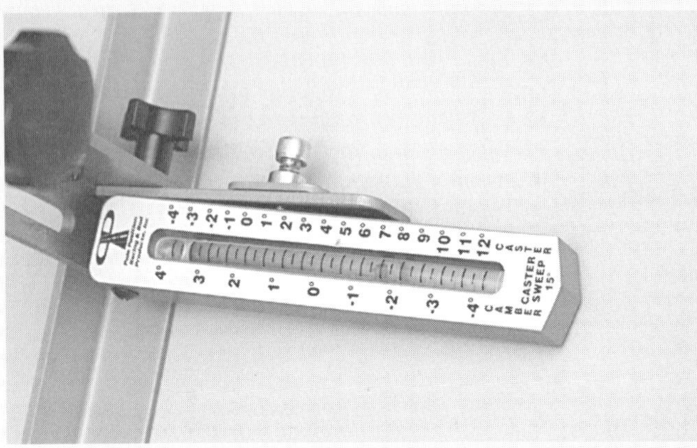

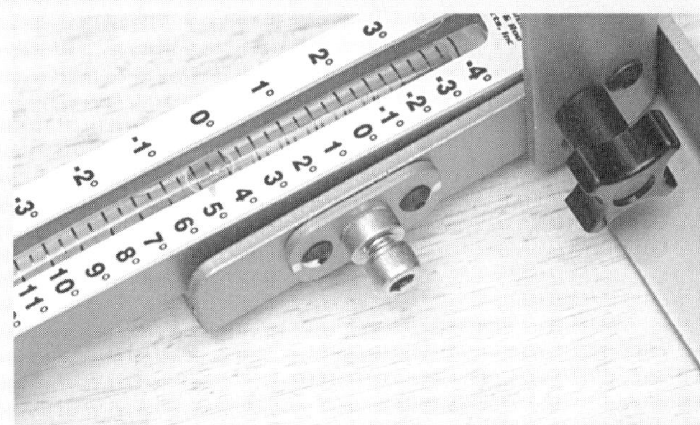

This is the same kind of bubble camber and caster gauges the pros use. We like this one because it's easy to read and easy to use. Lock it to your front wheels using the adjustable thumb wheel. Make sure you have your Mustang on a proven level surface, otherwise, readings will be erroneous.

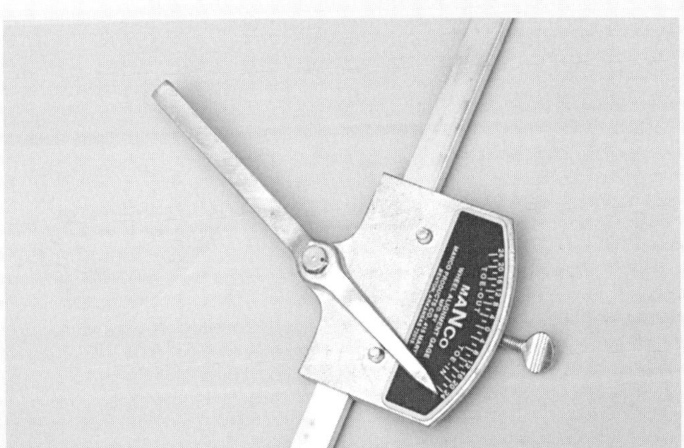

Here's the Manco alignment gauge for toe-in. With a helper close by, you can set toe-in in your garage just like professionals.

Chapter 18

Replacing a Battery Apron

Text and Photos by Jim Smart

Here's a quick fix for one of the Mustang's most common corrosion areas: the battery apron.

Classic Mustangs have always been cursed with rust and corrosion problems. The most notorious are cowl-vent leaks that result in floorpan and kick-panel rust-out, and the inner-fender apron under the hood where the battery sits. Because lead acid batteries contain sulfuric acid and emit corrosive fumes, they are especially hard on sheetmetal body parts like the inner fender and battery tray. In the past, we've shown how to replace the entire inner-fender assembly. But what if you don't want to replace the inner fender? It's best to keep the serial number intact, which is stamped into the inner fender. And sometimes it's just too much trouble to pull the fender and drill out all of those spot welds.

We visited the Northwest Pony Shoppe in Snohomish, Washington, just outside Seattle, to see how they correct battery-apron woes without major surgery. This is an easy fix for anyone who knows how to weld. And, if it's too much for your talents and abilities, this job can be done by your favorite autobody shop.

We're going to show you how to patch the inner fender. It won't look like a patch, and you'll save time and money in the process. For our repair, Mustangs Plus provided a new right-front inner fender, which is made of the same heavy-gauge steel as original equipment.

Sources

Mustangs Plus
www.mustangsplus.com

Northwest Pony Shoppe
northwestponies@yahoo.com

1. Here's a '67 Mustang battery apron and inner fender. At first glance, it doesn't look that bad.

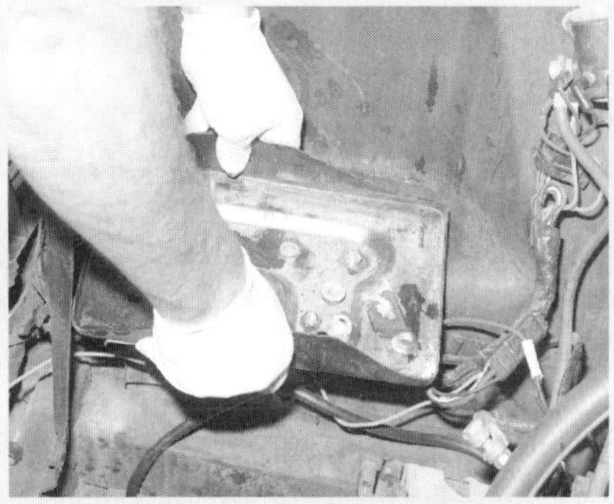

2. First, remove the battery tray by extracting the four retaining bolts. Depending on how your Mustang was assembled, you may have to remove the front bumper bracket to get to the bolt heads.

3. With the battery tray removed, it's apparent this inner-fender apron has been repaired before. Notice the sheetmetal patch that's been welded in place. This is not a solid repair. What's more, it looks awful.

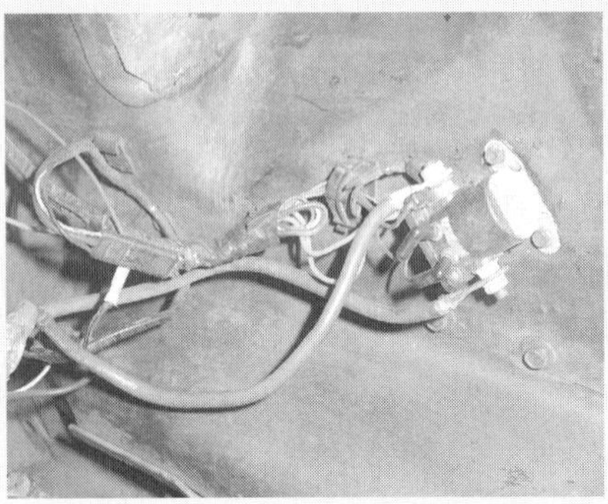

4. We suggest removing all electrical components in this area. No one wants to cut anything by mistake during inner-fender repair, especially the wiring harness.

5. Where you cut the inner-fender apron is a matter of preference. The Northwest Pony Shoppe cuts the battery apron only where necessary to eliminate the rust. The replacement patch from Mustangs Plus will fit this area nicely.

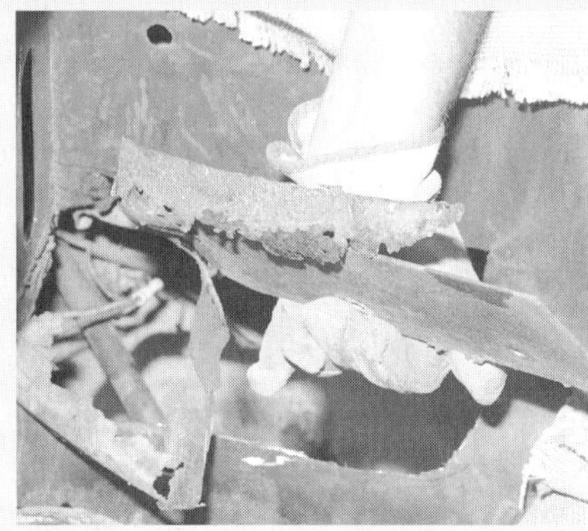

6. The corroded metal is removed as shown.

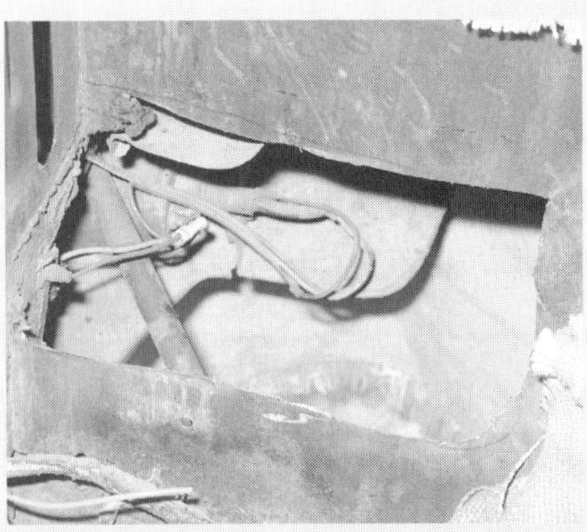

7. This is the hole after cutting. Don't panic! It will look much better when we're finished.

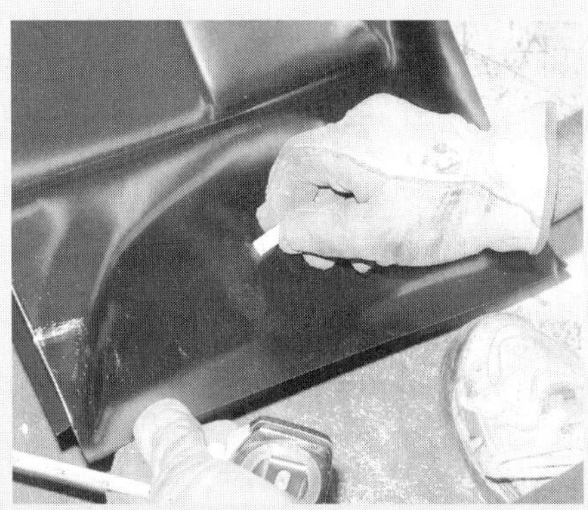

8. The huge inner-fender apron has to be cut down to size for proper fitment into the existing apron (left). Scott Robinson of the Northwest Pony Shoppe marks the new apron with chalk, then cuts it down to size (right).

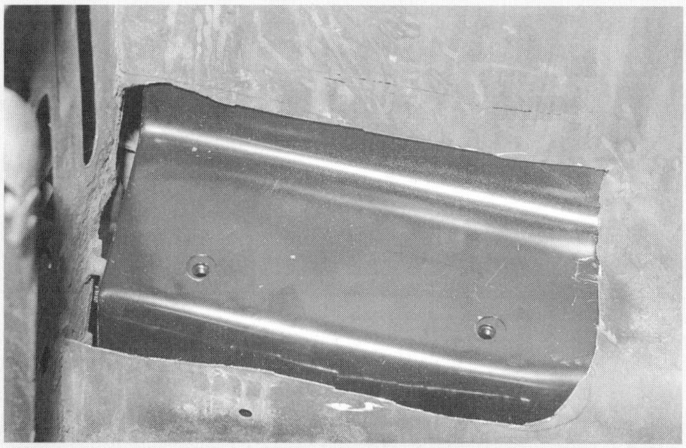

9. Scott fits the apron in place in order to mark the perimeter for trimming.

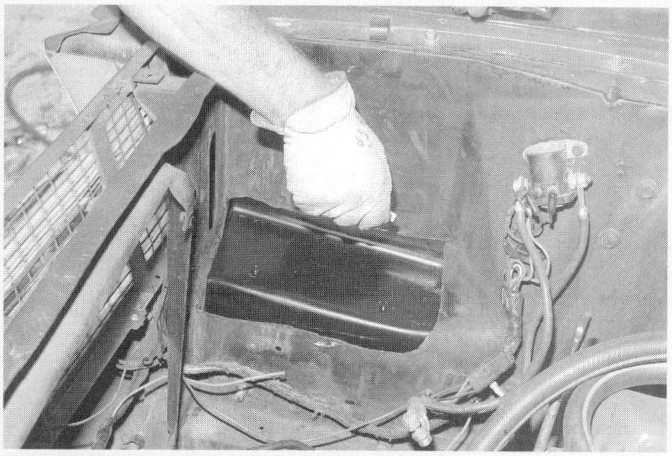

10. With the new inner-fender patch positioned correctly, Scott marks the perimeter with chalk for cutting and trimming.

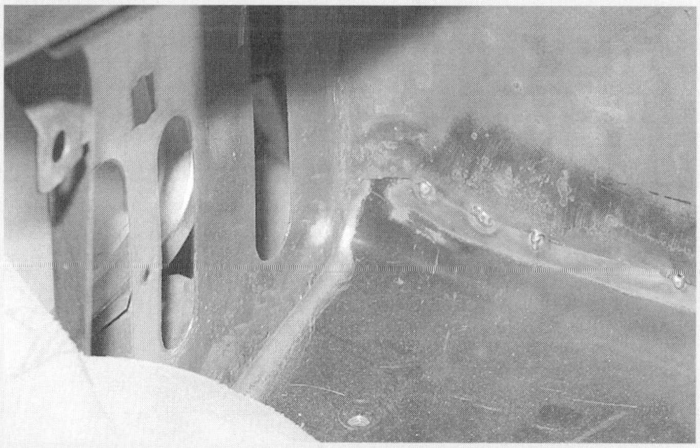

11. The inner-fender patch has been trimmed and fitted. Next, weld the patch in place. We're using a small, 110-volt wire feed welder from The Eastwood Company.

12. The welds are ground smooth in preparation for filling, priming, and painting.

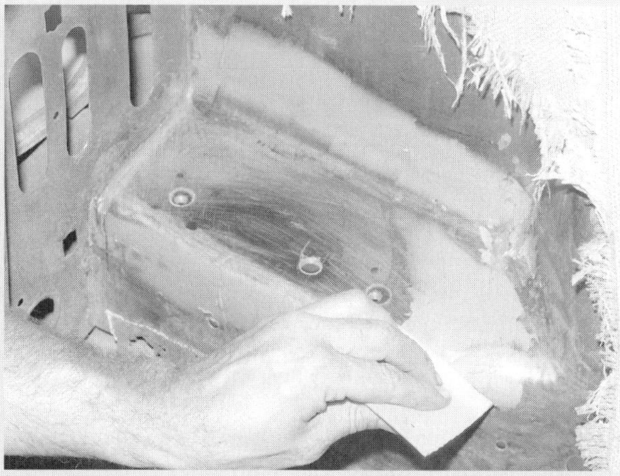

13. Body filler is used around the edges to fill imperfections. It will be sanded smooth, then primed with a self-etching primer/sealer.

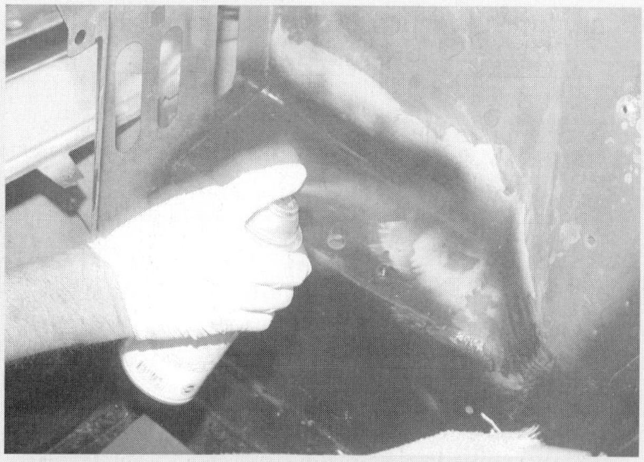

14. Primer and satin black paint are applied to finish the job. Installation of the battery tray is all that's left.

Chapter 19

Upgrading to Power Front Disc Brakes

Text and Photos by Miles Cook

At first we were just going to install a power booster, but after speaking with Master Power, we decided to upgrade to both a front disc brake package and the power-brake conversion. The brakes are relatively easy to install, and while the booster does take more effort, the results are worth it.

Installing front disc brakes ranks among the most popular upgrades for vintage Mustangs. We've covered the subject several times over the years and with good reason. From a purely safety standpoint, converting to front disc brakes is just about the single best improvement option.

This time we're looking in on a Master Power system as it's installed on a '67 Mustang. However, we're going beyond the installation of front disc brakes; we're detailing the conversion to power brakes as well. By adding Master Power's booster system, which includes the brake booster, a dual-reservoir master cylinder, and a power-brake pedal, we'll have better braking as well as a better feel.

While doing the whole setup at once is a great way to go, keep in mind that there are several options. If your '67–'73 Mustang already has manual front-disc brakes, you can convert to power brakes. Conversely, if you have manual drum brakes and want to do the power-brake conversion first, that's also possible. However, we recommend doing the disc brakes first and the power brakes later should you need to budget out the project over time. The power booster will work equally well with drum brakes or front discs. Master Power also offers complete power-disc kits for '65–'66s that came with manual drum, power drum, or manual disc.

We turned to our friends at Mustang Country International who supplied the installation know-how as well as the '67 Mustang subject car.

Source
Master Power Brakes
www.mpbrakes.com

Mustang Country International
www.mustangcountryintl.com

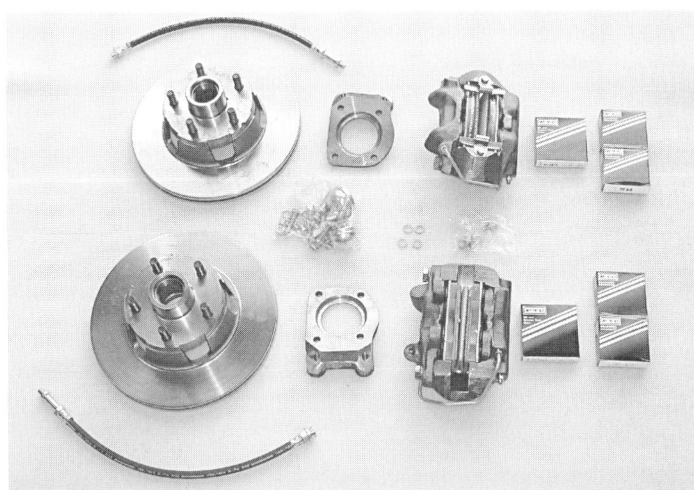

The Master Power front disc brake package is an ideal choice for a regularly driven early Mustang. The main bits include four-piston calipers, rotors, caliper-mounting brackets, brake lines, and wheel bearings. All related installation hardware is also included. The '65–'73 Mustang kits are similar to the '67 version with slight variations depending on the year of the car.

1. The only good thing about crusty, old drum brakes is how easy they are to remove. Each side only takes about 10 minutes. Begin by removing the dust cover, cotter pin, washer, and wheel bearing to allow removal of the brake drum.

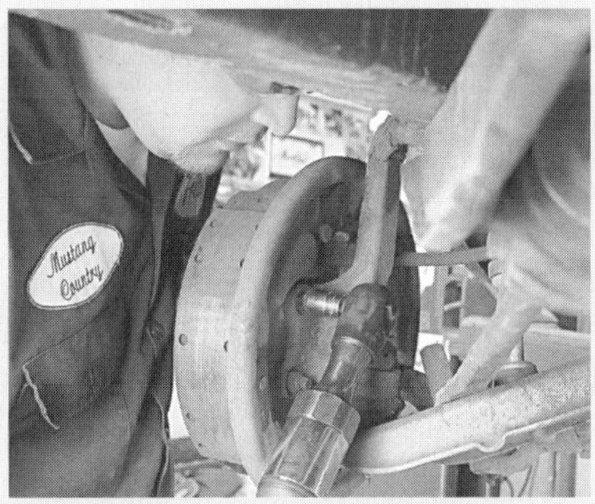

2. Once the brake drum is off, remove the backing plate by taking out the four nuts that hold it in place on the spindle. Access the nuts by reaching around to the backside of the plate. The plate, brake shoes, and all other related bits come off together. There's no need to disassemble the drum brakes prior to removing them.

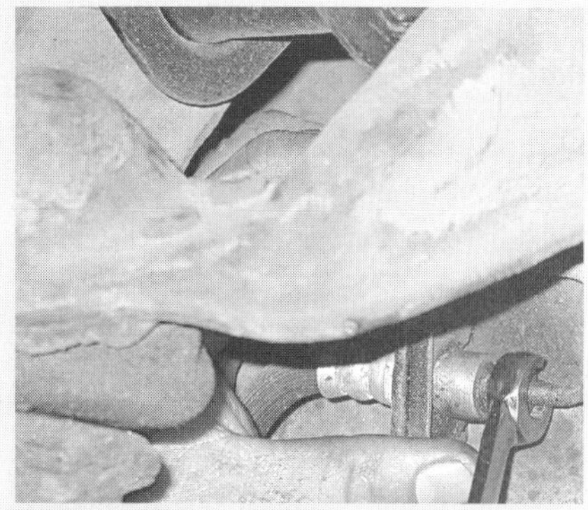

3. Next, unbolt the brake line from the drum-brake hose using a line wrench to ensure the nut on the line isn't rounded off. The line and the nut stay with the car and are reused with the new rubber hoses included with the Master Power brakes.

4. Remove the clip that holds the drum brake hose in place ...

5. ... to allow removal of the drum-brake backing plate. Then say goodbye to crummy drum brakes forever.

6. Clean the old wheel-bearing grease from the spindle so that it looks like the above. Then install the caliper-mounting bracket onto the spindle as shown. The two threaded holes should be mounted inward, toward the front of the car.

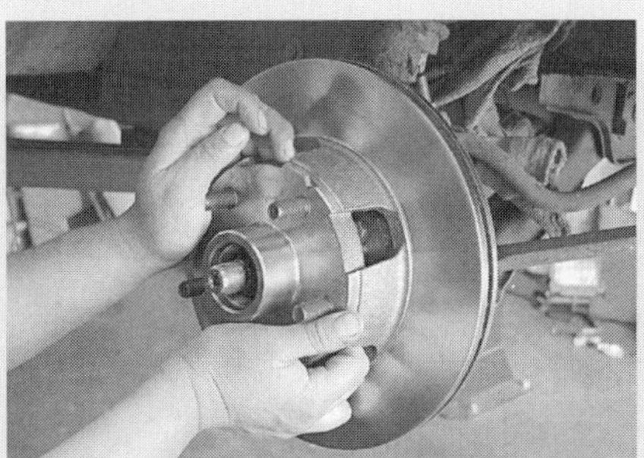

7. After the caliper-mounting bracket is installed, thoroughly coat the spindle with wheel-bearing grease. Next, grease the bearings, and install them into the brake rotors. Then install the brake rotors onto the spindles, and secure them in place with the new inner-wheel bearing along with the washer, castle nut, and cotter pin from the drum brake.

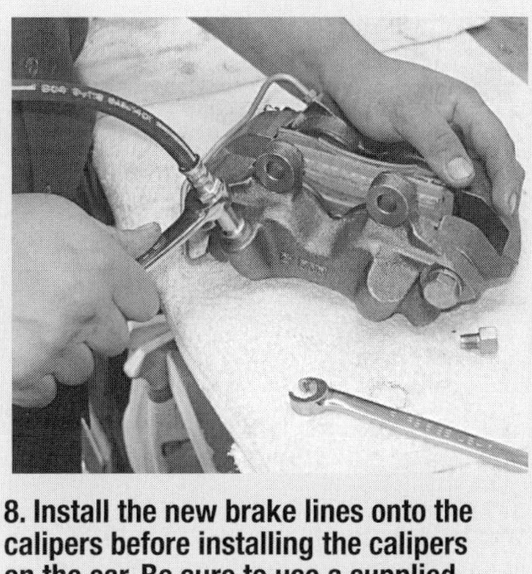

8. Install the new brake lines onto the calipers before installing the calipers on the car. Be sure to use a supplied washer between the line and the caliper to ensure there are no brake-fluid leaks.

9. Install the caliper by placing it over the rotor.

10. Using the supplied hardware, bolt the caliper into place by thoroughly tightening the bolts. Be sure to use the supplied lock washers along with the bolts.

11. We had to unbolt this brake-line bracket and slide the line forward slightly to allow it to reach the new brake line. This was a minor issue that took only a few minutes.

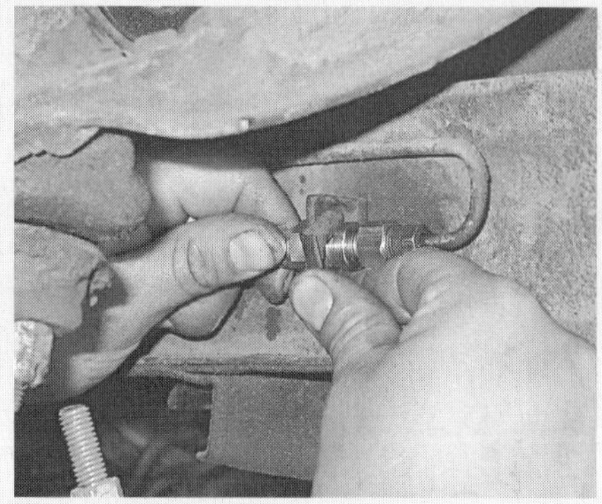

12. To complete the disc-brake installation, connect the new brake-line hose to the car's existing steel line. Then tighten it with a line wrench and reinstall the retaining clip as shown.

13. With the Master Power brake-booster package, the assembled master cylinder and booster are ready to install, and the vacuum-hose kit and power-brake pedal are also included. This package is considerably more difficult to install than the disc brake setup, but power brakes are a worthwhile upgrade on any vintage Mustang.

14. Remove the driver-side strut brace to access the master-cylinder area. This piece needs to come out so the new master and booster can be installed. It will also give you working room. Consider removing the hood as well to gain even more access to the firewall area. Remove the driver seat also, which will make it more comfortable to access the area underneath the instrument panel.

15. Unbolt the brake lines from the factory junction block, which will also be removed to make way for the block included with the new brake booster. Use line wrenches to be sure you don't damage the line nuts on the car's factory brake lines, which will be reused.

16. This cotter pin, located behind the instrument panel, secures the brake pushrod to the brake pedal. Remove it to take out the master cylinder. There isn't enough room underneath a car's dash for a tool, hand, and camera at the same time to show exactly what's happening, but a pair of needle-nose pliers is necessary to remove the pin, which will also release the brake light switch.

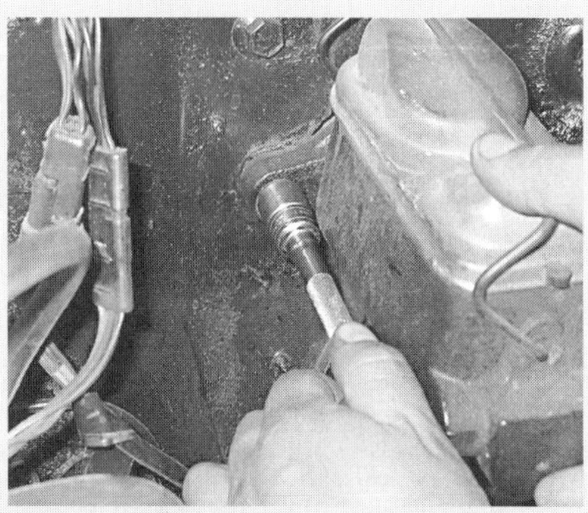

17. Back under the hood, the old master cylinder is easily removed by taking out the two bolts that hold it in place.

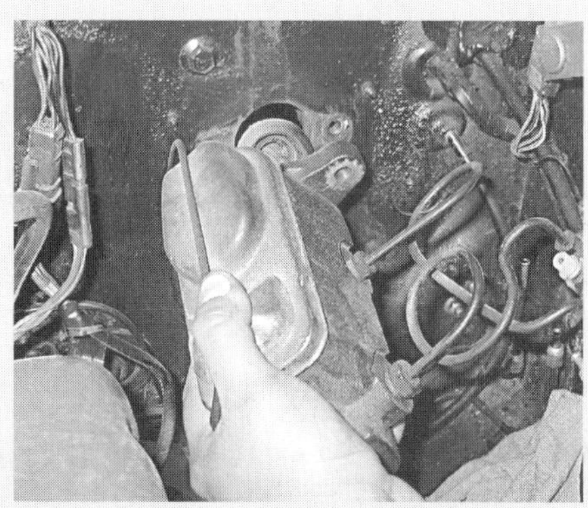

18. Carefully pull it away from the firewall while at the same time threading the pushrod out. Note also the stock junction block and brake lines coming out with the non-power master cylinder.

19. Back under the dash, there's another cotter pin that needs to be removed. This one holds the brake pedal in place, and once it's out, the brake pedal can be removed. On manual transmission cars, the clutch pedal also has to be removed to separate it from the brake pedal.

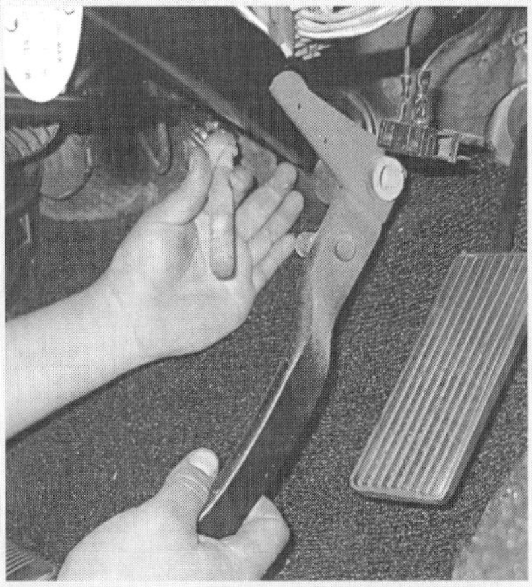

20. Once the cotter pin is removed, the pedal shaft and brake pedal will come out as shown above.

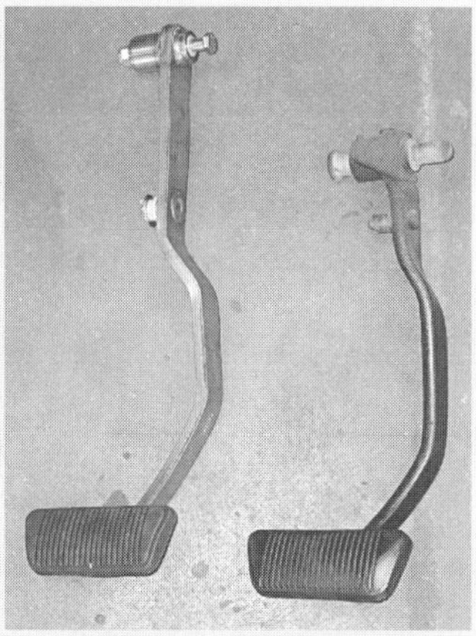

21. When upgrading to power brakes, replace the brake pedal with the supplied power-brake pedal. Due to the power-assist supplied by the booster, the power-brake pedal on the left is considerably longer and mounts to the pedal support higher up to provide less leverage. If a car has a manual transmission, the power-brake pedal installs the same way, and the clutch pedal ends up installing to the pedal support separate from the brake pedal.

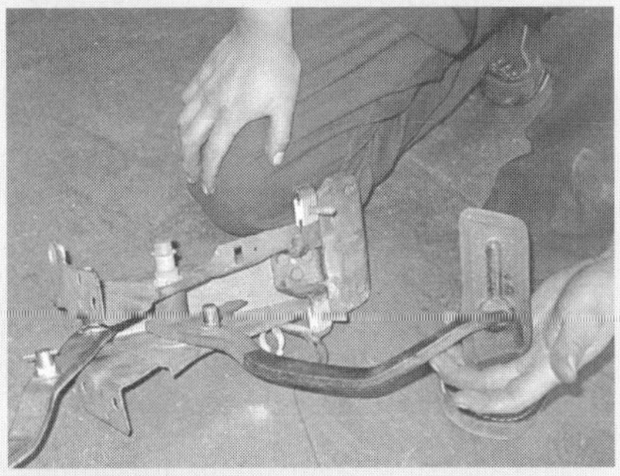

22. Here are both manual- and power-brake pedal attachments to illustrate how they install differently. While the power-brake pedal mounts higher onto the pedal support, it hangs down from the support the same distance to the car's floor due to its longer length.

23. In a supplied diagram, Master Power's instructions say to remove three of the four clinch nuts attached to the pedal support. This requires removal of the pedal support so the clinch nuts can be drilled out, which is shown here. The threaded clinch nuts must be removed so that separate nuts can be fitted to the studs attached to the new power-brake booster. These studs pass through the firewall and pedal support. To gain enough access to remove the pedal support, the steering column must be removed. To do this, we removed the four nuts on the firewall and the two nuts that hold it to the dashboard and unplugged the wiring harness. With the steering box's one-piece shaft still in place (later cars don't have a shaft in the way because the collapsible column and shaft come out together), undo the four bolts that hold the pedal support, and remove it past the steering shaft.

24. With the pedal support in a vice, the Mustang Country techs grind the clinch nuts down in preparation to knock them out.

25. Once the nuts are partially ground down, they are easily knocked out of the pedal support with a few good hits of a hammer and a punch.

26. Turning our attention to the firewall, several modifications are needed. First, drill a hole in the firewall directly below the large hole where the pushrod passes through. There is a dimple in the firewall indicating hole placement. The dimple was supplied for cars with original power brakes, but with manual brakes, the hole wasn't necessary.

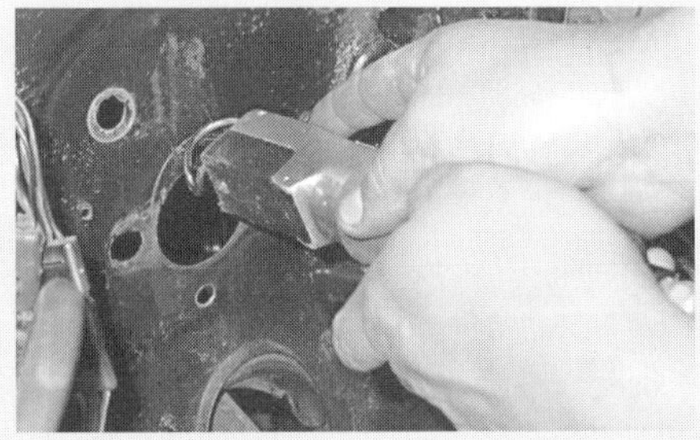

27. Open up the large hole at the top to allow the power booster to fit. Mustang Country used an air-powered tool with a small hacksaw-blade attachment. A hand hacksaw might also work since the appearance doesn't matter; the opening will be hidden by the brake booster. The directions and a supplied gasket or template illustrate how much material to remove. The directions also say to enlarge the two smaller holes on either side of the large hole to allow the brake booster to fit. By test fitting the booster/master, we discovered that we didn't have to do this on our car, but this will vary from car to car.

28. Now the new brake pedal and modified pedal support can be installed along with the steering column and steering wheel. Note the large holes in the pedal support where the clinch nuts used to reside. Those two holes are visible to the right of the brake pedal.

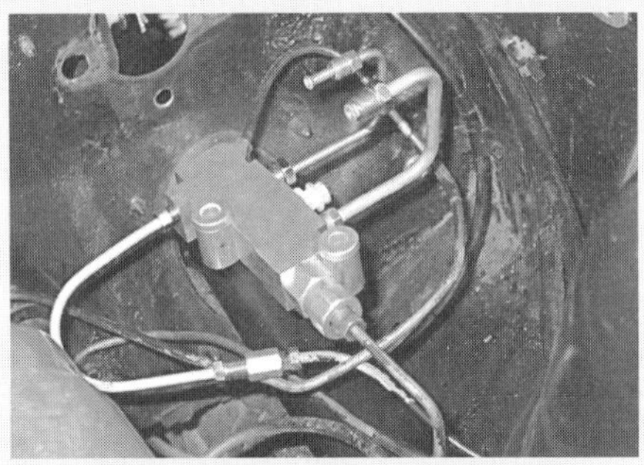

29. Before installing the brake booster and master, install the supplied junction block to the car's existing brake lines according to the Master Power instructions. The two front brake lines are joined at the rear of the block and the single rear line at the front. The master cylinder is then joined to the junction block with the two lines installed on the right side of the block as shown.

30. Next, the master cylinder/brake booster is installed and secured to the firewall using the supplied nuts. These are installed from inside the car underneath the instrument panel where the pedal support is located.

31. The brake booster requires a vacuum source from the engine. Master Power supplies a vacuum hose that can be cut to fit, and it's simple to run the hose from the booster to a fitting on the intake manifold. If no fitting is in place, Master Power supplies one for installation on the manifold.

32. Complete the installation by connecting the vacuum hose to the brake booster.

33. Here's the completed brake-booster/master-cylinder installation. The setup looks great, works perfectly, and has a cool OE-look to it. Bleed the car's entire brake system, and enjoy the huge upgrade from manual drum to power front disc brakes.

Now you can have the impressive stopping power of a Baer brake system on a vintage Mustang without having to increase wheel diameter.

Chapter 20

Installing Four-Wheel Disc Brakes

Text and Photos by Miles Cook

Upgrading to front or four-wheel disc brakes is one of the most popular ways to improve a vintage Mustang. While that's nothing new in a general sense, better brakes are and always will be a crucial way to make an early Mustang safer, more fun to drive, and better overall.

A well-respected and major player in aftermarket brake circles is Baer Brake Systems (formerly known as Baer Racing). Its systems and related components have been on the market for more than a decade, and the company consistently enlarges its catalog to include applications for almost any vehicle on the road today.

Hal Baer and Todd Gartshore started Baer Racing by offering four-wheel disc systems for Fox Mustangs in the early '90s—a setup that's still a hot seller today. By adding 12- or 13-inch rotors and the PBR calipers found on C4 Corvettes and SN-95 Mustang Cobras, a Fox 5.0 can stop with the best of them, including Porsches, Ferraris, BMWs, and Benzes.

Magazines have been doing brake-installation how-tos for years, and many of them are just that: installation how-tos. We're going beyond that in two areas. First, we have a complete set of before-and-after test numbers that show the phenomenal difference in performance offered by a set of four-wheel disc brakes on a typical vintage Mustang compared to the prehistoric drums many early Mustangs still have today.

Source
Baer Brake Systems
www.baer.com

We positively flogged our '69 SportsRoof with the new brakes in place, stopping back-to-back-to-back more than ten times from 60 mph to show the repeatability of the four-wheel disc package. We also did five stops from 100 mph to further illustrate stopping power. For comparison, of the four attempts we made from 100 mph with the drums, only the first two were anything resembling stops. The last two times, the car only coasted to a stop, as braking power was essentially nonexistent at that point.

There's also intriguing news for all vintage enthusiasts who want to retain a car's original wheels. Baer now offers the Serious Street front-disc system that utilizes the excellent PBR dual-piston caliper, newly designed caliper mounting brackets, and 11-inch one or two piece rotor assemblies that are functional and pleasing to the eye. The 11-inch rotor allows you to retain the stock wheels (almost any 15-inch and some 14-inch), thus eliminating the need to run a 16-inch wheel with a Baer package, which is required with a 12-inch rotor. In short, you can now have the quality of a Baer package but it can be almost completely hidden by a vintage car's stock wheels.

However, since our Mustang was already equipped with 16-inch wheels and we love the look of the trick PBR calipers and cross-drilled rotors showing through the wheels, we fitted front Sport and rear Touring systems that use 12-inch rotors. The installation sequence shows that system, and the test numbers that were generated with it.

For those interested in retaining the stock look of perhaps a Magnum 500 or Shelby-style wheel, see the sidebar, "Big Brakes for Small Wheels" on page 94. We detail the Serious Street Plus package that features the optional high-level 11-inch, two-piece rotor. The Sport and Serious Street systems use the equivalent PBR caliper; the installation procedure for both is basically the same.

1. Since our '69 fastback is equipped with Vintage Wheel Works 16-inch Vintage 45 wheels, we installed front and rear Baer systems that would spec out the car with a good size 12-inch rotor at each corner. The front package is known as a "Sport" system. If you're running 17-inch wheels, go with a 13-inch rotor, which would label the setup a "Track" system.

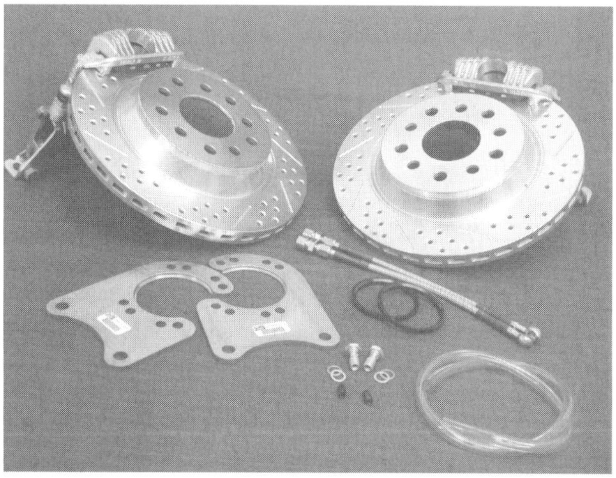

2. Out back, the "Touring" system also features a 12-inch rotor. A smaller rotor is available that fits behind most 15-inch wheels, which would make the rear assembly a "Rod & Drag" setup. In all cases, the completeness of the Baer systems is a hallmark of its product line. A clear rubber hose is used to connect to the caliper bleeder valves during the bleeding process. Included are the rotors, calipers, hubs, mounting brackets, master cylinder, stainless lines, and all installation hardware. The rear Touring and Rod & Drag systems, which use single-piston PBR calipers, are similarly outfitted.

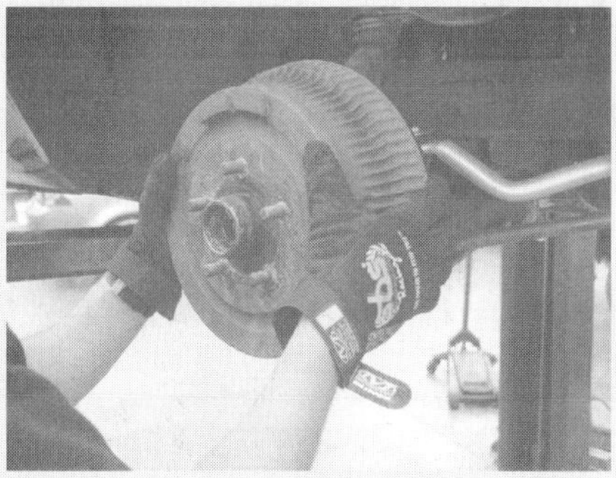

3. Good riddance to the crusty old drum brakes from more than three decades ago. Our test results show that, by today's standards, these things are downright unsafe in extreme braking situations. The drum comes off after removing the cotter pin, spindle nut, and washer.

4. While removing the brake shoes and other related pieces, unbolt the drum backing plate from the spindle as shown, and the whole thing can come off.

5. Once the line is unbolted from the car with a line wrench. You can opt to keep the drum brakes, but since our car isn't a big-dollar Mustang like some, we tossed them in the dumpster.

6. The car's factory drum brake spindle—all cleaned up and ready for the Baer hardware—looks like the above.

7. First up is the caliper mounting bracket, which attaches to the spindle as shown. Securely tighten the bracket to the spindle with the supplied hardware.

8. Install and seat the new hub. This is a snap because the races and studs are installed, and the hubs are also assembled with Timken bearings and already packed with Redline synthetic grease. Next, after the keyed washer (shown), install the castle nut and tighten to 10 ft-lb. Rotate the hub back and forth several times, loosen, then torque the castle nut again. Rotate the hub again, then loosen a second time and retighten just enough to take out any slack in the hub. Install the supplied cotter pin and dust cap.

9. ...placing the rotor onto the hub. Secure the rotor with one or two lug nuts to hold it in place, then install the caliper and anchor plate over the rotor as shown. Be sure you have the correct rotor and caliper on the correct side by confirming the rotor slots rotate in reverse, and the bleeder valve on the caliper is facing up. The rotors are also marked for the side on which they should be placed.

10. The Baer techs stress the importance of properly torquing the caliper anchor-plate bolts with a half-inch-drive torque wrench to 110 ft-lb; we're showing that here.

11. Attach the supplied stainless steel brake hose to the caliper using one of the supplied banjo bolts. Be sure there's a crush washer on either side of the banjo bolt before fitting it to the caliper. Cinch the bolt to hold in place, but don't tighten yet.

12. Route the other end of the hose through the existing hose-retainer bracket on the frame rail and attach the hard line to the bracket with the supplied clip. Connect the car's brake line to the hose as shown and tighten to 20 ft-lb. Route the hose so it doesn't interfere with any suspension components or the wheels/tires during full suspension travel and lock-to-lock steering. Tighten the banjo bolts with only enough force to seat them securely. Do not over-tighten.

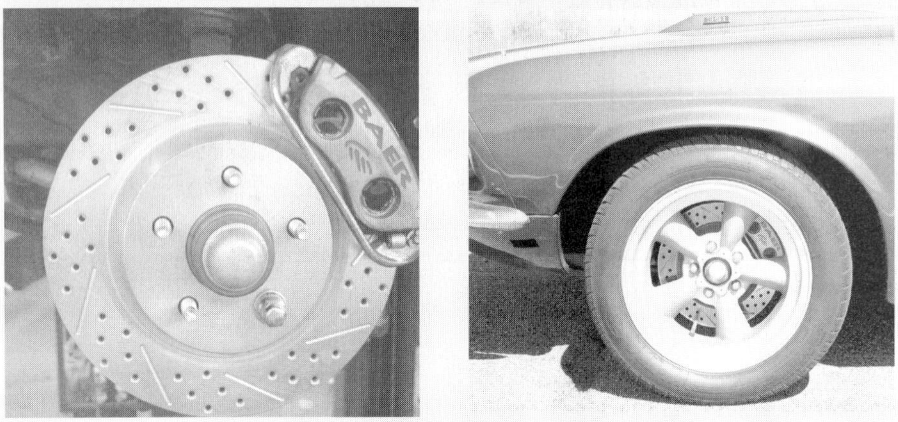

13. The installed Sport system looks like this on the driver side, with and without the 16x8-inch wheel and 225/50ZR16 tire combination.

14. Moving to the rear, begin by removing the drum and unbolting the axle backing plate.

15. Carefully remove the axle shaft from the rearend housing. This is a good time to inspect the bearings and bearing surfaces for damage or wear.

16. Disconnect the brake line and remove the drum backing plate from the rearend by pulling off the housing.

17. To better secure the rear brake lines to the rearend and protect the lines from damage while driving, Baer devised this simple and effective L-clip that attaches with a hose clamp. Fitting it involves carefully bending the factory hard line away from where it connected to the drum backing plate and pointing it down at approximately a 90-degree angle so it meets up with the L-clip. Then feed the end of the hose clamp through the opening in the supplied clip and install onto the axle as shown.

18. Carefully reinstall the axle into the rearend housing and install the combination caliper mounting bracket and axle-bearing retainer to the axle housing, using the original housing-to-backing-plate fasteners, torquing them to 40 ft-lb. With these brackets, there are eight possible caliper-mounting positions. The calipers and brackets can be swapped from side to side to achieve all eight positions. The calipers can be mounted in any of the positions illustrated in the instructions; however, they must always be oriented so the bleeder valve is on the top side. Use the mounting position that clears the leaf springs and shock mounts.

19. Install the supplied black centering rings over the axle hub.

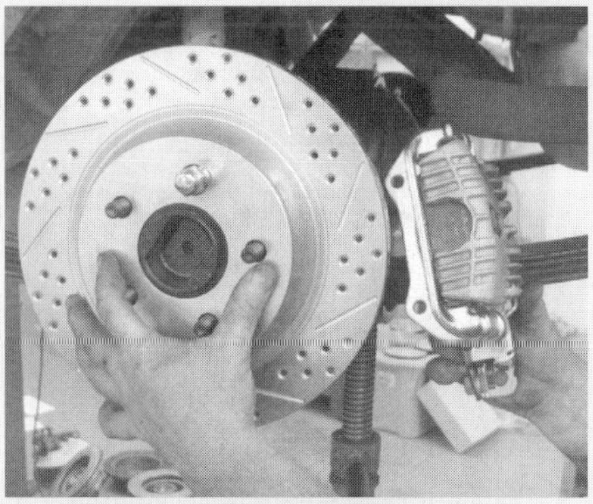

20. Install the rotor onto the axle and lugs. Use a lug nut or two to hold the rotor in place. Then install the caliper and caliper-mounting plate onto the rotor.

21. Torque the caliper mounting bolts to 85 ft-lb.

22. Install the stainless brake hose onto the caliper using a banjo bolt and two crush washers on either side of the banjo fitting. Because of the leaf spring in our car, we had to start the banjo bolt on the caliper before the caliper went into place. But once the caliper was on, we had enough room to get in a wrench to tighten the fitting.

23. Install the other end of the hose into the L-clip with a hose-lock clip.

24. Next connect the stainless hose to the axle's hard line as shown with a pair of correct-size wrenches. Torque the fittings to 18 ft-lb.

25. Here's the finished Touring system installed on the car's 9-inch rearend and ready for the 16-inch wheel and 245/50ZR16 tire. At this point, the supplied parking-brake cables are next to install, but we'll do that when we locate the required factory parts.

26. The new master cylinder installs onto the firewall using the same bolts that held the stock unit in place. This shows the configuration of the new master cylinder, lines, and proportioning valve. As shown in photo 1 of the front system on page 90, the supplied lines are flared only at one end. The other ends need to be flared to complete the installation of the master cylinder. Since installations vary from car to car (front disc only with no proportioning valve, front and rear disc with a proportioning valve, and so on), the lines that have to be flared (and equipped with new or existing flare nuts) will vary. With everything in place, bleeding the system is the final step.

Big Brakes for Small Wheels

The Serious Street system with its 11-inch rotors is the way to go with original 14- or 15-inch wheels.

The one-piece Sport 12-inch front rotor on the left went on our car for testing. The two-piece 11-inch rotor on the right is the upgrade on the Serious Street system, making it a Serious Street Plus package. The two-piece rotor precludes a separate hub that's required with a one-piece rotor, and the rotor/hub installs onto the spindle all at once.

In addition to the specially made rotors, the 11-inch-rotor-equipped Serious Street systems also use a unique caliper mounting bracket (right) to allow the rotors to fit inside wheels smaller than 16 inches in diameter. At left is the caliper mounting bracket for the Sport system that uses a 12-inch rotor.

Installation of the Serious Street package is the same process as other Baer systems. The caliper-mounting bracket is installed first...

...followed by the two-piece hub and rotor (or hub and one-piece rotor) that go on as a single unit.

Here's the Serious Street Plus system behind a 15x7-inch American Racing Torq-Thrust D wheel and 205/60R15 tire on the same '69 Mustang, before we fitted the front Sport and rear Touring systems for testing.

Test Results

We measured before-and-after brake performance by using Baer's Stalker radar gun along with a laptop computer and custom software that measures and records braking distances. Testing was done from two speeds: 60 and 100 mph.

The first trials were the baseline 60–0 runs. Then we drove the car for about 8–10 miles to cool the brakes before performing the four baseline runs from 100 mph. After the 1,200-foot-plus stop, we knew the drum brakes were toast. The same process was done with the disc brakes in place. While the overall distance reductions and averages were dramatic, most astonishing is how consistent and repeatable the disc brakes are. Quite simply, no stock drum brake system on any vintage car can hold a candle to the Baer components.

Finally, the stops are listed in the order they were achieved, which further shows how the stock brakes fade into oblivion and the Baer brakes stand up to repeated hammering with virtually no fade. You can expect the Serious Street 11-inch system to perform fairly closely to the Sport/Touring 12-inch system we tested. Of course, tires also play a key role in overall stopping distances. Stickier tires help produce shorter stops.

Stopping distances from 60-0 mph in feet

Drum	Disc
215.77	147.11
182.76	146.23
168.55	144.61
199.79	143.33
201.35	135.74
248.38	139.73
246.98	147.71
260.12	137.79
272.37	146.94
306.69	145.71
N/A	150.62
Average:	**Average:**
230.28	144.13

Stopping distances from 100-0 mph in feet

Drum	Disc
636.03	382.62
665.12	361.27
844.86	389.58
1247.91	381.89
N/A	351.45
Average:	**Average:**
848.48	373.36

Chapter 21

Replacing the Clutch & Brake Pedal Assembly

Text and Photos by Miles Cook

The clutch and brake pedal assembly on early Mustangs can wear out. Here's how to replace it.

Several readers have written in the past wondering how to replace a manual transmission clutch spring in a vintage Mustang. While it's a fairly esoteric subject, we still wanted to investigate the issue. In the meantime, we ran across a car that needed a complete pedal-assembly replacement, which in turn also gave us the chance to examine the issue of where the clutch spring is and what needs to be done to get to it.

The result is we're killing two birds with one stone. That is, we're doing quite a bit more than originally intended, which we hope will be more helpful than if we replaced only a clutch-pedal return spring.

The main problem we've seen on some vintage Mustangs is the way the pedals hang in the footwell. They're slanted at the wrong angle, and clutch-pedal effort is much harder than it should be. We knew that in the case of our subject car, the situation would grow worse every time the clutch was used, since it was already grinding.

On vintage Mustangs, the clutch and brake pedal operation and design is pretty much the same, so the problem on this '69 Mustang isn't confined to any model year. The issue is more related to how much the car has been driven and for how long maintenance of the clutch linkage has been overlooked. After 37 years of service, this car's pedal assembly needed serious help, so we decided to show how the entire assembly is configured, at the same time illustrating the clutch-spring location if its replacement is necessary.

Source
Mustangs Etc.
www.mustangsetc.com

1. After disconnecting the battery, we looked up underneath the dash with a flashlight and determined right away that the whole pedal assembly could be jiggled around. It would have to come out of the car for inspection. Since the heavy bracket that supports the pedals mounts to the firewall at a point above the steering column, the steering column had to be removed first. Under the hood, we disconnected the column from the steering box at the rag joint. Fasteners at the base of the column tube going into the firewall were then removed. The underdash support for the column was undone, and we disconnected the electrical wiring. On some cars, the turn signals and horn have two separate plugs to disconnect before pulling the column. Other cars have a single plug found above the column, directly behind the speedometer.

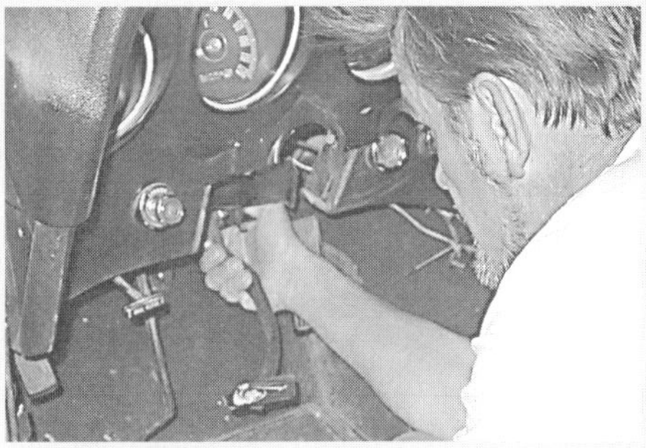

2. After removing the driver seat to ease underdash access and disconnecting the clutch pushrod at the equalizer bar, we removed the brake-light switch and master-cylinder pushrod at the brake-pedal arm. After removing four bolts at the firewall and two under the dash that retain the pedal-support bracket, we removed the pedal assembly from the car.

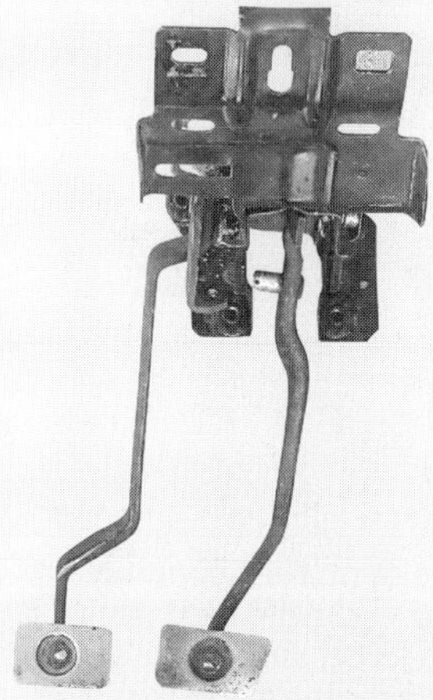

3. Here's the pedal assembly on the ground. The offset brake pedal doesn't look quite right, but we also discovered problems in several other areas.

4. This close-up of the pedal assembly shows the badly misaligned pedals. The pin on the brake-pedal arm that joins the master-cylinder pushrod slants off at an odd angle, and metal shavings are evident.

5. Disassembly of the pedals begins by removing the large clip used to retain the pedal cross-shaft. Look closely and you can see the remains of a nylon bushing.

6. Considerable damage is evident on the other side of the pedal assembly. The clutch and brake-pedal cross-shaft has worked its way through a nylon sleeve and the support bushing. As a result, the cross-shaft was operating against the bracket itself, further wearing away with each cycle of operation. Before long there would've been complete failure of the linkage, resulting in no clutch operation. Looking at these parts after nearly four decades of wear, it's easy to understand why the pedals were misaligned and the pedal-effort level was so high.

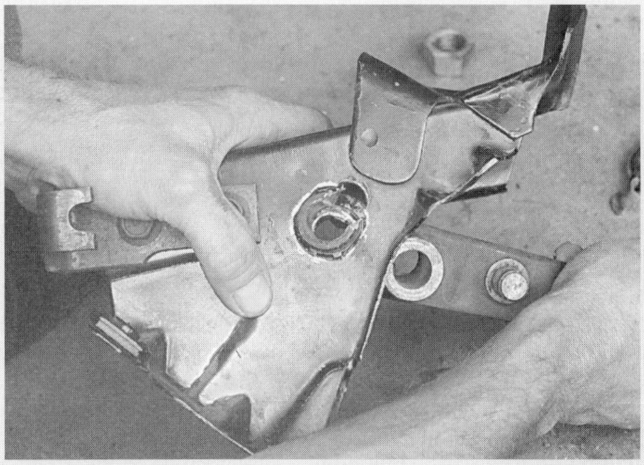

7. As we removed the brake pedal, the old bushings fell out. The required nylon sleeve inserts are destroyed or gone, and the brake-pedal arm seems like the only salvageable part so far.

8. The other side of the bracket shows how the bushing-support opening looked before the damaging wear began. Two types of pedal-mounting brackets were used by Ford. This type has the flat-sided openings shown here. The other type has a simple round hole, which can easily be repaired with a commonly available rebuild kit that includes a new shaft and bushings. There is no repair kit for our type of bracket, so we contacted Mustangs Etc., a well-stocked shop with a large inventory of used parts. The damage to our pedal mechanism was so widespread that we needed to replace the entire assembly. The main bracket we need went out of production long ago.

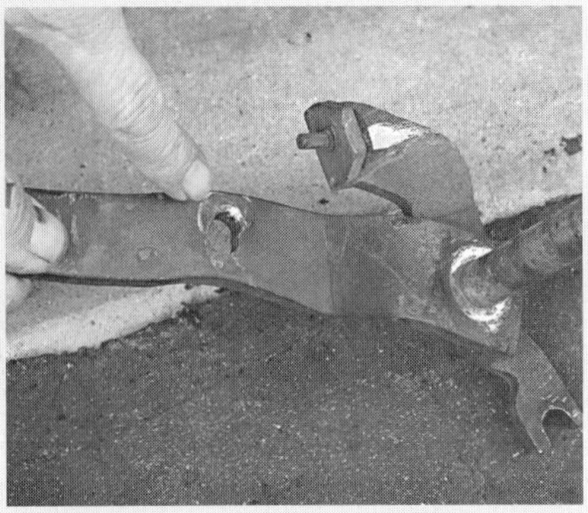

9. This clutch-pedal arm can't be reused either. The hole in the pedal arm for the clutch pushrod has worn oblong to the extent that the nylon bushing would no longer fit securely.

10. The result of metal-to-metal contact during operation can also be seen on the end of the pushrod itself. This pushrod is clearly ready for the scrap heap. A new pushrod is shown at right.

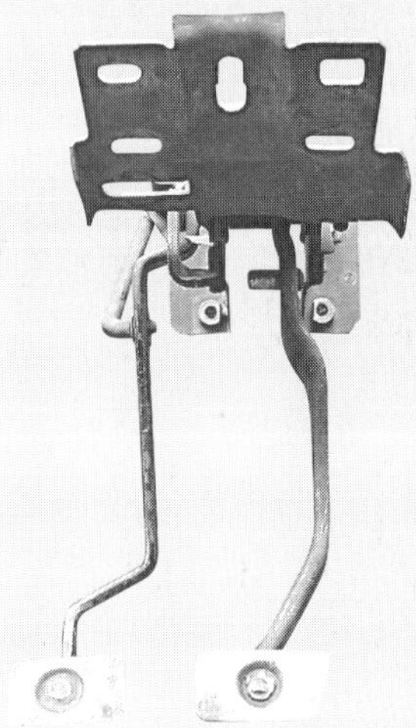

11. Here's the used manual-transmission pedal assembly—complete and in good working order—we obtained from Mustangs Etc. They also had a N.O.S. bracket. While some parts might not be available all the time (as is the case for any used components), Mustangs Etc. or other used-parts houses around the nation are all possible sources. This assembly differs from the original in that it has the round-style cross-shaft bushings. Regional differences in local suppliers to the three different Mustang assembly plants (Dearborn, San Jose, and Metuchen) could account for the existence of the two styles. Also note that if Mustangs Etc. has a pedal assembly in stock, you'll need to exchange your old one as a core.

12. Before we installed the replacement pedal assembly, we installed new nylon sleeves, or bushings, that we also obtained from Mustangs Etc. White grease was used to lubricate the bushings during assembly.

13. Even though this nylon pushrod bushing was in one piece, we replaced it before assembly. We also installed a new bushing at the equalizer end.

14. Before installing the replacement assembly, we made sure the clutch-return spring at the pedal was properly installed and lubricated. Once inside the car, pedal travel is limited and installing the spring is more difficult. This is what the clutch spring looks like for '69–'70 Mustangs. It's slightly different for '65–'66 and '67–'68. The clutch spring can be replaced without removing the pedal assembly, but does require that you reach up under the dash and remove it from there. At this point, we also made sure everything was correctly set up under the hood. Luckily, the equalizer bar between the framerail and engine block was not damaged or worn at its pivot points, and it was cleaned up and reinstalled.

15. The replacement pedal assembly is now ready to be installed. You'll need another person under the hood to support the master cylinder and start the fasteners while the pedal bracket is held in position inside. It's also easier to reconnect the brakes at the pedal while the steering column is still out of the car.

16. Reinstall the steering column while another person guides the rag joint back onto the steering box. First install the underdash fasteners to support the column in position while completing other connections.

17. Here we're getting the last of the dash trim back into place.

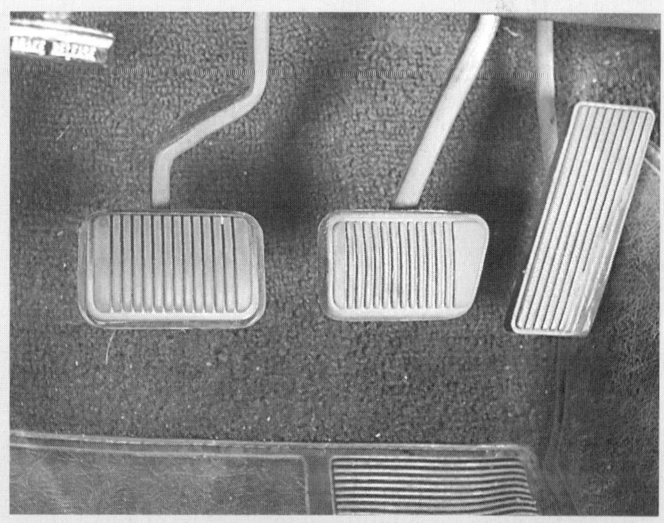

18. A pair of new pads had the pedals looking good. The clutch linkage had to be readjusted by a considerable amount as it had been lengthened to take up all of the slop in the linkage. Pedal effort was cut in half and the car is now much more enjoyable to drive.

Chapter 22
Installing a New Fuel Tank

Text and Photos by Miles Cook

A new fuel tank does wonders for the appearance of the trunk. We freshened up the inside with some spackle paint from the local auto parts store. By no means a show-quality restoration, the car's trunk looks much better now.

After 35 years, some parts just wear out. It's likely Ford never intended for most '69 Mustangs to last upwards of four decades, but many of them have because we love our vintage Mustangs so much.

It was clearly time to replace the fuel tank in this '69 fastback. There were small pinholes in the top of the tank that caused leakage (this is also the trunk floor), and the fuel smell inside the car became stifling after a short period of time. Furthermore, after discovering the holes, we couldn't fill the tank because the fuel ended up in the trunk.

The obvious solution was to replace the tank, which would eliminate the smell and provide the capacity for more fuel by switching to a larger one. The 22-gallon tank is made by Spectra Premium Industries and is available from Laurel Mountain Mustang and other Mustang mail-order vendors. Installing the tank is a basic R&R procedure outlined in the following photos and captions. Details on the different-size tanks available from SPI follows.

Sources

Laurel Mountain Mustang
www.laurelmountainmustang.com

Spectra Premium Industries
www.spectrapremium.com

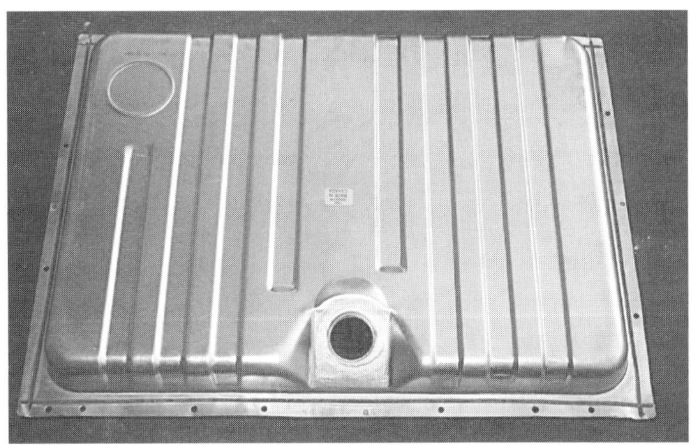

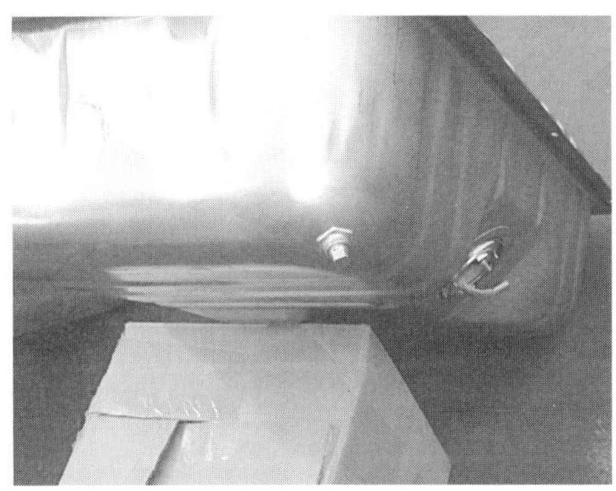

A faithful reproduction of the original tank, Spectra's '69–'70 tanks include a drain plug at the right-front corner.

You Can Go Bigger

Many of you vintage Mustang experts likely know about the various fuel-tank sizes that came in the '65–'73 cars. But it's interesting information for those of you who might be unfamiliar with the particulars.

The '65–'68 Mustangs came with a 16-gallon tank. Ford nicely upped the capacity to a healthy 20 gallons for the bigger '69s, like our subject car. It was enlarged again for the '70s, as those cars came with a 22-gallon tank. For the '71–'73 big-body cars, a 20-gallon tank was secured to the trunk floor from underneath the car with retaining straps.

The good news is you have several options in tank sizes for '65-'70 Mustangs. You can go with the size the car came with originally, a likely option for a restored show car. But for a driver, going with the largest possible tank is the hot ticket. In other words, a 22-gallon tank, as found in '70 Mustangs, can be installed in a '69 or '65–'68 cars. But the filler neck and hose might need to be modified if you're installing a 22-gallon tank in a smaller '65-'66 car. We were able to reuse our hose without cutting it and slightly bending the filler neck to make it work because the 22-gallon tank is a bit taller than the 20-gallon unit.

For a '65–'68 Mustang, adding a 22-gallon tank is a great way to dramatically increase range, more than 100 miles in some cases, depending on your car's gas mileage. Finally, if you make the move to a larger tank, use the correct corresponding sending unit to retain gas-gauge accuracy.

1. Here's our starting point. Nice, eh? The stain in the middle is from fuel leaking out of the tank when it was filled up. We can only guess how many years this had been going on. One day we filled the tank to the top, then pressed lightly on it, and fuel squirted out through several pinholes. We realized it was time for a new tank.

2. Removal of the old tank is easy, and takes about 20 minutes. First remove the bolts that hold it in place inside the trunk. There are 11 bolts. Of course, you want as little fuel as possible inside the old tank when you remove it.

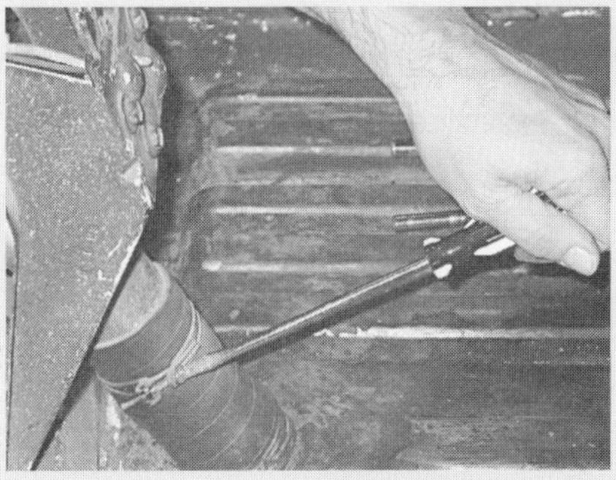

3. Unscrew the filler hose so it can be removed from the filler neck and tank. These factory clamps were in pretty good shape, even after 35 years, so we reused them after a cleanup with spray lubricant.

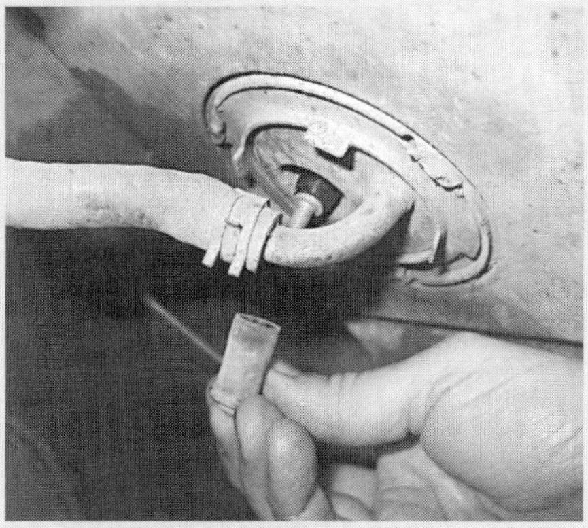

4. Underneath the car, unplug the sending wire for the fuel gauge...

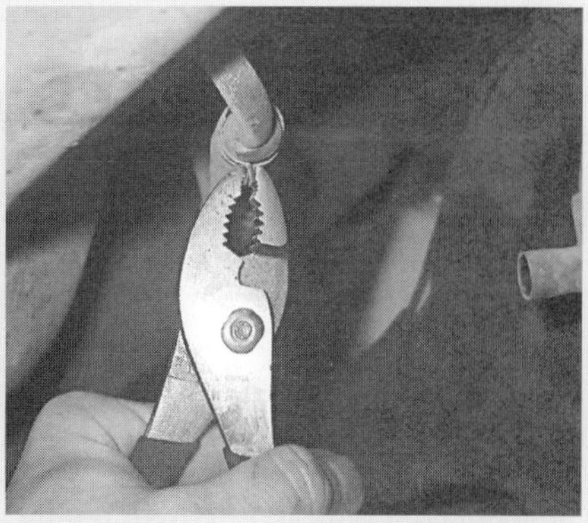

5. ...and remove the fuel-line clamps with a pair of pliers. When removing the fuel line, be sure not to spill fuel as some will likely come out when you disconnect the line from the tank. Have a gas can handy to catch any fuel that leaks from the fuel line.

6. With bolts, filler hose, fuel line, and gauge sending wire unplugged, the tank is lifted out through the trunk opening. Drain any excess fuel into a gas can, then run water through the old tank to eliminate the danger of an explosion from fumes in the tank ignited by an outside source.

7. You can reuse the existing sending unit or order a new one from Laurel Mountain Mustang. If you use the same-size tank, the original sending unit can be reused. We're using a larger '70 tank (see sidebar on page 101), so we replaced the original with the correct version to retain fuel-gauge accuracy. To remove the sending unit, push the retainer ring with a screwdriver to disengage it from the retaining clips and remove the ring; the sending unit lifts out of the tank.

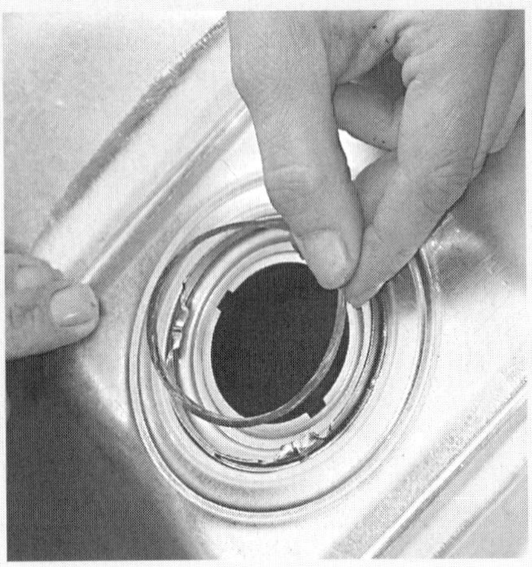

8. A new gasket is supplied with the new tank. Set it in place...

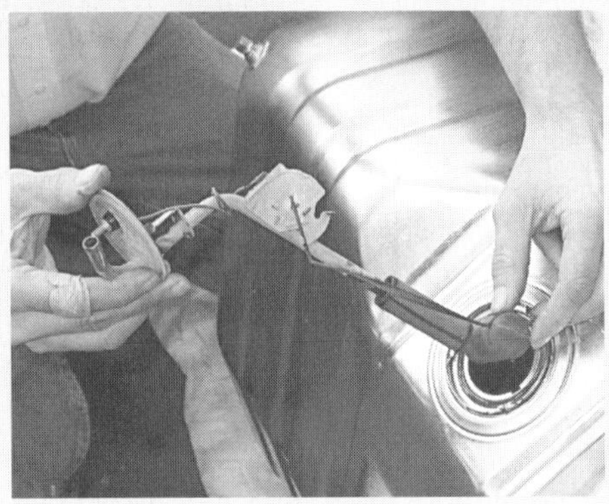

9. ...and install either your existing sending unit or a new one into the tank. Tabs in the sending unit and notches in the tank locate the unit exactly where it needs to be so the float will send an accurate reading to the fuel gauge.

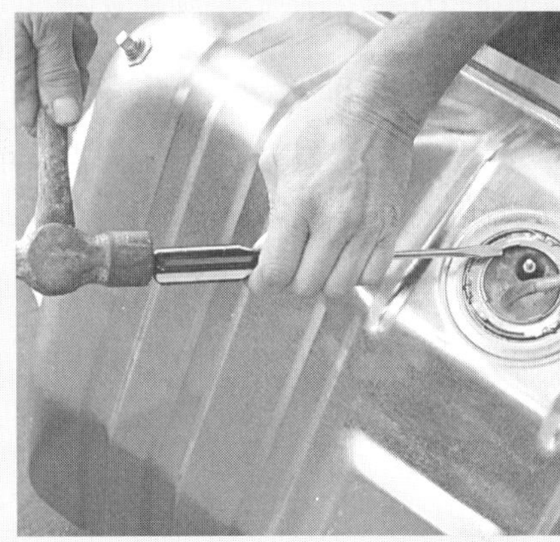

10. A new retainer is supplied with the new tank. Use it to secure the sending unit in place inside the new tank.

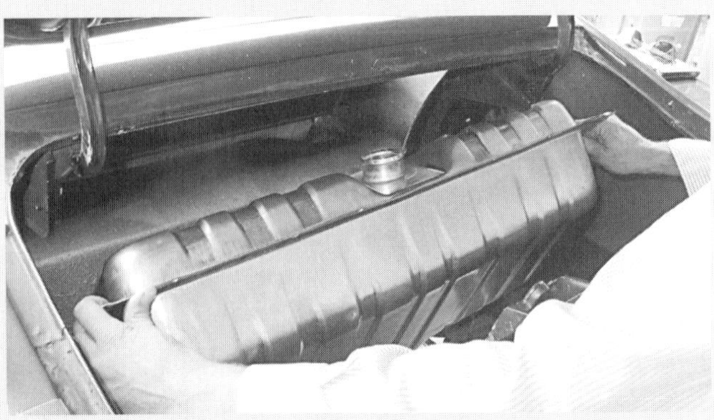

11. After some cleanup inside the trunk, we're ready to install the new tank—an easy procedure that's just a reversal of the removal process. Set the new tank in place inside the trunk.

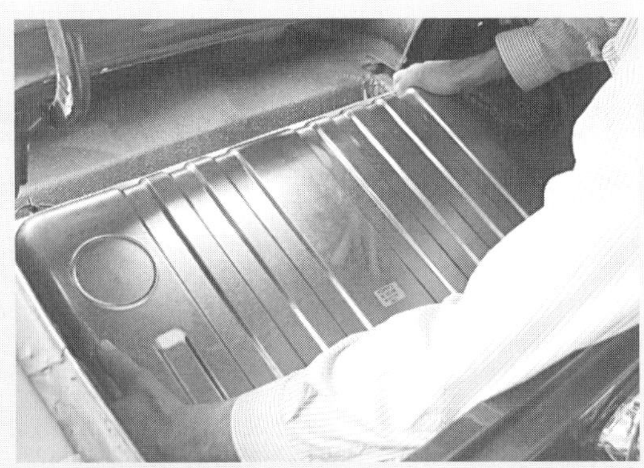

12. Once inside the trunk, put the rear of the tank in place first to clear the filler neck, then set the front of the tank in place afterward. Since this car is a fastback, the trunk-lid opening is considerably smaller than a coupe or convertible, so a little jockeying is required to get the tank into place. Of course, with the bigger trunk in a coupe or convertible, placing the tank will be easier.

13. With our larger 22-gallon tank, it took some elbow grease to refit the filler hose. It involved loosening the filler neck and bending it slightly, but we were able to make it work. Be sure to have the clamps back into position before reinstalling the hose to avoid having to take it all apart again.

14. Once the hose is back into position, tighten the hose clamps securely. After cleaning with spray lubricant and leaving them in the lubricant for a while, the old clamps work as good as new.

15. Retighten the filler-neck bolts that attach it to the taillight panel.

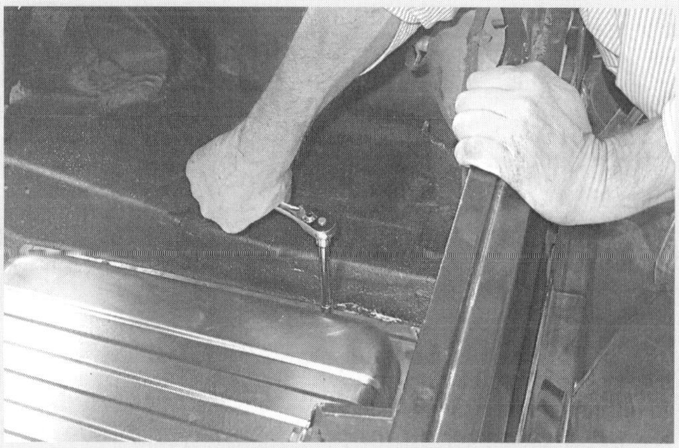

16. To complete the installation, reinstall the bolts that hold the tank in place inside the trunk.

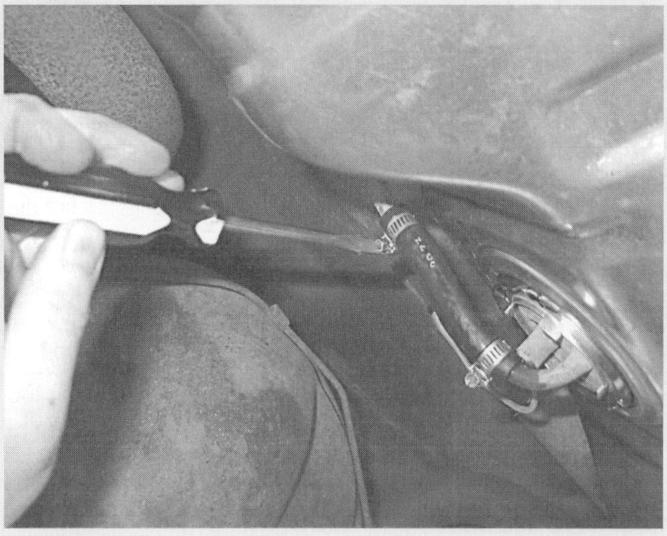

17. Underneath the car, plug the sending wire back in and reinstall the fuel-line hose, which is about 4 inches long. Replace the old hose with a fresh piece and a pair of new clamps. Standard rubber fuel line is available at any local auto parts store.

Chapter 23

Installing a New Fuel Sending Unit

Text and Photos by Miles Cook

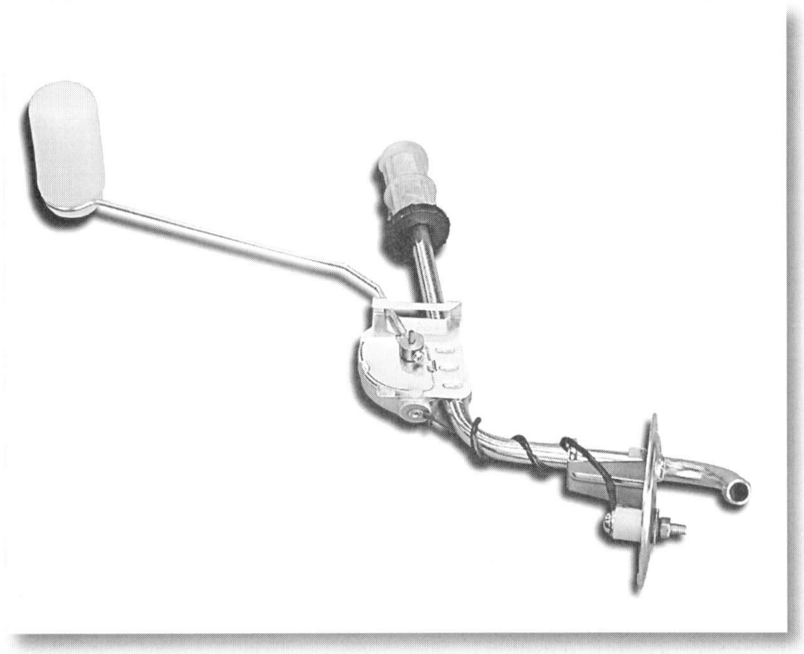

Is your fuel gauge inaccurate or not working at all? A new sending unit might fix the problem.

Most cars, including vintage Mustangs, have a fuel-tank sending unit that conveys to the fuel gauge the amount of gasoline in the tank. Sometimes they quit working and need replacement. No need to worry though, as the R&R procedure on a vintage Mustang is quite easy.

The following photos and captions show exactly how we replaced the sending unit in our '69 Mustang project car. Our friends at Mustang Country International had our car up on the shop hoist, the fuel tank drained of about six gallons of gas, and the new sending unit installed in our car in about a half hour. We acquired a sending unit from Mustangs Plus, a Northern California vintage-Mustang parts house whose print catalog and Web site offer just about any part you can think of for '65–'73 Mustangs, as well as a few late-model bits, too.

While the sending unit in our car still worked fine, we wanted to show you how to do this basic but still fundamentally important repair.

Sources

Mustang Country International
www.mustangcountryintl.com

Mustangs Plus
www.mustangsplus.com

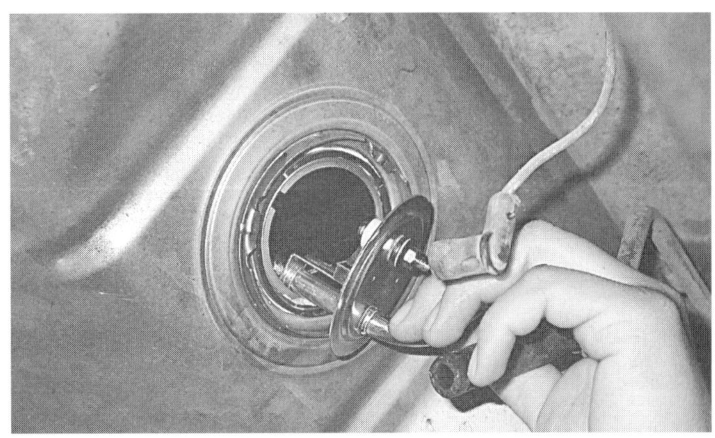

This is the new sending unit we installed in our car. Available from Mustangs Plus, it's a direct replacement for the original. Basically, '65–'73 Mustangs came with three fuel tank sizes: '65–'68s had a 16-gallon tank, '69s had a 20-gallon, '70s had a 22-gallon, and '71–'73s went back to a 20. We installed a larger 22-gallon tank in our '69 awhile back, but at the time, we reused the car's original 20-gallon sending unit, which affected gauge accuracy. We wanted to install a sending unit for use with our 22-gallon tank. Also note that it is possible to install the 22-gallon tank in any '65–'70 Mustang. The '71–'73 tanks install from underneath the car so the upgrade to a 22 isn't possible.

1. Have as little fuel in the tank as possible before starting. We only had about six gallons, which wasn't too difficult to handle. If your tank doesn't have a drain plug, siphon the gas directly into a gas can. We drained the gasoline into this plastic bucket, and then immediately poured the gas from the bucket into a plastic six-gallon can with a funnel so it could be poured back into the tank after the installation.

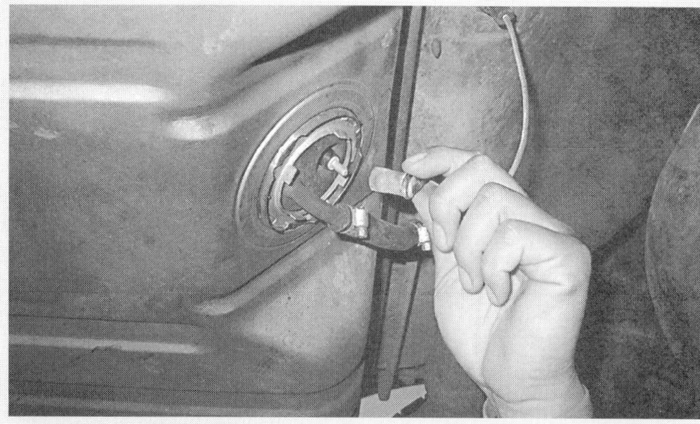

2. Once the tank is drained, removing the old sending unit is easy. Access the sender's location underneath the tank just behind the car's rear axle. Begin by unplugging the wire for the gauge from the sender. A gentle tug with your fingers should facilitate its removal.

3. Next remove the fuel line. Once the small hose clamp is loosened with a screwdriver, the hose can be pulled off. Have an absorbent towel or rag handy to catch any excess fuel still in the line.

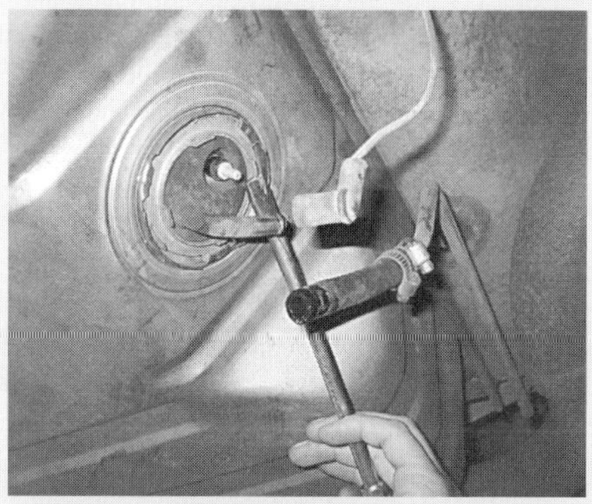

4. With the hose and wire removed, use a large flat-head screwdriver and a hammer to carefully tap the sending-unit retaining ring counterclockwise to loosen it from the tank. Position the screwdriver as shown here …

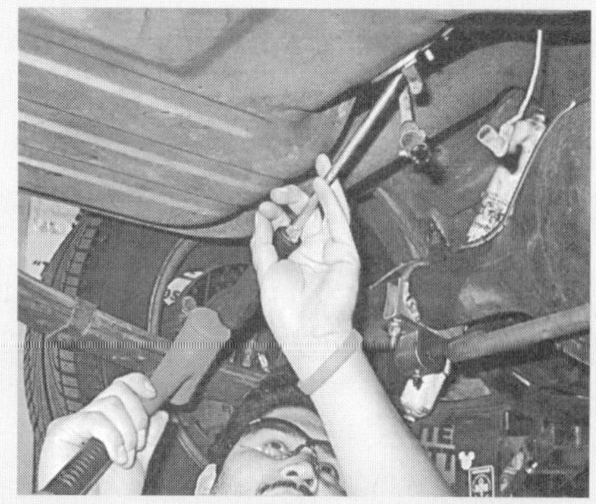

5. … and tap the handle with a hammer to rotate and loosen the ring.

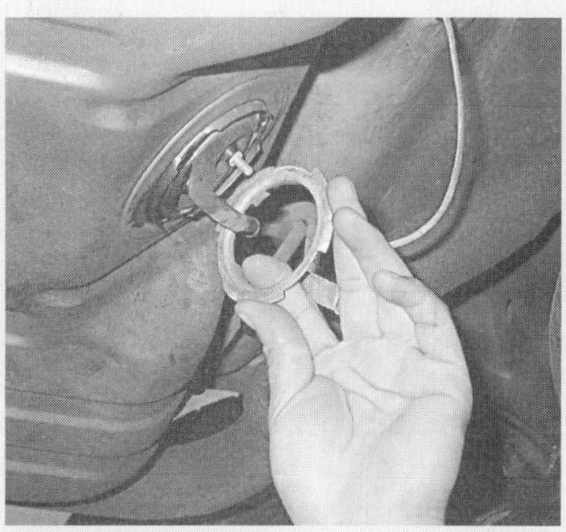

6. Once the ring is loose, it will come off the rest of the way, separating from the tank as shown.

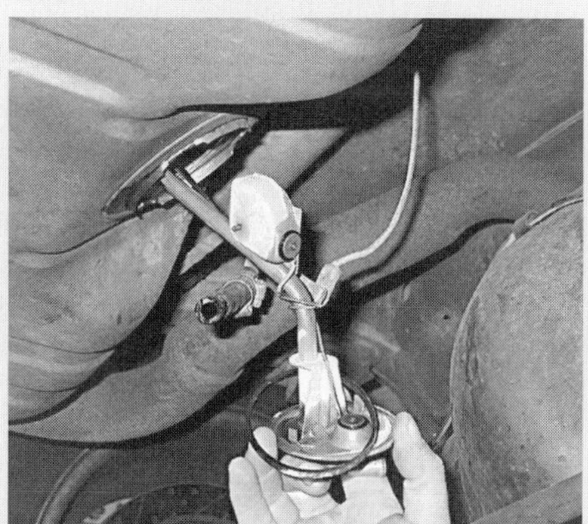

7. You may have to carefully jiggle the old sender to get it out, but it will likely drop out of the tank once you gently pull.

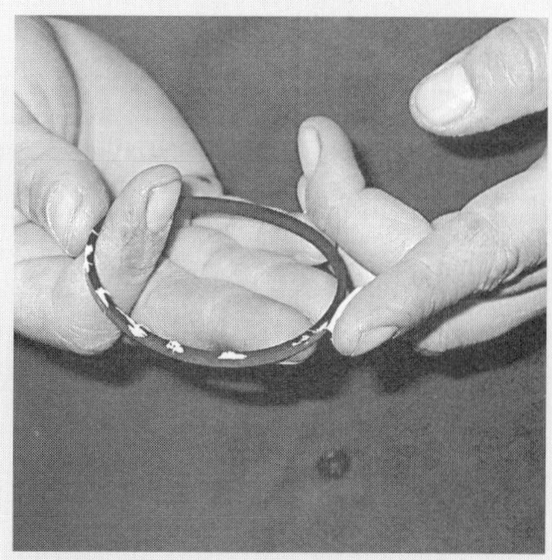

8. Mustang Country likes to apply a small amount of grease on the supplied new gasket. This keeps it from getting crushed and causing leaks once the new sender is positioned.

9. Position the gasket in place in the tank. There is a small groove for the gasket.

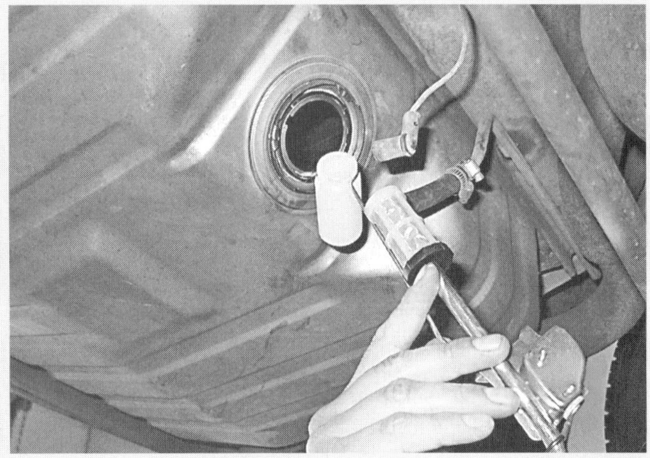

10. The new sender (with a slightly longer float arm compared to one for a 20-gallon tank) simply installs in place of the old unit. Carefully place it inside the tank, and line up the unit's tabs with the notches in the tank. There are two notches that accept the tabs on the sender to correctly orient it in the tank. This ensures the float is in the right location and the fuel gauge is accurate. Be sure the sending unit is completely seated in the tank before reinstalling the retaining ring.

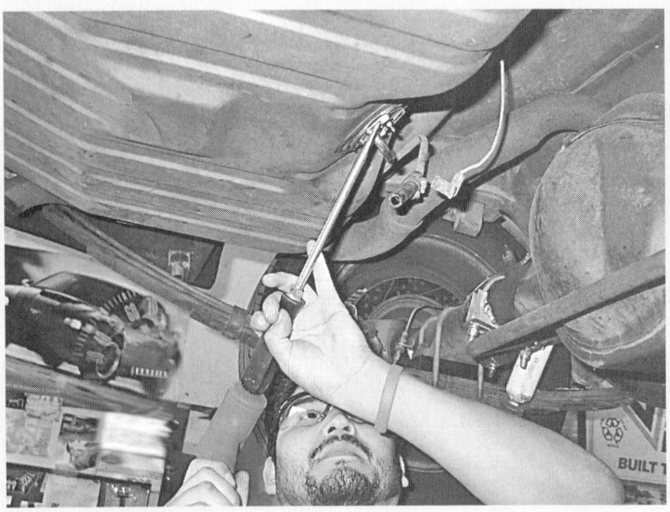

11. Place the retaining ring back on the tank, and tap it into place with a screwdriver and a hammer.

12. The retaining ring has tabs on it that allow it to seat into place. Try to seat it all the way to be sure it doesn't come out. Complete the procedure by reconnecting the fuel line and plugging in the gauge wire. You might have to slightly trim the wire boot to allow the wire to plug into the new sending unit. We did this on our car; a minor issue that we solved in a few seconds.

Chapter 24

Installing a High-Performance Front Suspension

Text and Photos by Miles Cook

Here are the new pieces installed on our '69 Mustang project car. While not as radical as a coilover setup, the Blue Moon and Global West pieces are a huge step up from the factory suspension bits. If you want to be on the path to a good-handling vintage Mustang street car, these tough-as-nails components are about as good as it gets, short of going to a more expensive coilover system.

For years, various manufacturers have been producing suspension parts for vintage Mustangs that are much better than the original components. Several areas have been addressed, including strength, suspension geometry, adjustability, and more.

Best of all, many of these items are virtually invisible to the casual observer, which is the case with the front upper and lower control arms we're featuring here. Unless you peer inside the fender or have the car on a hoist, no one will know you've updated a daily driven vintage Mustang with suspension bits that take advantage of modern technology. We're all for anything that makes an old Mustang more enjoyable to drive.

With that in mind, we checked out a set of tubular upper and lower control arms being installed on our '69 fastback project car. A pair of adjustable strut rods is also part of the mix. These parts are a major step up in performance front-suspension components that retain the original architecture of the '65–'73 Mustang front suspension. The next step would be to go to a full coilover system, which is slick but considerably more expensive than what we're showing here.

The Control Freak upper and lower control arms are made by Blue Moon Motorsports, and the strut rods are produced by Global West Suspension Systems. These suspension goodies and numerous other vintage Mustang parts are available from National Parts Depot, which has warehouses in California, Florida, Michigan, and North Carolina.

We've also established a new working relationship with another shop in the Los Angeles area that, among other things, specializes in vintage Mustang repair and upgrade work. If you're in the area, Don & Harold's Automotive Center in Long Beach, California, is a worthy resource for anyone interested in professional help with early Mustangs.

Source
National Parts Depot
www.nationalpartsdepot.com

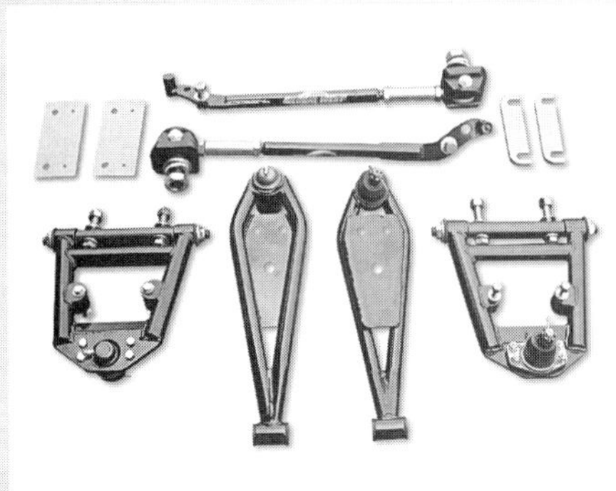

1. Available from any of National Parts Depot's four warehouses, the Blue Moon Motorsports upper and lower control arms are built like Fort Knox. Marketed under the Control Freak name, several applications cover '65–'66, '67, and '68–'73 Mustangs. They're also available in the black powdercoat finish shown here or a polished stainless steel finish. The arms include Grade 8 hardware and are made from 1.050 x 0.154-inch-wall seamless tubing. Equally durable are Global West's tubular adjustable strut rods (top) that allow for more caster adjustment than stock strut rods.

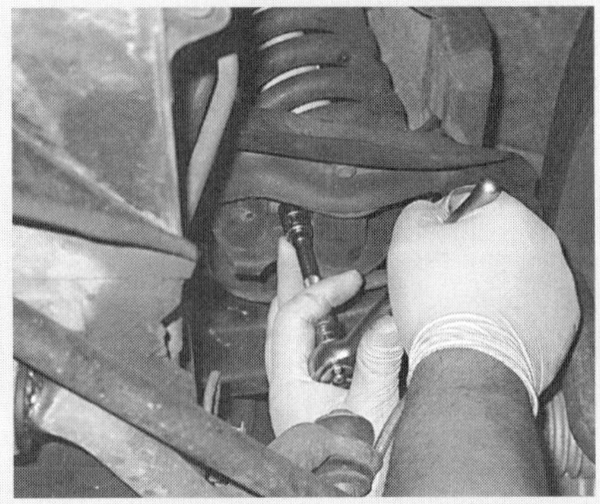

2. We arrived at Don & Harold's ready to begin the installation and discovered the shop has a fully computerized alignment rack, which meant a trip to an alignment shop wouldn't be necessary. We began by unbolting the shock from the spring perch and upper control arm before raising the front wheels off the ramps of the alignment rack/vehicle hoist. This avoids having to contend with the shock being extended before separating it from the spring perch.

3. Unbolt the shock from the top of the tower by removing the three bolts. The shock then lifts out of the tower as shown.

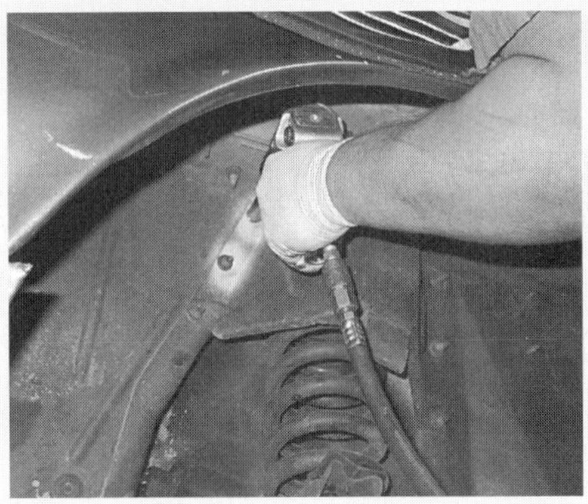

4. Next, remove the dust shield inside the fenderwell, which is held in place with six bolts. This allows better access to the coil spring to facilitate its removal.

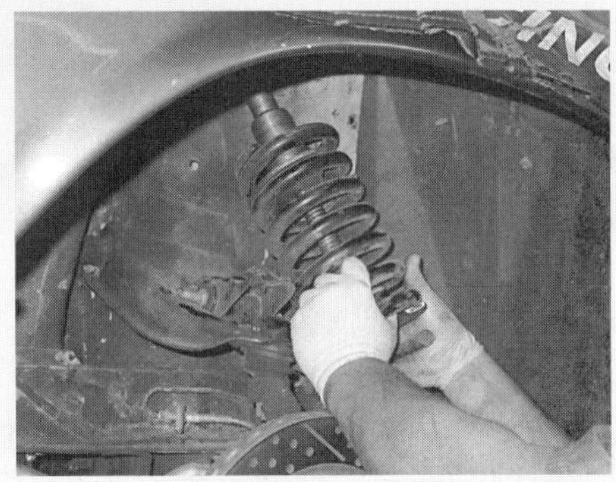

5. A spring compressor is required to remove the spring. You must know what you're doing here because the spring must be compressed enough to get it off the spring perch and out of the shock tower. The force of the compressed spring can be deadly if mishandled. While we've tackled this type of procedure before, it was nice to have seasoned experts at Don & Harold's handle it safely. Once the spring is adequately squeezed together with the compressor, it can be safely removed.

6. With the spring out, removal of the stock parts gets underway, and we begin by unbolting the upper control arm from the spindle as shown. Don & Harold's has compressed air, which is another advantage. Although the arms can be installed with handtools, the air-powered tools make the job move along faster.

7. After several strong hits on the spindle with a hammer, the spindle separates from the ball joint of the stock upper arm.

8. With the upper arm separated from the spindle, it can be unbolted from the shock tower. This involves getting a wrench on the nut on the underhood side of the shock tower and a socket inside the fender. It's a big help to have two people for this step—one to hold the wrench under the hood and one to unbolt the arm from inside the fender.

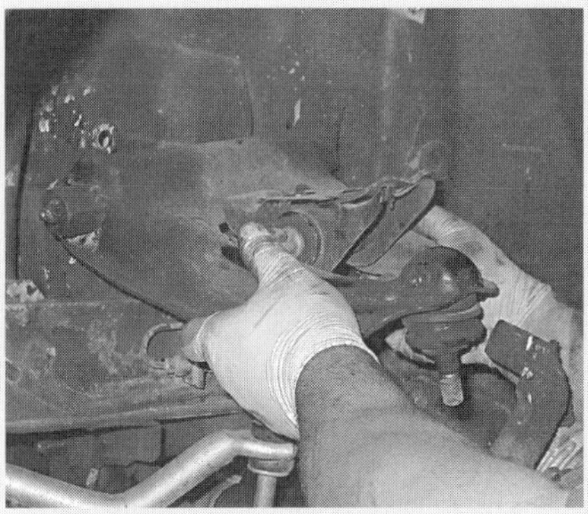

9. With all fasteners out, the upper control arm is removed from the car.

10. You can go two ways here. Either use new spring perches if required or reuse the ones on the stock arms. Ours were recently replaced, so they were in good enough shape to be reused. They were therefore removed from the stock arms as shown.

11. The Blue Moon upper control arms come with a pair of machined spacer plates that must be used with the new arms. They slip into place like this before the arm is installed in the car to prevent the tubular arm from hitting the shock tower.

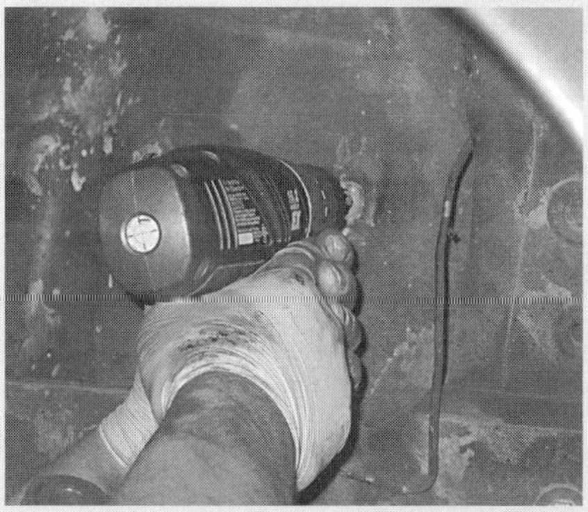

12. As was done on '65 Shelby GT350s, the Blue Moon instructions state to drill new holes in the shock towers to relocate (lower) the upper arms in the shock tower using supplied templates as guides. Although Blue Moon recommends relocating the upper arm for optimum performance and geometry, we decided to forgo that for now and install the arms in the stock holes, which Blue Moon also said is adequate for a street car. But we had to enlarge the stock holes to accept Blue Moon's 9⁄16-inch hardware.

13. Whether or not you drill new holes, the remaining installation procedure is the same. The new arms install into the shock tower using the supplied hardware. The Blue Moon arm is a direct replacement for the stock arm and reinstalls as such. The bolts go through the shock tower (with the spacer plate between the arm and shock tower) and attach with the supplied nuts on the other side of the tower.

14. With the arm installed, reconnect the spindle to the new arm with the supplied hardware. Be sure to install the supplied cotter pin once the castle nut is tightened.

15. With the upper arm installed, the techs at Don & Harold's use a block of wood to keep the upper arm and spindle up and out of the way while they remove the lower arm and strut rod. In this photo, the spindle has already been disconnected from the lower arm.

16. The next step is to remove the strut rod, which is a fairly easy task. Unbolt the large nut at the front of the rod and remove the rod from the lower control arm by loosening and removing the two bolts.

17. After disconnecting the antisway bar from the lower arm and removing the bolt that holds the other end of the lower arm in place near the oil pan, the lower arm is removed from the car.

18. The Blue Moon lower arm installs in place of the stock arm. You might have to spread the mounting ears slightly where the lower arm attaches to the car near the oil pan; otherwise, the Blue Moon arm is a direct replacement. Use the car's existing hardware (be sure to include the alignment eccentric) to secure the lower arm in place.

19. As with the other parts, the Global West strut rods are a direct replacement and come with all required hardware. Attach the front of the strut rod and secure into place with the supplied hardware. The Global West rods also replace the stock rubber bushings where the rod attaches at the front of the car. Tighten the large nut as shown.

20. Here, the anti-sway bar has already been reattached to the Blue Moon lower arm using the existing hardware. Next, attach the strut rod to the lower control arm with the supplied bolt and one bolt from the stock strut rod. This bolt installs next to the steering stop on the Global West strut rod. Securely tighten the bolts and nuts to complete the installation of the strut rod.

21. Next, the lower arm and spindle are connected and secured together with the supplied castle nut and cotter pin.

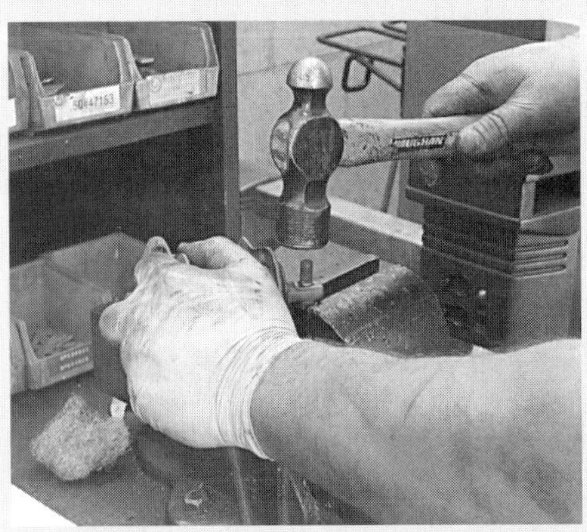

22. Remove the stock bolts from the spring perch by tapping them out with a hammer. This is best achieved while the perch is placed in a bench vise.

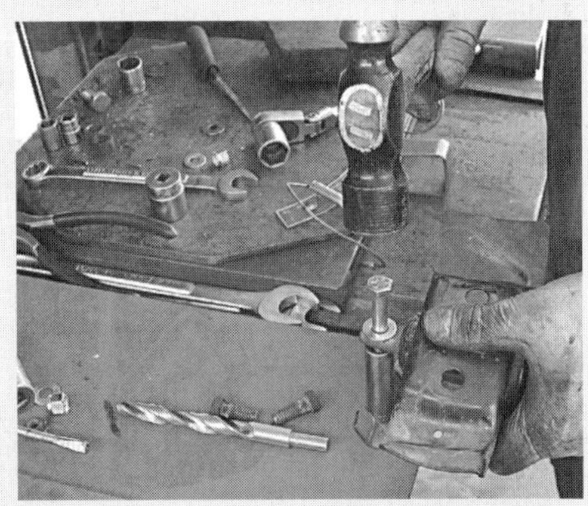

23. Remove the supplied hardware on the Blue Moon upper arm and install the longer bolts into the spring perch. The longer bolts are required to run through the thicker perch-mounts on the Blue Moon upper arm.

24. According to the Blue Moon instructions, some spring perches must be trimmed for clearance on the upper arm. We determined our perches would need trimming to clear the tubular construction of the Blue Moon arm when the suspension is in motion. We took off about 1⁄8 inch of material with a die grinder.

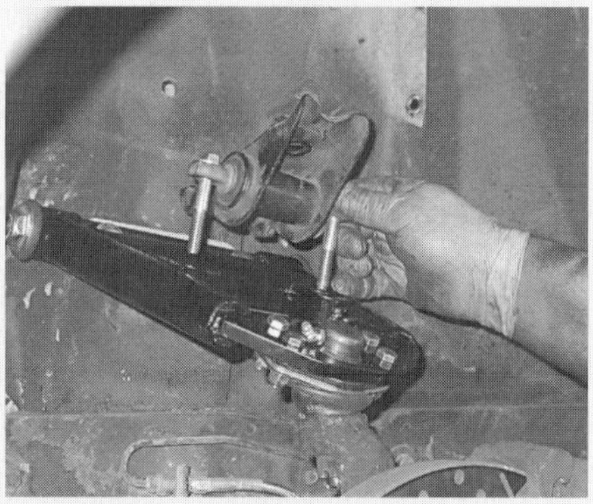

25. The perch is installed onto the upper control arm…

26. …and secured into place with the supplied hardware.

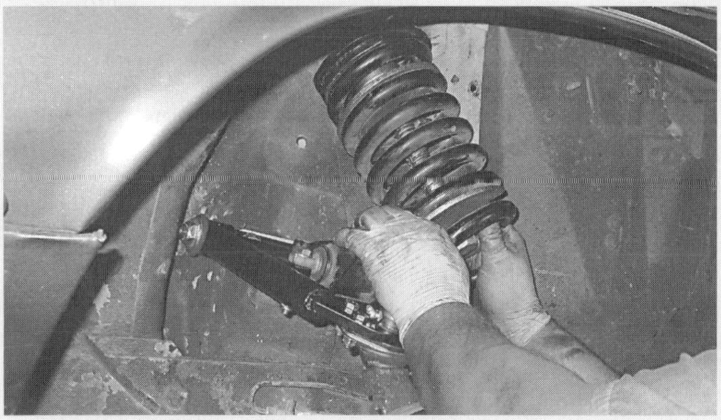

27. By reinstalling the spring, installation is almost complete.

28. Reinstall the shock, and lubricate the upper and lower ball joints with grease.

29. Have the car aligned to complete the installation. Don & Harold's has a fully computerized Hunter alignment rack and did it on the spot. Caster was set at about 2 degrees positive (tops of spindles tilted toward the back of the car), camber at about 3⁄4 degree negative (tops of wheels tilting in toward the center of the car), and toe-in at 1⁄16 of an inch—all fairly typical street-oriented settings. For more sporting open-track activity, set camber to around 2–3 degrees negative, caster to around 3 degrees positive, and toe-in about the same or slightly less at 1⁄32 inch. If ride height changes and you want to retain the height you had before installation, the coil springs might need to be trimmed. Each car is different. We trimmed three quarters of a coil from our front springs with a cutoff wheel to get the height we wanted, which was slightly lower than where we began.

Chapter 25

Fitting Larger Wheels & Tires

Text and Photos by Miles Cook

Running larger wheels and tires is a great way to enhance and individualize the look of vintage Mustangs.

Over the years we've received numerous inquires about installing larger wheels and tires on vintage Mustangs. Are you worried about backspacing, tires rubbing fenders, and tire sizes that may or may not look right on your car? We'll answer these questions and more to steer you in the right direction for fitting the ideal wheel-and-tire package to your vintage Mustang.

While we have plenty of photos to help you decide what you want, this isn't a wheel-styling guide, but rather a look at what fits in terms of wheel diameter, width, backspacing, and tire size.

In general, installing larger wheels and tires on vintage Mustangs falls into two categories. The '65–'66s are in a separate group. They have a few limitations because they're a bit smaller in every dimension than the later cars and have smaller fenderwells. For example, we've seen 18-inch wheels on these early cars, and they don't look right.

The second category is the '67–'73. There are more choices for what will fit in the fenders and look good. Any '67–'73 Mustang with the right 16- or 17-inch-diameter wheel and tire filling the wheelwell provides that cool open-track, road-racy look many owners desire.

We're using a '65 to discuss the small cars, and our reliable '69 fastback for the '67–'73 cars. We'd like to thank the folks at Vintage Wheels Works (VWW) and Islas Tires for helping us get the wheels and tires mounted.

Sources

Islas Tires
800/941-1797

Mustangs Plus
www.mustangsplus.com

Vintage Wheel Works
www.vintagewheelworks.com

Our '69 fastback test car is a textbook example of any '67–'73 Mustang in terms of wheelwell room and wheel/tire fitment. We believe the current wheels and tires on this car are perfect because they're big enough to look cool, but don't cause interference with fenders or suspension pieces. Remember—we're discussing dimensions only and aren't recommending a specific make or model of wheel or tire. We love VWW wheels in a big way, but are leaving the design and style choices up to you.

The wheels are 16x8-inch Vintage 45s all around, which are the original wheels VWW began offering about eight years ago. With 4 1/2 inches of backspacing, there's plenty of clearance from all obstacles and they fit exactly where you want in a '67–'73 wheelwell.

The tires are staggered sizes, with 225/50R16s in front and slightly larger 245/50R16s out back. You could also run the same size tires all around (a 245/50R16 at each corner on a '67–73), but we like the look the staggering provides. We think 225/50R16s all around might be too small for the rear wheelwell of a '67–'73 Mustang. You could also run a 255/50R16 tire out back, which we've seen on other '69–'70 Mustangs, with a 245/50 or a 225/50 up front. That would depend on how much size difference you want with the front and rear tires.

For a compromise between looks and ease of fitment, a 16-inch wheel is the perfect diameter for a '67–'73, but that's not to say the right 17-inch wheel isn't just as cool. The 8-inch width of the Vintage 45s is also ideal for this or any aftermarket wheel. Seven-inch-wide wheels are also good for whichever style you choose.

At Vintage Wheels Works, we mounted a set of 17s on our '69 to see how they looked. Cool, eh? They fit as well as the 16s, but are more radical with respect to a restomod look. They provide more clearance up front in relation to the front suspension's upper control arm. A staggered setup all the way, these two-piece 17-inch Vintage 45s differ from the 16-inch wheel's one-piece construction. We had VWW mount a 245/45R17 tire to a 17x8-inch wheel up front—which would also look good out back and would allow tire rotation. But the car here has a big 17x9.5-inch Vintage 45 out back with a 275/40R17 tire. VWW feels safer recommending a 17x8 for the rear, but the bigger wheels fit well enough to run them with no worries. Some inner-fender rolling at the rear might be needed with the bigger wheels, which will vary on a car-to-car basis. The difference in looks between 16s and 17s is evident in our photos. Though we love the 17s, we were content to leave VWW's shop with our 16s still in place; they've been on this car for nearly two years. As for tire sizes on 17-inch wheels, 245/45s up front or all around are ideal on any car, though we love the bigger 275/40s out back. The biggest tire we'd run at the front would be a 255/40R17, although we'd likely be satisfied with what you see here.

It is possible to fit 18-inch wheels and tires on a vintage Mustang, but that's beyond the scope of our discussion. We're sticking to 15-, 16-, and 17-inch wheels, since the 17s already provide a low-profile look most of us might seek on an early Mustang.

Using the '69 as the guinea pig, we'll show what 15-, 16-, and 17-inch wheels look like on the front of a typical vintage Mustang. Using VWW wheels of various styles, we begin with a 15x7-inch Vintage 50 with a 3 7/8-inch backspacing. Essentially a Shelby 10-spoke copy, it looks great with the 225/60R15 tire (about the best size for this wheel). However, a 235/60R15 tire would also fit well. You could also stagger tire sizes by running a 225/60 or 235/60 up front and fit a 245/60 on the back. Although the 245/60 might be a little big for a wheel that's only 7 inches wide, it will fit and would look good. The Vintage 50 is also available in a 16x8-inch size with a 4 1/4-inch backspacing and would work great with the 16-inch-diameter tires mentioned earlier.

A 16x8-inch wheel with a 4 1⁄2-inch backspacing wrapped with a 225/50R16 (shown here) or 245/50R16 tire is a true no-hassle fit on the front of any '67–'73 Mustang. Use these dimensions with just about any make or model of wheel and tire and you can't go wrong. This one-piece 16x8-inch Vintage 45 wheel started it all for Vintage Wheel Works.

Now we're going big and it's really cool. This two-piece Vintage 45 in a 17x8-inch version with a 245/45R17 tire is the de facto setup to use in terms of size should you desire 17-inch wheels of any make and style. Fitment is perfect, fenderwell clearance isn't an issue, and you have plenty of rubber on the road—much more than any Mustang came with from the factory. Backspacing is 4 1⁄2 inches. A 255/45 or 255/40 tire will also fit here with the same size out back, or a larger 275/40 if you want to stagger tire sizes.

Moving to the back of the car, our current setup is the same 16x8-inch wheel as the 16-inch wheel shown earlier up front, but with a slightly larger 245/50R16 tire. This ideal combo looks great and won't cause any problems. On some cars, slight fender rolling might be required because some 245/50 tires are fairly big. You could have four tires that are the same size, but are different brands, so the tires will measure differently. For example, we've seen 245/50R16s on the back of two different '69 Mustangs and the tires were from different makers. One brand was clearly taller than the other, even though the size was labeled the same on each tire. Another option is the 255/50R16, which we've recently seen on the back of a '69 Mach 1. They are big, however, so some fender rolling will probably be required compared to running 245/50s.

If you need anything bigger than this out back, you can do it, but you'll likely have to contend with custom backspacing, which VWW offers. This 17x9.5-inch Vintage 45 wheel with a 5 1/2-inch backspacing and a 275/40R17 tire is downright huge. It's a nice fit, but VWW recommends staying with 17x8s out back. Outer fender clearance will be an issue on some cars, since all cars are different in this area. VWW offers several 17-inch wheel setups, including 17 x 7s, 17x8s, and these 17x9.5s. Custom backspacing is available on all sizes if needed. This package is about as far as we'd go. Beyond this, you'll likely get into custom sizing and backspacing issues, as well as tires rubbing on outer fender lips without considerable fender rolling.

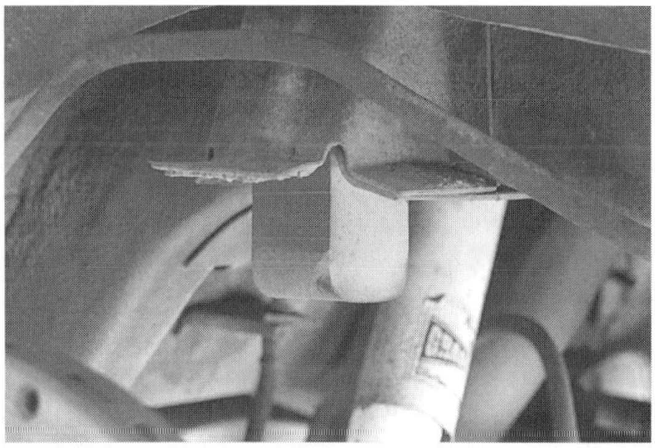

Although the 17x9.5-inch wheels fit on the back of our car, they got pretty close to the rubber bumpstops for the rearend. If we ran the 17s, we'd trim the metal edge on the part of the bumpstop mount that's closest to the wheel, since removing the bumpstops altogether on a regularly driven street car isn't a good idea.

Going Big on Smaller '65–'66

Although there are many similarities when it comes to fitting larger wheels on '65–'66 Mustangs as compared to the bigger '67–'73 cars, special considerations are necessary because the earlier cars are smaller in most dimensions. Although they might fit, some believe a fairly large 17-inch wheel-and-tire combo doesn't look right on an early car. In some circumstances, we'd agree.

We have some recommendations for those with a '65 or '66 Mustang. Factory 14-inch wheel-and-tire combinations that came on most '65–'66 cars are fine, but we're going beyond that here. While some might consider it conservative, we think the right 15-inch wheel and tire looks perfect on these smaller Mustangs. For example, the '65 fastback below rides on 15x7-inch Magnum 500 wheels with 225/50R15 tires all around. Now that VWW offers several 15-inch wheel options, you have an ideal choice for a Vintage 45 in a 15-inch size. Several Mustang parts houses also offer a factory Styled Steel wheel in a 15-inch size that looks cool on a '65 or '66 Mustang. Other tire options include taller 225/60R15s all around or a staggered arrangement with a 215/60R15 up front and 225/60s out back.

As for running 16-inch wheels on '65–'66 Mustangs, you can go with 16x8-inch wheels with 4 1/2-inch backspacing and 225/50R16 tires all around, or maybe 245/50s in back. That's as big as we'd go, and it's the ideal 16-inch setup for these smaller cars. But fenderwell clearance can be an issue with some cars when trying to run 16-inch wheels in front. For no-hassle fitment on a '65–'66 Mustang, a 15x7-inch wheel with one of the above-mentioned tire sizes is an ideal setup that will look great and won't cause any problems.

Finally, while you can run 17-inch wheels on a '65 or '66 Mustang (and there are many options), be careful because there isn't as much room as on the larger '67–'73 cars. Shop carefully and seek out the advice of a tire/wheel shop or the wheel manufacturer you're working with before going big on an early small Mustang. While it's not impossible, it isn't the norm, so be careful before buying 17-inch wheels and tires for a '65–'66, installing the tires on the wheels, and not being able to return them before realizing they won't fit. Test fit before you mount 17-inch wheels and tires on a '65–'66 Mustang.

Chapter 26

Installing an Export Brace & Monte Carlo Bar

Text and Photos by Miles Cook

Making the front end of a vintage Mustang stronger is an important step to better handling.

Adding an export brace and a Monte Carlo bar to a vintage Mustang is an excellent way to improve the overall rigidity of the car's structure. It's something Carroll Shelby incorporated on his GT350s, and enthusiasts have been following Shelby's lead ever since. By replacing the two thin braces that connect the shock towers to the cowl with a much thicker single brace, then adding the Monte Carlo bar as well, the front of the car is dramatically firmed up. It's something we were able to feel on the road during all driving conditions.

A one-piece export brace, available from National Parts Depot, attaches both shock towers and the cowl together, thus making the front of the car more rigid. After driving our '69 SportsRoof for well over a year without a brace, we were amazed at the difference the brace made in the car's overall feel. Even during smooth highway driving, it was noticeably more solid.

However, while test-fitting the brace on our SportsRoof and checking it on several other cars, we knew it wouldn't simply drop into place. The shock towers on all the cars we checked had settled inward to the point they needed to be spread apart to get the factory-spec brace to fit over the towers. In our car, the discrepancy was more than an inch.

To make the brace fit, we enlisted Marlo's Frame & Alignment in Chatsworth, California. They were able to move the towers outward by jacking up the car and placing the jack under the lower control arms' attaching point. In addition to raising the car at those points, we pulled the brace inward with a come-along cable and trimmed the brace slightly where it attaches to the cowl. These procedures combined to make the brace fit well, and it all went back together without a fuss.

Sources

National Parts Depot
www.npdlink.com

Marlo's Frame & Alignment
www.fly-ford.com

1. Available either in black or chrome, reproduction export braces are available from National Parts Depot. At about $40 for the brace ($50 for chrome), it's amazing how effectively it firms up the front end of a vintage Mustang.

2. Here's our engine compartment before the installation of the export brace and Monte Carlo bar. Like most early Mustangs, the usual arrangement is visible here: two thin braces that attach to the cowl and each shock tower. They're not nearly as strong as the one-piece, thicker export brace.

3. We began by unbolting the shock and letting it rise up and out of the way of its mount. Then we unbolted the shock mount from the tower.

4. With the braces free from the shock tower, each one was unbolted from the cowl.

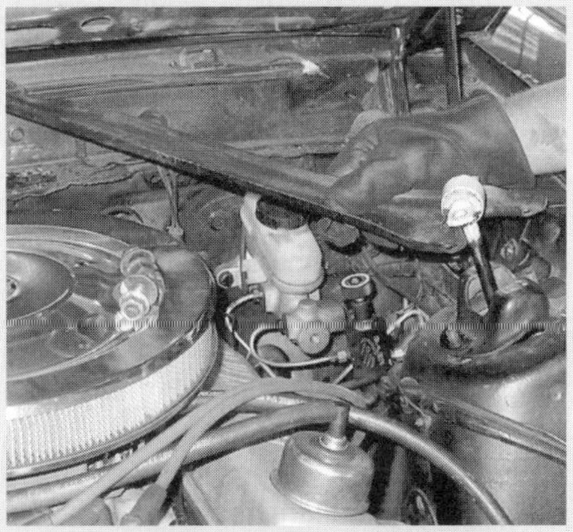

5. Once unbolted, the flimsy factory braces could be removed. You may want to hang on to them, especially if they're original.

6. Next, the shock mount was removed by lifting it up over the shock and off the car.

7. With all parts removed (except for the protruding shocks, which won't get in the way), we test-fit the brace for the first time. Keep in mind that all Mustangs will have a different set of circumstances. After more than 35 years with a heavy 351W up front, the towers on our '69 had settled inward more than a full inch. Using our car as a reference, we think some Mustangs will be better in this area, while others might be worse, especially if there has been any front-end damage that wasn't repaired properly.

8. Here you can see how far off the shock-tower mounting studs were compared to the holes in the brace. At first we thought something was wrong with the brace, but National Parts Depot set us straight. We were assured the export braces are made to the same dimensions as the originals and the issue is different for each car. In our case, the towers had simply settled. In some cases (ours included) the shock towers and front of the car have to be manipulated to get the brace to fit. Based on NPD's advice and our experience, at least some movement of the towers will be required for an export brace to drop onto any '65-'70 Mustang that has never had one.

9. After some head scratching, we came up with a plan to spread the towers apart and make the weight of the car work for us. First, we did some comparative measuring with this toe-check tool to see where the towers were. The front outside shock-mount studs were our reference point.

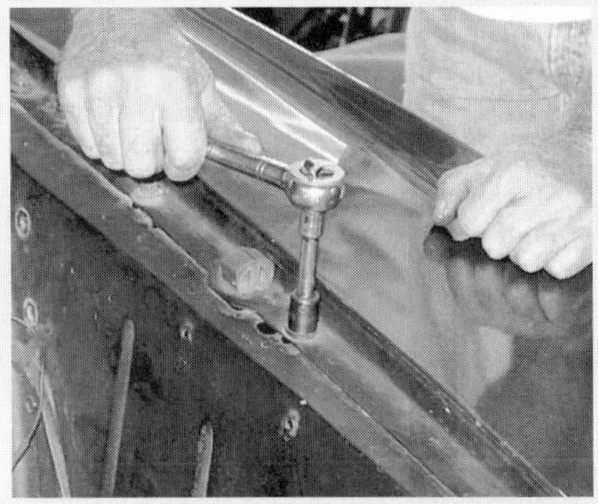

10. Next, we loosened the six bolts on the tops of each front fender in case a slight amount of movement for the inner fenders was needed. The bolts don't have to be removed, just loosened to allow some movement.

11. At an alignment shop like Marlo's that specializes in vintage Mustang front-end work, it helps to have an alignment rack and these burly multi-ton bottle jacks. This shows our first trick. Jacking up the front where the lower control arms connect to the towers gave us a large portion of the movement we needed to get the shock towers to move farther apart. We left the car raised in this position until the brace was completely bolted down and installed.

12. This shows how much movement we got from the towers by jacking up the car at the lower control arms. With the toe-check tool still showing the same measurement before we raised the car, you can see about half an inch difference between the shock-mount stud and the tool. We were definitely on our way to making the brace fit.

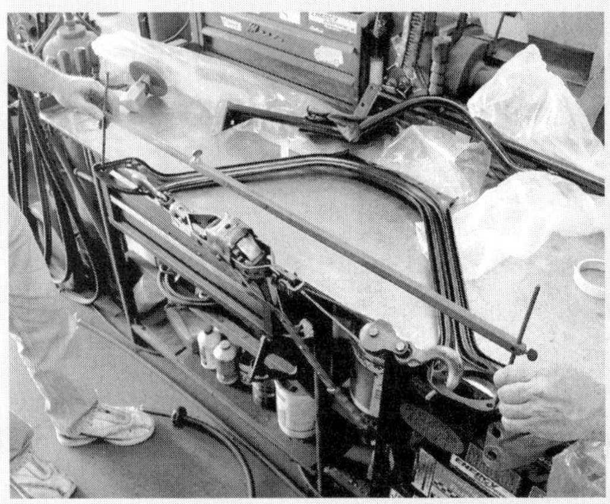

13. The car can be jacked up so the front subframe droops, which gives more movement to the front-end components. This wasn't necessary in our case. Slightly pulling the brace together on the bench with a come-along tool made the brace fit perfectly over the shock towers.

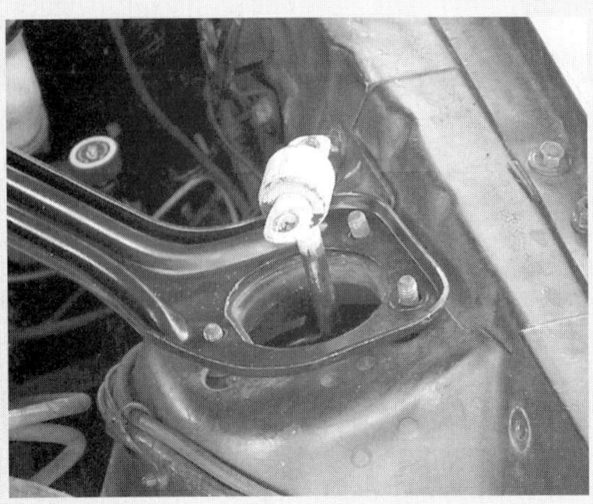

14. We pulled the brace inward only a little, making sure it didn't bow at the center.

15. To make the brace fit, jacking up the car so the front droops is an option. We didn't do that, but we ran into another issue. The brace was a bit too long where it mounts to the cowl, so we trimmed it slightly on the bench with a cutoff wheel. Note the piece of tape on the brace to show how much (or how little) material needs to be ground off. For us, it was only 1⁄4 inch.

16. Slightly enlarging the holes at the back of the brace and cowl helped ease the final fitment of the brace using the car's original mounting bolts.

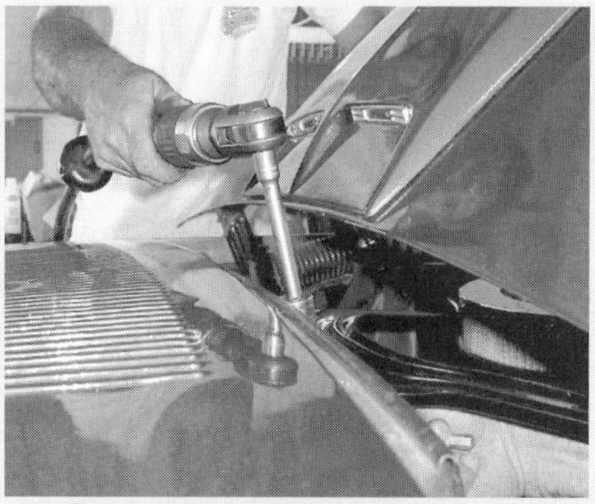

17. With fitment complete, the brace was secured to the cowl using the original hardware. It's best to have two people for this, one holding the nut in place underhood with a wrench and another tightening the bolt from outside the car between the hood and cowl.

18. With the brace in position, the shock mounts and shocks were reinstalled.

19. Tighten the shocks in place to complete the installation of the export brace. Be sure to retighten the fender bolts (six on each side) that were loosened previously. When installation of the brace is complete, recheck the car's front alignment specs and adjust them as needed. We checked our car and found all measurements were still within spec.

20. We also installed a Monte Carlo bar. Two types are available. The curved one shown above is designed to clear distributor caps and large (e.g., stock-type) air cleaners. If you have room, go with the straight bar, which is stronger.

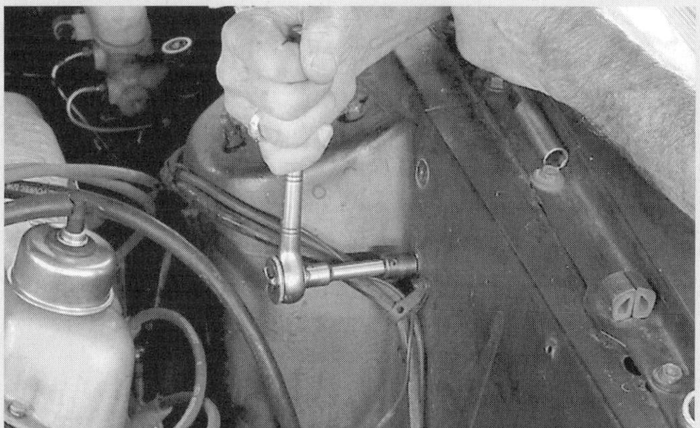

21. To test-fit the bar, remove the bolts that will hold the bar in place on the side of the inner fender.

22. While the car was still raised up with the bottle jacks for the installation of the export brace, we test-fit the Monte Carlo bar to see if its holes lined up on the sides of the inner fenders. They didn't.

23. The holes had to be elongated to fit a '69 Mustang. It took only a few minutes. Every car will be different; since Monte Carlo bars were originally designed for '65–'66 Mustangs, fitting one to a dimensionally different (and larger) '67–'70 Mustang usually requires minor alterations.

24. Once the holes were lined up, the bar was installed using the car's original hardware on the side of the inner fender.

25. Using the partially installed bar as a template, we drilled the four required holes in the top of the inner fender to complete the bar's installation.

26. NPD has a hardware kit for the Monte Carlo bar's additional holes, or you can procure the nuts, bolts, and washers from a hardware store. Tighten these bolts to complete the installation.

BODYWORK

Chapter 27

How to Replace the Cowl Panel

Text and Photos by Donald Farr

Dynacorn's new reproduction cowl panel is a blessing for '67–'68 Mustang owners with rusty, leaking cowls. All brackets are included, so there's no need to cut them out of the old panels for transfer to the new panels. When welded together, the upper and lower panels create a brand-new cowl-panel assembly.

The days of cutting, welding, and otherwise repairing rusty, old cowl panels are over. Well, at least for '67–'68 Mustang owners. Dynacorn International has released brand-new sheetmetal reproductions of the upper and lower cowl panels for '67–'68 Mustangs, which means you can trash the old leaky panels and replace them with Dynacorn's fresh sheetmetal, available from National Parts Depot.

Cowl-panel repair has been the bane of vintage Mustang owners for decades. Leaves, pine needles, and other trash can enter the cowl through the exterior vents and accumulate around the water dams at the corner vents inside. Rain water adds moisture to the inaccessible mess, resulting in rust, especially at the dams, which eventually allows water to enter the interior through the fresh-air vents. Many a Mustang owner has been surprised by wet carpet after a rainstorm or even a car wash.

Because the cowl panel is an enclosed assembly, repairs are difficult. Temporary repairs like plastic cowl-vent caps and fiberglass replacement dams have been available for years. More permanent repairs have meant cutting open the cowl panel to replace the dams and surrounding sheetmetal. According to Matt Simmons from Classic Creations of Central Florida, he can spend up to a week repairing a rusty original cowl.

Of course, even with completely new panels, the replacement is not a job for the typical weekend mechanic. Removing the original cowl requires removing fenders, cutting out spot welds to remove the old cowl panel, and welding the reproduction panels together before the final alignment and welding. You'll also have to paint the upper panel to match the car's exterior because the cowl forms the sheetmetal between the hood and windshield. It's a big job if you're performing a cowl-panel R&R on a complete car, but less time-consuming and labor-intensive if you do it during the course of a restoration.

As you've likely heard, Dynacorn is working on a complete '67 Mustang body, which explains why the '67–'68 cowl panels are available before '65–'66. According to National Parts Depot and confirmed by Classic Creations of Central Florida, Dynacorn's lower panel can be adapted to '65–'66 Mustangs, although the shape is not quite correct because it's made to clear the '67–'68 Mustang's underdash pedal support. However, that area is not visible when the panels are welded together as an assembly and installed.

Sources

Dynacorn International
www.dynacorn.com

National Parts Depot
www.npdlink.com

1. To remove the old cowl-panel assembly, the fenders must be removed and the original spot welds cut out. You'll need to support the car's frame underneath the cowl area, especially on convertibles, because otherwise only the firewall is supporting the car. Without support underneath, the car could sag in the middle.

2. With the old cowl panel flipped over, you can see the hidden rust damage. In fact, this cowl had been repaired previously, and even the repair sheetmetal was rusted through.

3. Before welding the upper and lower panels together, they must be aligned to drill the holes needed for the spot welds. Matt secures the panels with clamps before drilling.

4. With the panels aligned and secured together with clamps around the edges, drill small pilot holes through both panels around the edges.

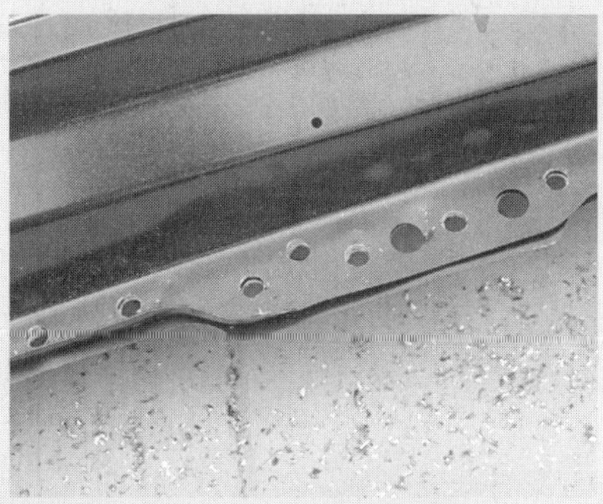

5. Next, open up the pilot holes to 3/8 inch. The final pattern should look something like this. The two large holes are part of the panel stampings, used for alignment purposes.

6. The holes along the top of the upper panel, where the panel will be welded underneath the windshield, can be made with a sheetmetal hole punch, which Matt says is available from tool companies for around $20.

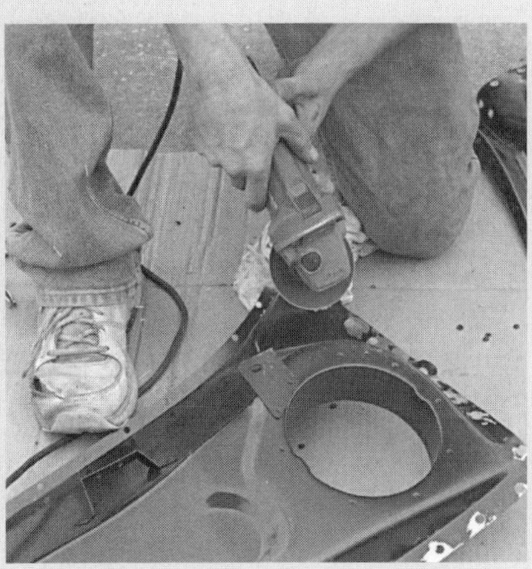

7. Use a sander to remove the burrs from the freshly drilled holes.

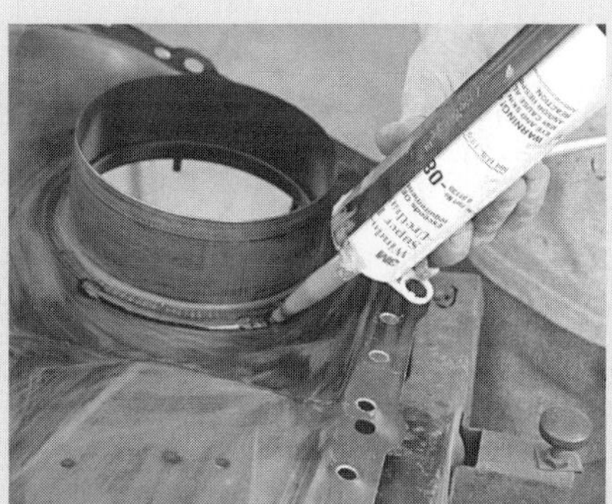

8. For durability, it's a good idea to apply a bead of sealer around the water dams before welding the panels together. This will provide extra protection from the moisture collected by leaves, pine needles, and other debris.

9. After scuffing in preparation, paint the inner panel surfaces body color because you will be able to see into the cowl through the upper vents.

10. After the paint dries, remove the paint on the areas to be spot-welded.

11. Once again, clamp the upper and lower panels together in preparation for final welding.

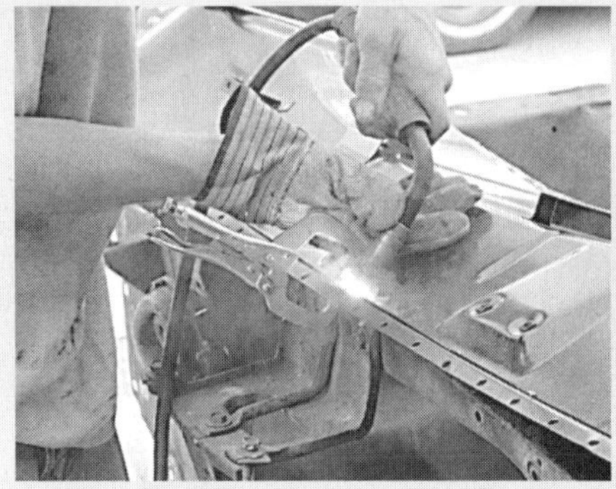

12. Use a spot-welder to weld the panels together at each drilled hole where the panels meet at the windshield area. Matt welds from underneath the top panel, which reduces the amount of final grinding and finish work, making for a cleaner appearance on the top panel.

13. Here's a good look at Matt's spot-welds where the lower panel meets the upper panel at the windshield area.

14. Place the semi-welded cowl-panel assembly into place on the car body. It's a good idea to cover the vents with tape to prevent debris from accidentally falling in. Once the panels are welded together, you won't be able to access the interior area.

15. Use an awl to align the cowl panel to the body. There are holes, four in front and one at each side, in both the cowl panel and the body for this purpose. With the cowl aligned, clamp it into place before welding.

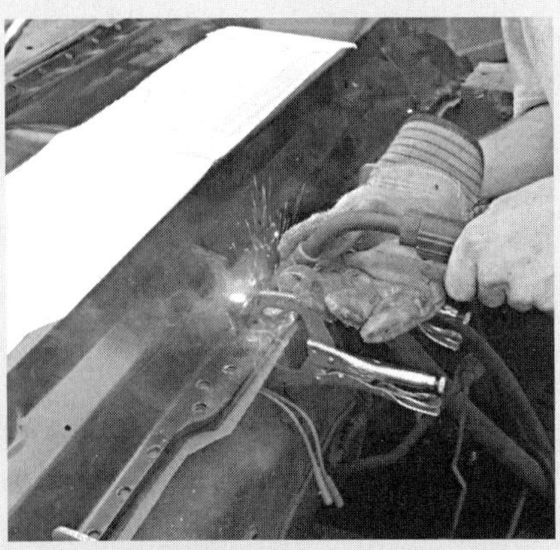

16. Now comes the tedious part: Spot-weld the cowl panel into place at all of the nearly 200 holes.

17. Before final welding, it's a good idea to lay the fenders into place to check the fit. It's easier to remove a few welds for adjustment now instead of having to cut out all the welds later.

18. As you see, the fender and cowl fit perfectly, so Matt continued with the welding to complete the installation. Both Matt and Merv from Classic Creations of Central Florida agreed that the Dynacorn reproduction cowl panel is a quality piece.

With fresh lenses for the side-marker lamps, we could see a difference in the overall brightness of the lights. Our car didn't have any lenses up front so the difference is even more dramatic in the dark.

Chapter 28

How to Replace Side Marker Light Lenses

Text and Photos by Miles Cook

By now, most of you are familiar with our '69 Mustang fastback that has turned into a project car. Although not a project in an official sense, we've used this car for several technical how-to projects, showing the installation and testing of various parts.

Since early 2003, we've installed and extensively tested a Baer brake system and a Flaming River rack-and-pinion steering system. We've swapped a T5 five-speed into the car, and recently installed and dyno-tested a new 351W short-block that was detailed in our sister magazine, *Mustang & Fords*.

While our main purpose is to show the mechanical side of vintage Mustang how-to tech, most of us want our cars to look good too. With that in mind, we decided to address our car's lack of side-marker lights. We'll show you how to replace side-marker lenses at each corner as well, with a trick or two we learned along the way. The rear lights were a direct R&R procedure, but the fronts were a new installation because our car didn't have them.

We'll show you how to replace the side-marker lenses and surrounding bezels, even though we didn't do that at all four corners the same way.

The parts shown here are available from Mustang Country International, which has new and used vintage Mustang parts as well as an in-house shop that does various restoration and upgrade work on vintage Mustangs.

Source
Mustang Country International
www.mustangcountryintl.com

Here are the new lenses in place. Even in full daylight, the clear, bright light provided by the new lenses is evident.

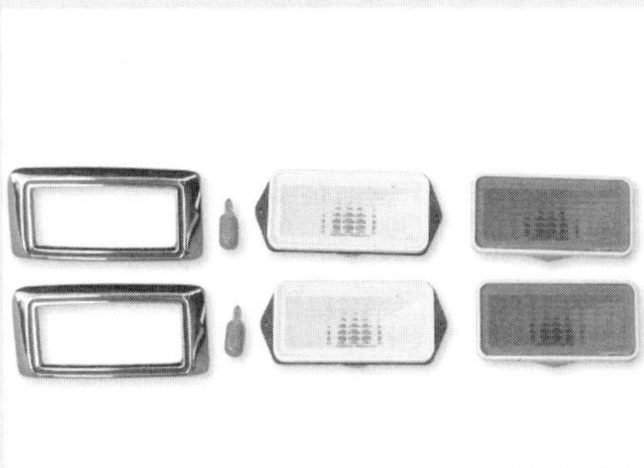

1. Mustang Country International stocks the lenses and bezels. Although '69 and '70 Mustangs are similar in many ways, the side-marker lamps are different. The red lenses are for the rear quarter-panels, while the white ones are for the front fenders on our '69. The yellow bulbs in the white lenses produce the correct pale-yellow glow with the side markers on at night. Mustang Country has an endless array of vintage Mustang parts, but had only two new bezels on hand so we reused our well-worn but still serviceable bezels at the rear of the car.

2. Our rear side markers were in fair shape, but the original lenses were a bit faded and the bezels were pitted. We reused the rear bezels and saved the new ones for up front.

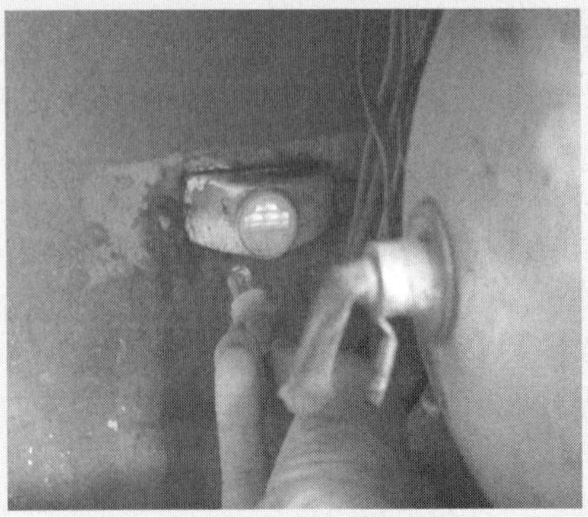

3. Removing and replacing these components requires simply unplugging the bulbs and removing the lenses. Begin inside the trunk by rotating the light socket until it can be removed from the lens housing as shown.

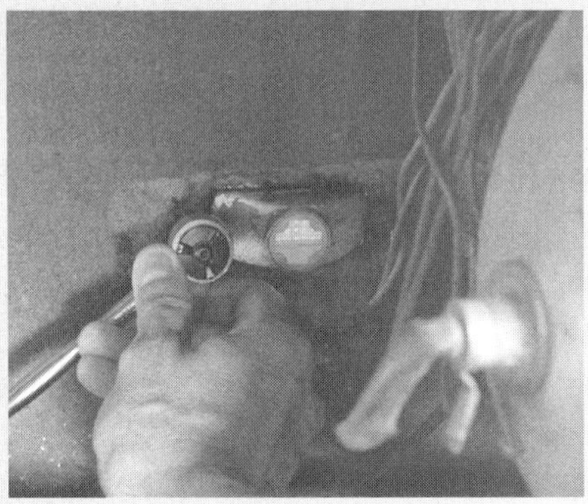

4. Next, the bezel that holds the lens in place is unbolted from the quarter-panel. After the two speed nuts are removed, the whole assembly comes out of the quarter-panel.

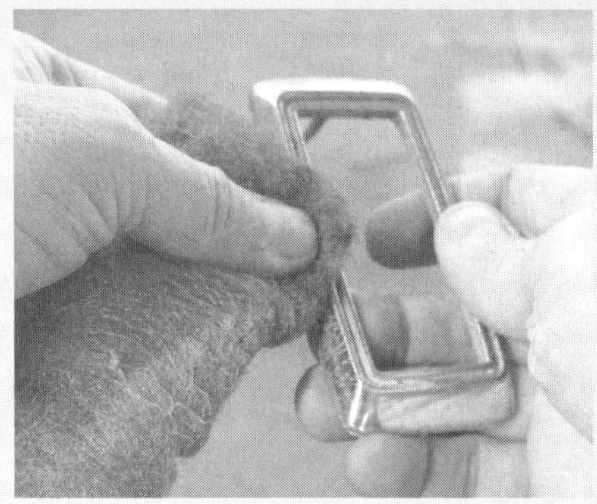

5. Cleaning the rear bezels with steel wool helps a little, but they don't look as good as the new ones.

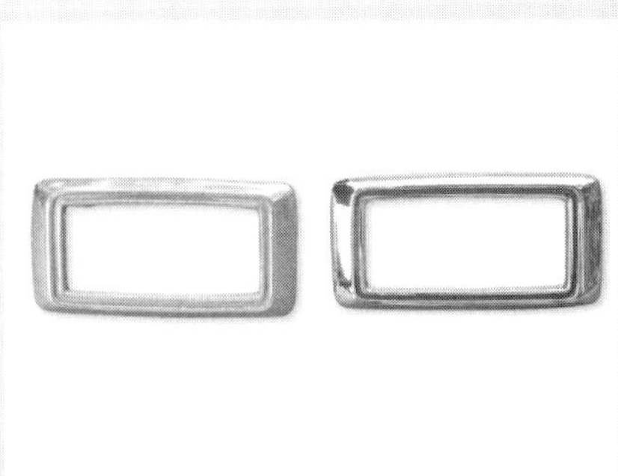

6. The rear bezel we reused is on the left, and the new one that went up front is at right. We have another pair on order from Mustang Country, but the original pieces are serviceable for now.

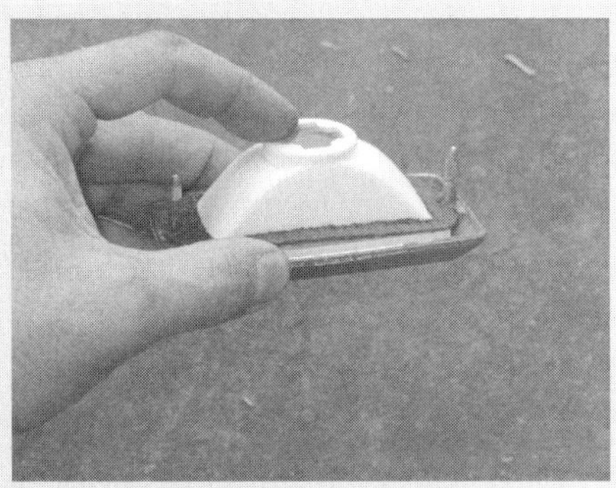

7. The gaskets on the rear bezels are in fairly rough shape, but they'll suffice until we get new ones. This shows the correct order of assembly. The new lens drops onto the bezel and the gasket fits over the lens and sits in place with the bezel mounting studs holding the gasket in position.

8. Reassembling the rear side markers on either side of the car is easy. Simply reinstall the bezel and lens assembly back into the quarter-panel, secure it in place with two speed nuts, and plug the bulb-and-socket back in place. With the process complete at the rear, we'll move to the front.

9. As mentioned, our car didn't have any front side markers in place, so this will be a nice upgrade. Were they there, removal is the same as the rear: unplug the socket from the lens and remove the assembly by unbolting the speed nuts from inside the valance.

10. At one point we thought the original front side-marker wiring harnesses were gone, but we found them tucked away behind the front bumper area. Did they work? To our surprise, we removed the old bulbs, plugged in the new yellow-colored ones on each side, and they both worked.

11. While we only replaced the lenses at the back of the car, we have all new components for the front, including bezels and lenses with gaskets attached to the lenses. The lens and bezel go together as shown with the lens fitting right behind the bezel.

12. The foam gaskets have holes through which the bezel studs pass, as shown here. Carefully fit the studs through the small holes in the gasket and push the lens all the way in until it fits flush against the bezel.

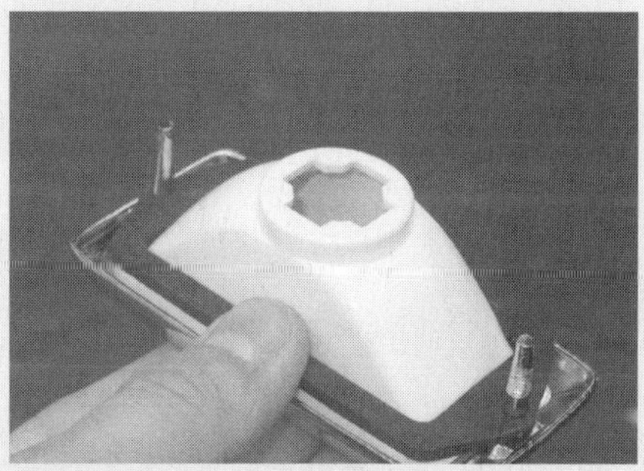

13. The correctly assembled bezel, gasket, and lens look like this.

14. Mustang Country has new speed nuts that are used to secure the bezel and lens assembly to the lower front valance panel under the front fenders.

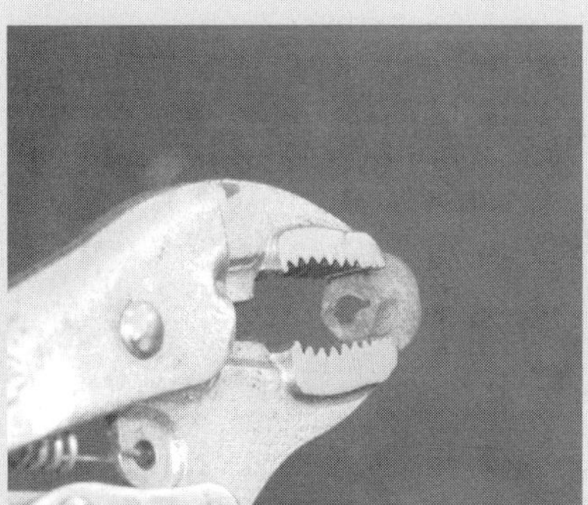

15. If you're replacing the lenses and don't get new speed nuts, the original ones can be flattened out a bit with a pair of Vise-Grips so they grab onto the bezel stud.

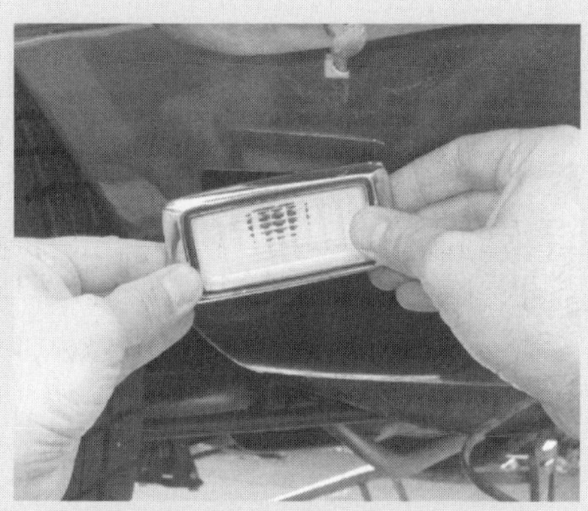

16. With everything assembled and with either new or resurrected speed nuts, the assembly installs into place as shown.

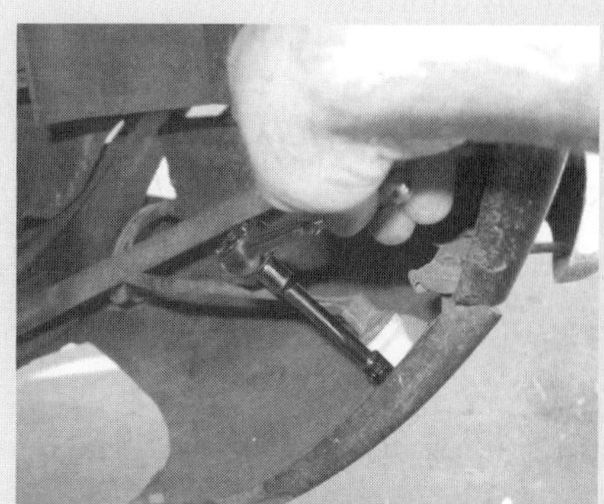

17. Tighten the speed nuts from inside the front valance panel with a 1/4-inch-drive ratchet and correct-size socket. Plug the light/socket assembly into the new lens, turn on your parking lights at night, and look at the improved outside lighting on a '68-or-newer vintage Mustang.

Chapter 29

How to Install New Bumpers

Text and Photos by Miles Cook

New bumpers at the front and rear make a noticeable difference in the overall appearance of our '69 Mustang subject car (above and below).

Source
Mustang Country International
www.mustangcountryintl.com

Back in olden days of restoring vintage Mustangs, one often had to spend additional time, money, and effort on the resurrection of a car's existing parts. Replacement components weren't available, which is still the case in some instances.

However, bumpers for '65–'73 Mustangs are a good example of what's available with respect to replacing a part rather than having it restored and waiting for it to be returned. While having the original bumpers rechromed on a rare and valuable car is a viable option that might be the best choice in some cases, replacing them on a less collectible car is a worthy alternative.

Our trusty '69 fastback subject car is such an example. Although it's a fastback and is now a presentable-looking and good-running daily driver, it isn't a rare car like a Boss, a Mach 1, or a Cobra Jet. In short, this car is the perfect candidate for a fresh set of replacement bumpers, available from Mustang Country International. As it turned out, the original bumpers had already been replaced before we owned the car.

In Chapter 28, we showed you how to replace the car's side-marker lights, and thought new bumpers would improve the car's looks since it leads a tough life as a heavily used test mule.

Whether you decide to replace ratty-looking bumpers or remove your car's existing ones to have them rechromed and reinstalled, the following remove-and-replace sequence will be of interest. Although we used a '69, bumper replacement for all '65–'70 Mustangs is similar.

1. Begin with the rear bumper. First disconnect the license plate light by unplugging it from the car's wiring harness. Then, from inside the trunk, unbolt the bumper. Using a half-inch-drive ratchet and the correct-size socket, remove the four bolts (two on each side) that allow the bumper to be taken off.

2. With the hardware removed, the bumper separates easily from the car. If possible, have another person hold the bumper in place while it's being unbolted so it doesn't fall off once you've removed the four bolts.

3. Two small bolts hold the license-plate light to the rear bumper. Remove them with a wrench to separate the light from the bumper. Save these two bolts, as you'll reuse them to reinstall the light on the new bumper.

4. Regardless of whether you replace the rear bumper or have it rechromed, you must remove the mounting brackets as shown. Each is held in place with two nuts and bolts. Removal is best accomplished with a half-inch-drive ratchet and socket.

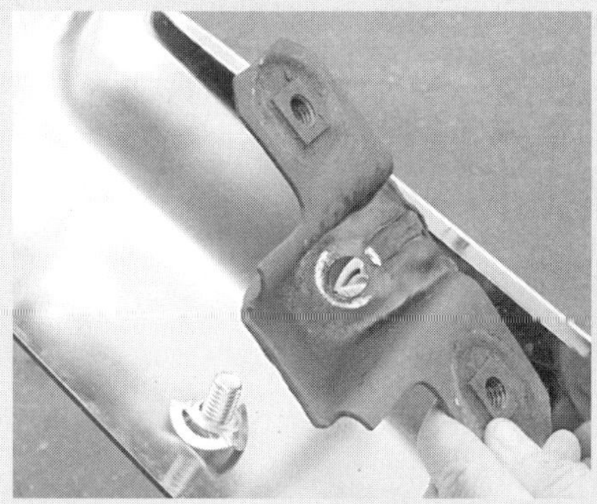

5. We used new bumper-mounting hardware because the bolts are visible with the bumper in place. On a new bumper, the existing bracket drops into place. Note the special curved washer, which comes with the new hardware. The washers provide space for the square part of the bumper bolt to fit through the bumper and to prevent dimpling and damage when the nut is tightened to hold the bracket in place.

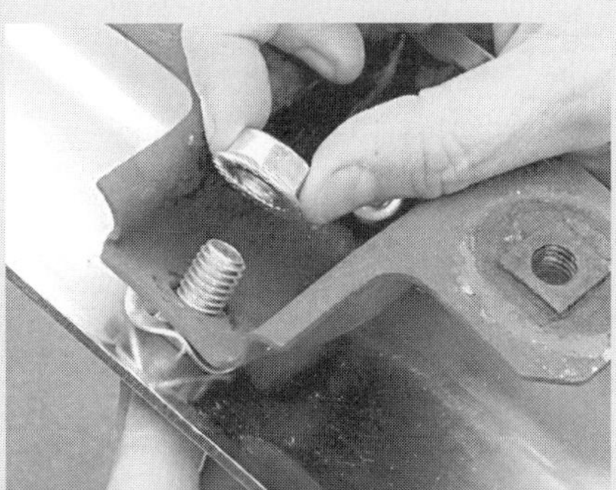

6. With the bumper bolts and bracket in place, the nut fits over the bracket and tightens with a ratchet and socket. Tighten securely, but be careful not to damage the bumper by overtightening the two nuts and bolts that hold the bracket in place.

7. Reinstall the license plate light on the new bumper using the same two bolts that held it in place on the old bumper.

8. Although it can be done alone as we did here, it's easier to have another person hold the bumper in place while you start to tighten the four bolts inside the trunk. Be sure to start all four bolts before final tightening. Once all four are started, some adjustment will be available to center the bumper on the car. When you're satisfied with the placement of the bumper, final-tighten the bolts and plug the license plate light back into the car's wiring harness to complete the installation.

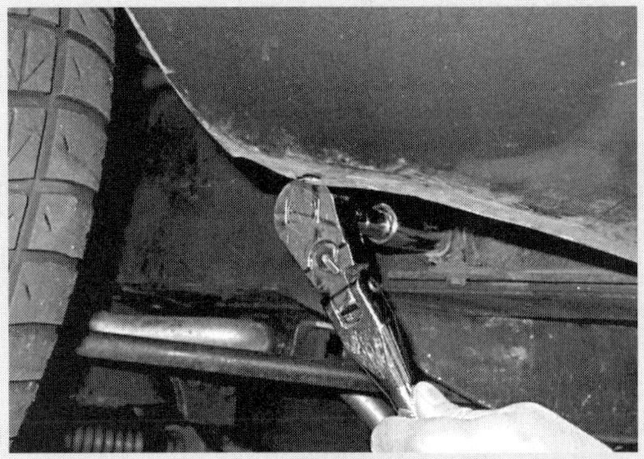

9. On the front of the car, removal of the bumper is similar to the rear, as there are two bolts on each side that hold it in position. The bolts are inside the front valance panel behind the front wheels, and are accessed with a half-inch-drive ratchet and socket. Remove the four bolts and set them aside, as they'll be reused to install the new bumper.

10. A great example of a dumb idea are these bent-to-fit lag bolts holding the bumper ends in place. The correct brackets are available from most Mustang parts houses. We unbolted them from inside the front fender and threw them in the dumpster.

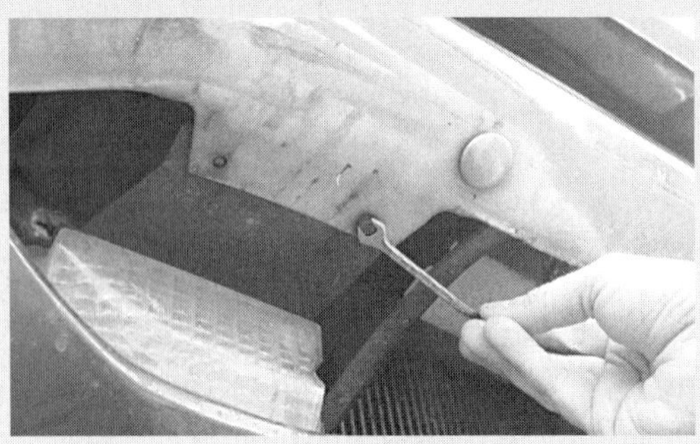

11. On a '69 Mustang, unbolt the parking-light housings from either side of the front bumper as shown. These small bolts are the same as those used on the rear bumper's license-plate light and take the same-size 1⁄4-inch wrench.

12. With the four main bolts and parking lights removed, and those hokey lag bolts out of the way, the front bumper and its mounting brackets are removed all at once by pulling the assembly forward.

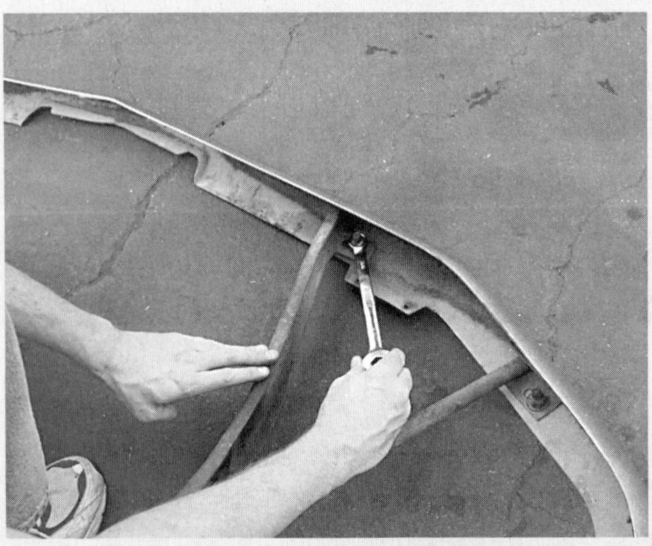

13. Like the rear bumper, the mounting brackets also need to be removed from the front bumper, which is easy with the correct-size wrench.

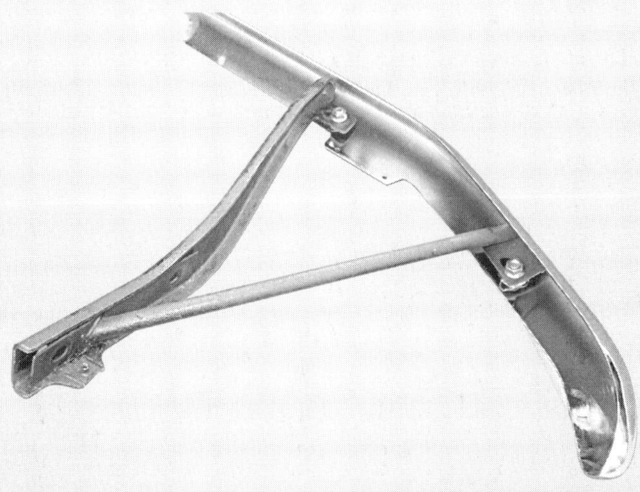

14. We did some quick resto work with a can of flat-black spray paint on the beaten-up bumper mounting brackets. After they dried, we reinstalled them on the new bumper as shown. There are two brackets for each side of the bumper (passenger side shown), which come together at a point where both are affixed to the car with the two main mounting bolts. Like the rear bumper, we used new hardware and the same special washers to attach the brackets to the bumper, again being careful not to overtighten them and damage the new bumper.

15. This is the correct hardware for the ends of the front bumper. The bumper-bracket bolts from AMK Products are in the plastic bag, and the small bumper bolts up top came with the new bumper hardware kit. The used brackets complete the picture for attaching the front bumper the right way. All of these pieces are available from Mustang parts houses.

16. While the used brackets were serviceable, Mustang Country located a pair of new ones, so we used them instead. This shows how they're attached to the bumper before installing it on the car. The bracket fits inside the bumper, with the bumper bolt going through the bracket, held in place with a flat washer and nut.

17. With the brackets and new hardware on the bumper, it's ready to be installed on the car. Carefully fit the brackets through the front valance to place them in their original positions. The parking lights are hanging from the car so they won't be in the way while putting the bumper back in place.

18. This is the correct AMK hardware used to attach the new brackets on each end of the bumper.

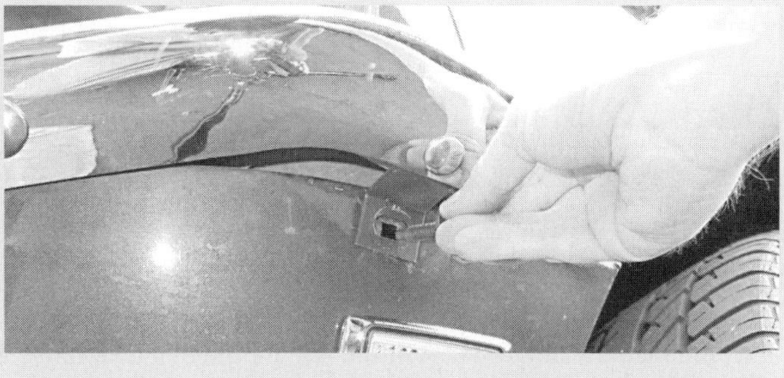

19. Since the valance panel doesn't fit well, we were surprised the holes lined up closely when we installed the end hardware with the new bumper in place. The bolt installs through the bracket and valance panel, and the retaining nut installs on the backside of the valance panel. Tighten the nut securely with a 1⁄4-inch-drive ratchet and pliers on the inside as shown, but not enough to damage the rubber grommet. At this point, the bumper is installed, and all that's left is to reinstall the parking lights.

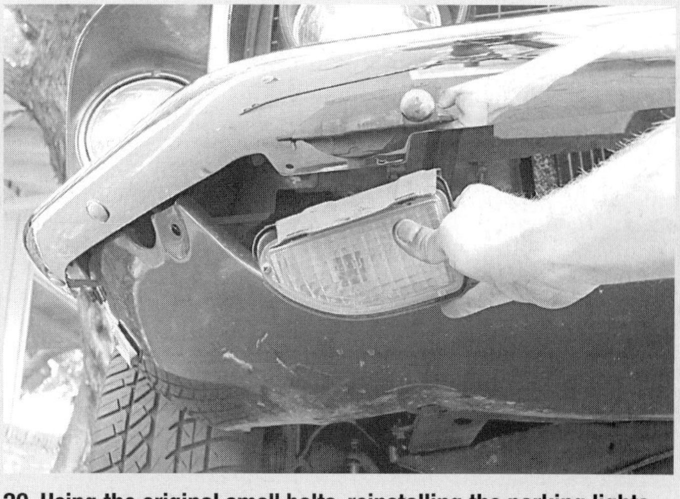

20. Using the original small bolts, reinstalling the parking lights onto the new bumper is easy. Set it into place and tighten the bolts securely with a 1⁄4-inch wrench as shown.

Chapter 30

How to Install New Lock Sets

Text and Photos by Miles Cook

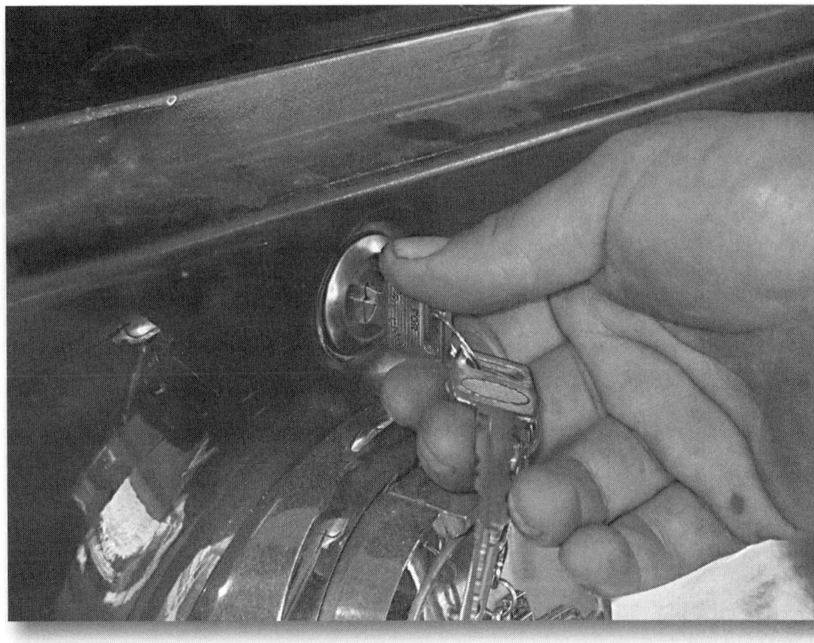

When it comes to fully restored Mustangs, what's wrong with this picture? The square key doesn't fit in the trunk lock, but a new kit sold by Mustang Country eliminates the need for two keys. Like a late-model Mustang (and most newer cars), you'll need only one key for the whole car. If you decide to go back to a two-key arrangement, simply swap out the trunk lock.

Our '69 Mustang fastback has been in the *Mustang Monthly* project garage for more than two years now, and we're just getting around to this relatively easy project. We've been concentrating on bigger items like a disc-brake conversion and the installation of a T5 five-speed transmission, and with a new lock set we'll be able to secure the car for the first time since we've owned it.

Why wouldn't it lock? A new ignition lock was installed before we purchased the car, but the door locks weren't changed at that time. So we had a square ignition key (and a round trunk key) that didn't work in the doors.

We originally planned to install a new ignition lock and a pair of door locks, while keeping the car's existing trunk lock. But after arriving at Mustang Country International for the installation, we were shown a complete kit made by Scott Drake Mustang Parts. It includes two door locks, an ignition lock, and a trunk lock set up to work with a single key—an ideal arrangement for a daily driver.

Of course, you can also keep the original system for vintage Mustangs, which means a separate key for the trunk lock. This may be a more desirable way to go for a restored show car that will face the scrutiny of a detail-oriented judge. Whether you want one key or two, the installation procedure is the same.

Source
Mustang Country International
www.mustangcountryintl.com

This is the complete one-key lock kit for '67–'69 Mustangs as offered by Mustang Country. At the top are the two door locks and their retaining clips. From left to right in the second row are the retaining clip for the trunk lock, trunk lock, ignition-switch lock, and three gaskets for the door and trunk locks. Below that is the rod that connects the trunk lock to the trunk latch and the two supplied keys. At this point, the single-key system is available only for '67–'69 cars; the '65–'66 Mustangs use a different single-sided key, and '70–'73s have the ignition switch in the steering column. Mustang Country stocks lock sets for all vintage Mustangs, and has an extensive selection of obsolete used parts yet to be reproduced.

For those who prefer the purist route, a standard two-key system can also be installed, such as the packaged door-and-ignition-switch kit shown at left. To the right are a trunk lock and its standard round key. Whichever you choose, the installation we're showing is the same.

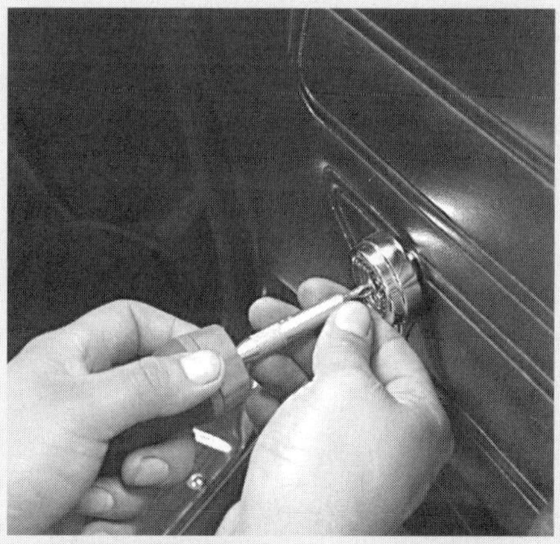

1. Starting with the doors, first remove the door panels to access inside the door. This involves removing the window crank and armrest pad.

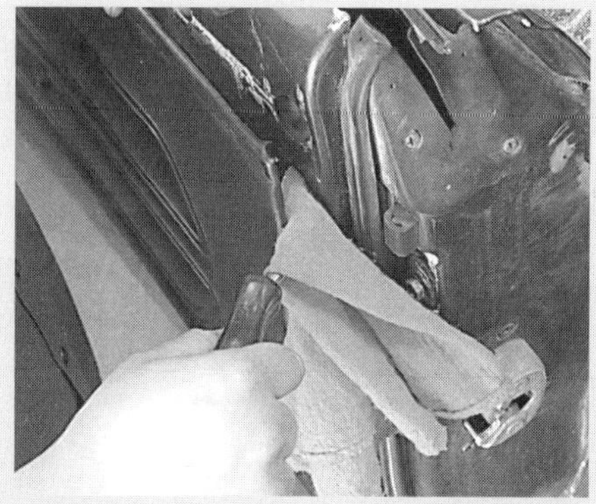

2. With the window crank and armrest pad removed, the panel can be carefully pried from the door. Be sure not to tear the door panel from the mounting clips, which happens because the clips are stronger than the cardboard in the panel. Pry on the clip, not the panel, to avoid damage. Cover the prying tool with a rag to prevent damage to the vinyl on the panel.

3. With the panel off, you have access to the door lock. This shows how the lock is held to the door with the U-shaped clip and attached to the lock-rod. We couldn't get a hand and a camera inside the door at the same time, but the following two photos gives you an idea of how to remove the lock from the door.

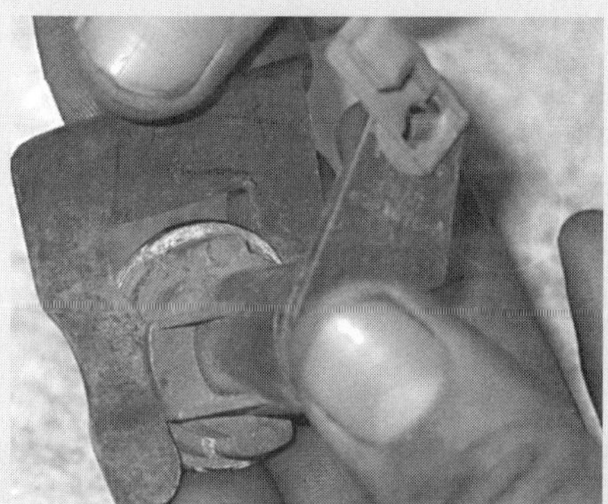

4. Although outside the door, the clip slides over the lock as shown. Each car is different; if you can't R&R these clips by hand, you may need pliers to pull off the clip to remove the old lock and install the new one.

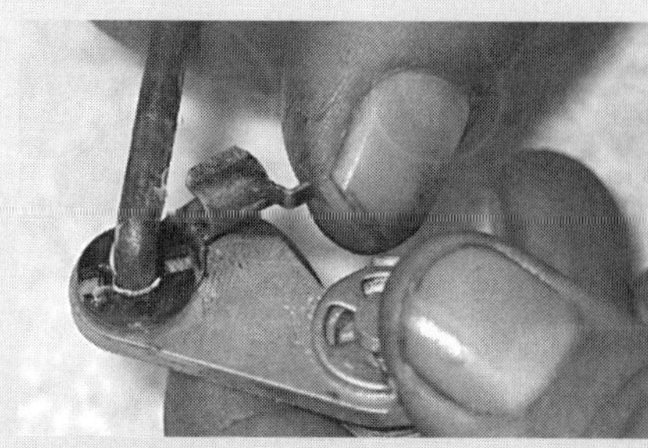

5. Again outside the door, this shows how the lock-rod clip attaches to the rod. It can be removed with your thumb or a small screwdriver. This is a simple R&R procedure, as the rod and the clip are reused with the new lock.

6. A comparison of the old and new door locks shows that the part that goes through the door-lock hole on the new lock isn't tapered as it is with the old lock. We had to trim this area slightly with a die-grinder to make them fit. This won't be an issue with every car, and we solved it in a minute or two.

7. Making sure the new gasket is in place and the keyway of the lock cylinder is at the top, the new lock installs into the door. The locks are side-specific, and the arm that connects to the lock-rod in the door faces toward the front of the car on each side. Once the lock is in place in the door, final installation is completed from inside the door by installing the retaining clip and reconnecting the lock-rod to the arm of the new lock.

8. Once the locks are installed, test them with the new key to make sure they work properly before putting the door panels back on. To complete the installation, reinstall the door panels by pushing the clips on the panel back in place in the door and reinstalling the armrest pad and window crank.

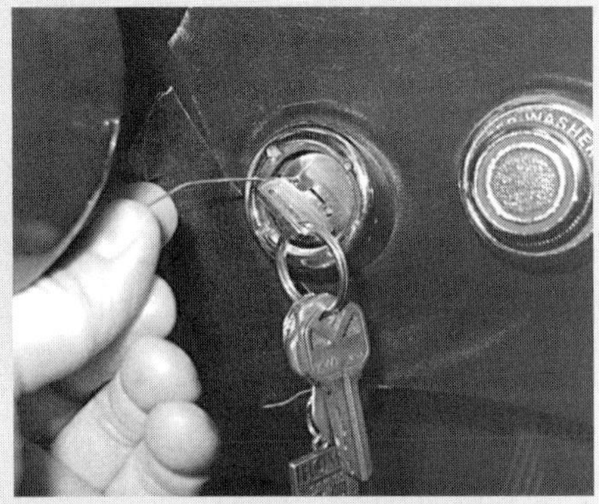

9. Moving to the ignition switch, you'll need its key and a straightened paper clip or similar tool to remove the lock cylinder from the switch. Turn the key to the Accessory position (toward the driver side) and insert the clip to release the lock...

10. ...that allows the lock to be turned beyond the Accessory position. Once the key is turned past Accessory, it can be removed from the ignition switch as shown.

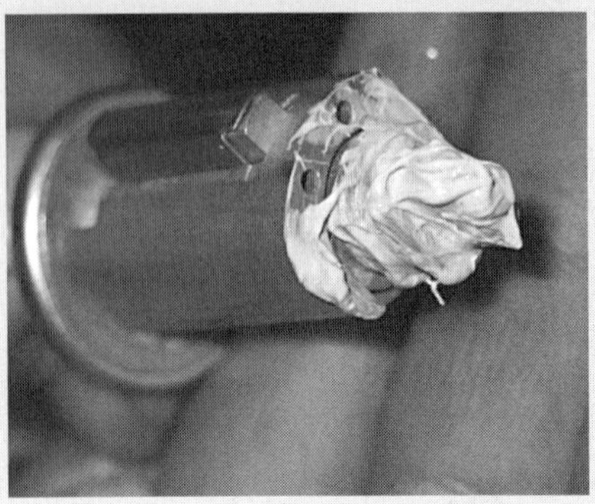

11. Before installing the new lock into the ignition switch, apply some grease on the end to make the new lock work smoothly for the long haul.

12. The new lock goes back in the place of the old one. Insert the new key in the lock, insert the lock into the ignition switch past the Accessory position (the ignition is in the Accessory position while the lock is removed), turn the key to Accessory, then to the Off position. At this point, you're back to where you started. The new lock and key should now go to the Run and Start positions.

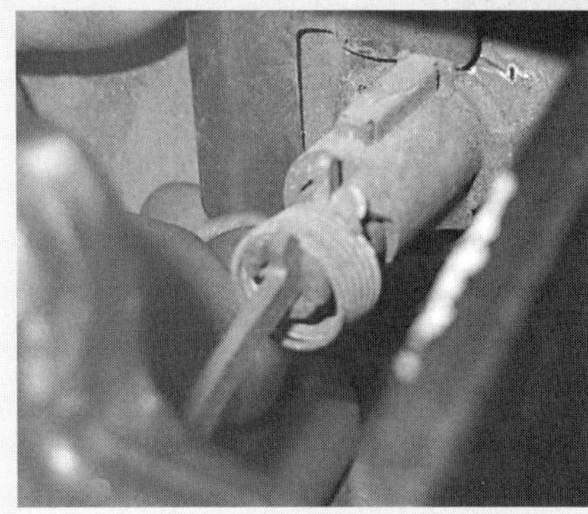

13. Removing the trunk lock is the same as with the doors—an identical clip holds the lock in place inside the trunk. You might need pliers to pull out the clip, depending on how tight it is. With the clip removed, the lock and lock-rod come out of the rear panel.

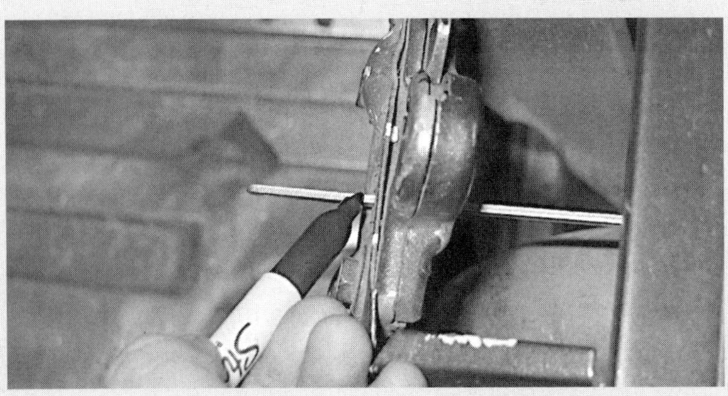

14. Test-fit the lock and rod to see if they work properly. We used the new rod and trimmed the excess length with a bench-mounted cutoff wheel. If you want to forgo this because the new rod is too long, the car's original one should work fine with the new lock.

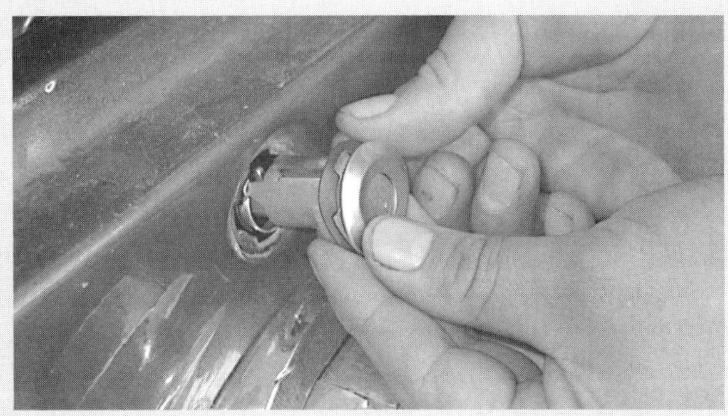

15. Install the new lock, making sure the new gasket is in position. Once in place, reinstall the retaining clip (either the supplied new one or the car's original one) to complete the installation of the lock package.

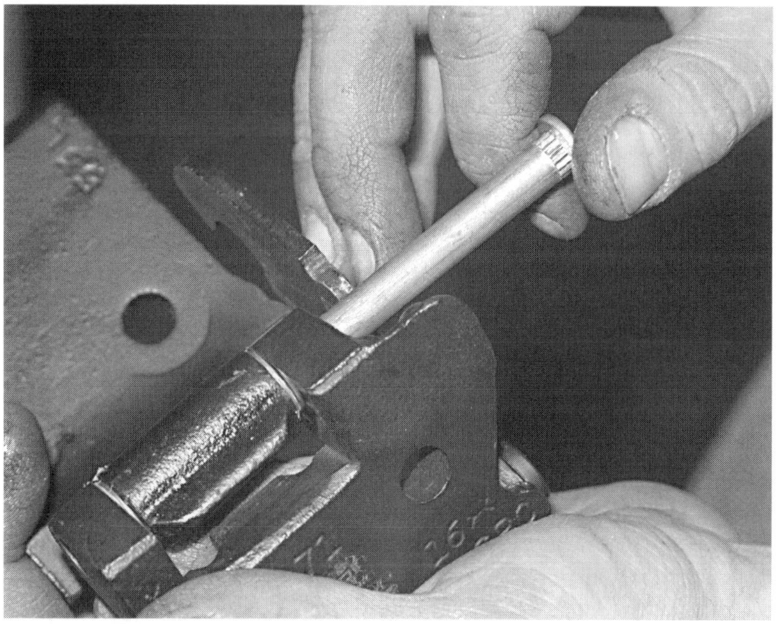

Good hinges are the key to proper door operation and fit.

Chapter 31

How to Rebuild, Replace & Adjust Door Hinges

Text and Photos by Jim Smart

Sources
Mustangs Etc.
www.mustangsetc.com

National Parts Depot
www.npdlink.com

Rustbusters Automotive Restorations
313/255-5990

A Mustang's integrity can be summed up by how the doors feel and sound when opening and closing. Does the door rattle? Does it echo? Doors that sound awful when closed say a lot about your Mustang. They also demonstrate your abilities as a car restorer and whether you sweat the attention to detail.

Door feel and function begin with good components and solid integrity. Door hinges with worn bushings and pins will never give you smooth operation because sagging doors won't work consistently no matter how many times you adjust them. They close smoothly one time and sloppy the next. To get doors working consistently, you need hinges and bushings with zero play, and strikers and lube-coated latches free from wear. Door-latch mechanisms, rods, linkages, and lock cylinders also need plenty of lubrication.

Building Better Hinges

Ford used two types of door hinges on '65–'78 Mustangs. From '65–'67, Ford used cast-iron hinges with steel pins, bronze bushings, and heavy-duty steel roller checks with coil springs. Toward the end of the '67 model year, Ford switched to stamped-steel upper hinges while retaining the cast-iron lowers. As the '68 model year unfolded, Ford went to stamped-steel hinges at top and bottom, with twin detent roller checks against a formed-spring steel rod. Like the iron

This is a '65 Mustang cast-iron lower door hinge. Media-blast, prime, and repaint the check spring before reusing it. Reproduction springs aren't strong enough to hold the door open on a hill.

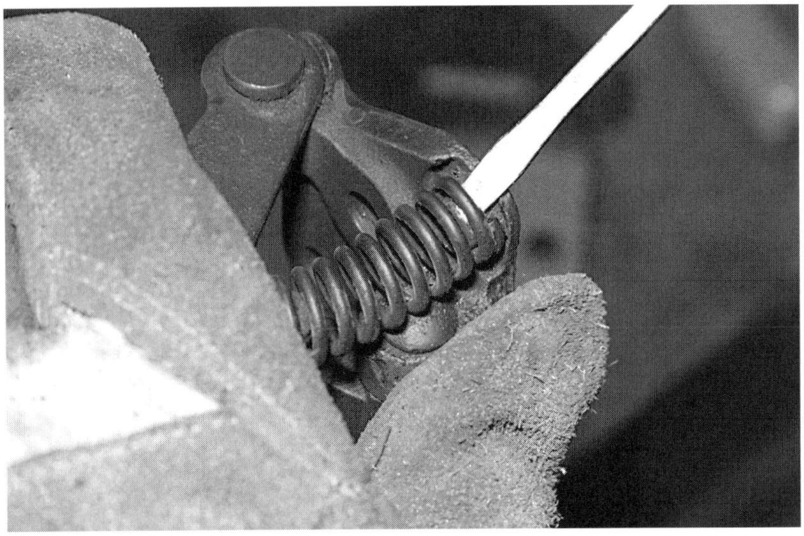

Spring removal isn't half as fun as spring installation. Wear eye and face protection, shroud the spring with your hand protected inside a glove, and pop the spring out as shown.

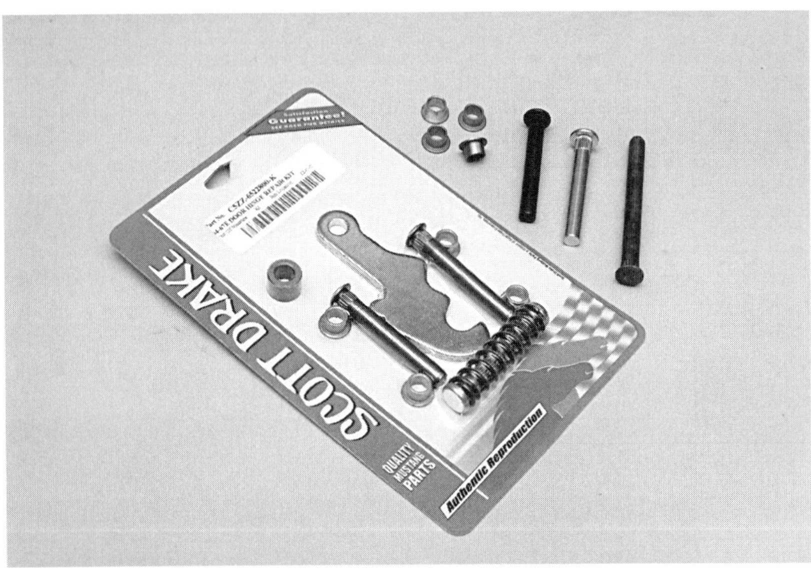

This is what you can expect from Scott Drake Reproductions, available from National Parts Depot. Pins, bushings, springs, and even check-detent mechanism parts are available in these kits. You can buy kits or obtain the components separately, depending on need. The nice thing about these kits is the virtually perfect fit without having to trim pins and make modifications.

hinges, the steel hinges had bronze bushings and tool steel pins. Stamped-steel hinges were used through 1978.

Stamped-steel hinges were lighter and less expensive to produce, hence the transition to them late in 1967. No matter how you slice or dice a stamped-steel hinge, it isn't as solid as a cast-iron hinge. If you're restoring a late '67 or '68 Mustang and aren't concerned about spot-on originality, use cast-iron hinges on top and bottom.

In all cases, rebuilding door hinges involves replacing pins and bushings. It also involves getting check mechanisms to work properly if you want the doors to stay open, especially when parked on an incline. There are a couple avenues you can take regarding service parts for Mustang door hinges. To tell the truth, some reproduction hinge-pin/bushing kits aren't the same caliber as original Ford parts, and much of what was available from Ford isn't available now.

Door Adjustment and Function

Solid door-hinge construction is the base on which door adjustment begins. But there's another angle: body dimension. If a Mustang has had even the lightest impact in any location, door opening dimensions could be off, making it impossible to get the gaps right. For example, a minor rearend collision can move B-pillars significantly, affecting door adjustment. Adjustment begins at the B-pillar where the door striker is located. Movement of even 1⁄16 inch can make proper adjustment impossible. Gaps should be 3⁄16 inch at the leading and trailing edges and larger at the bottom. A buckled rocker panel can affect bottom gap, which needs to be approximately 1⁄4–5⁄16 inch.

If you're unsure about a body's collision history, check the measurements all around. If a Mustang's body isn't "plumb" per factory specifications, body panel and door adjustment is virtually impossible. If you have a body that has suffered any kind of collision, ranging from a parking lot smack to a hard frontal or rearend impact, dimensions can be out of spec enough to give you headaches when it's time to adjust doors, fenders, the hood, and the decklid.

As stated earlier, door adjustment always begins at the B-pillar. Once the beltlines and door striker are lined up, you're ready to adjust front fenders. Door adjustment has a series of priorities to follow. Because this is what you see first, the most important alignment point is the door crown where it meets the quarter-panel and front fender. Beltlines are next, beginning with the top, then the bottom beltline. Ideally, both will line up along with the door crown.

Also important is window adjustment, which goes hand in hand with door adjustment. The wing-window assembly should seat firmly in the perimeter weatherstrip without hanging up when the door is opened. Once the wing window is locked in place, close the door and adjust the window travel and the seating against the weatherstrip. The same rule applies: It can't hang up on the weatherstrip.

Door adjustment takes place at the hinges at the A-pillar. Hinges adjust in two places: the A-pillar and door fulcrum. A-pillar locations control door angle and gap. Fulcrum locations control door angle in relation to beltlines, the B-pillar, and the door crown. The first things to adjust are the A-pillar–hinge attachment points to control gaps, beltline alignment, and the crown. Then fine-tune fulcrum attachment points to adjust the door angle. Snug the bolts, but don't tighten. Slowly close the door, and check gaps and alignment. Gaps need to be 3⁄16-inch (thickness of a wood paint-stirrer) at the B-pillar and front fender. The bottom gap needs to be 1⁄4–5⁄16 inch. What matters most is crown alignment with the fender and quarter-panel at the top. Align that part, then work your way down. Because stampings weren't consistent during the '60s, getting all of it to line up is challenging.

Another important point is the door striker at the B-pillar. Strikers are adjustable in any direction. Move them toward the door with shims. Loosen the screws, and move them carefully up, down, in, or out to control door position. Raise or lower the striker to fine-tune the closed door positioning. Move the striker in or out to control door depth when closed. Move the striker too high, low, or far in, and the door will be hard to open and close. Adjust and shim the strikers to get door adjustment dead-on. Just remember how important striker location is to ease of door operation.

The difference between reproduction door check springs and original equipment is obvious. On the left is a reproduction spring with a smaller coil wire and less tension. It isn't as strong as the original equipment, but it will hold a door open on level surfaces. The problems with original-equipment springs are damage, corrosion, and less tension due to time and use. Inspect both springs to determine which to use.

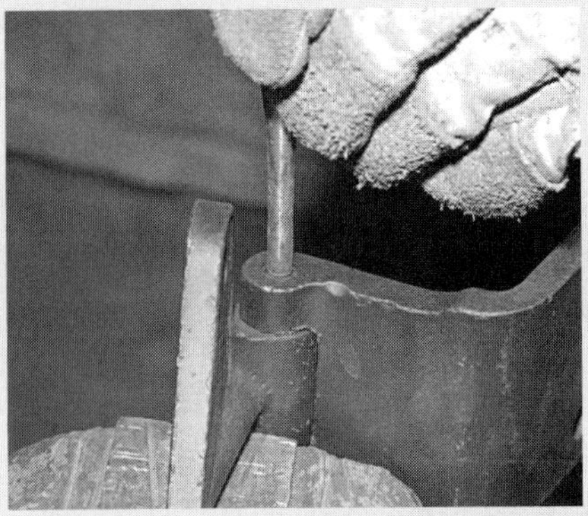

1. Begin rebuilding by driving the old pins out. This is an upper hinge without the check, but the basic principles are the same.

2. Check out this 40-year-old bronze bushing. It's worn completely through and needs replaced.

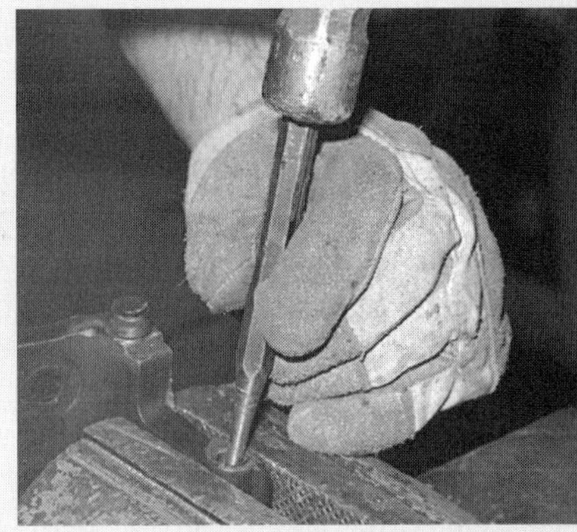

3. To replace the bushings, Rustbusters starts by driving out the old bushings and inspecting the castings.

4. On the right are original bushings, which have more bearing surface area for superior pin support. Reproduction bushings offer less support area, but they are the only choice if you need bushings right away. One option is to check with bushing/bearing suppliers to see who has exact dimension bushings for your hinge project. Aerospace hardware companies are one option. They produce all kinds of bronze bushings for aircraft applications that may also fit your hinges. Ideally, you want longer bushings on a par with original equipment if you're going to use your Mustang daily.

5. The bushings are fitted to the hinge and pressed in as shown.

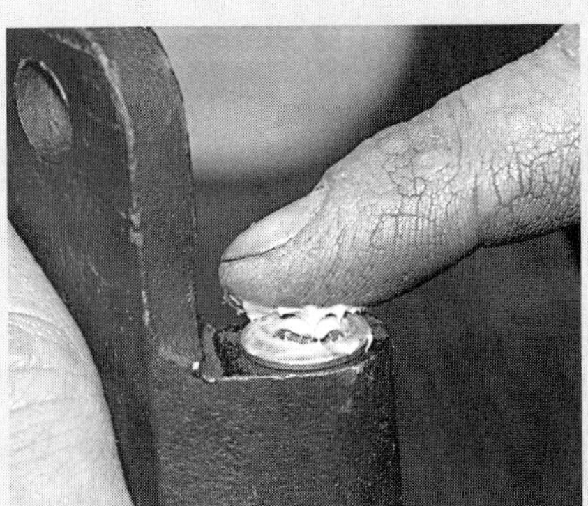

6. White lithium grease has plenty of staying power and provides smooth operation. Coat the bushings and the pins with plenty of it.

7. Drive the hinge pin in place until the head is firmly seated. If necessary, tack weld the head and grind it smooth. Sometimes pins work their way out, which is why we suggest a small tack weld.

8. Check-spring installation is never easy. Rustbusters works the spring as shown, seating it in the lower hinge like this. Again, use eye and face protection along with heavy gloves to protect your hands. A small engine valvespring compressor also works well for this step. See Briggs & Stratton or Tecumseh for this tool.

9. Rustbusters does a nice job on its hinge rebuilds. Note the taper pin tip and fresh bushings ready for operation.

10. When you're shopping for hinges and parts, pins like this are available from Mustang-parts vendors and body shop suppliers everywhere.

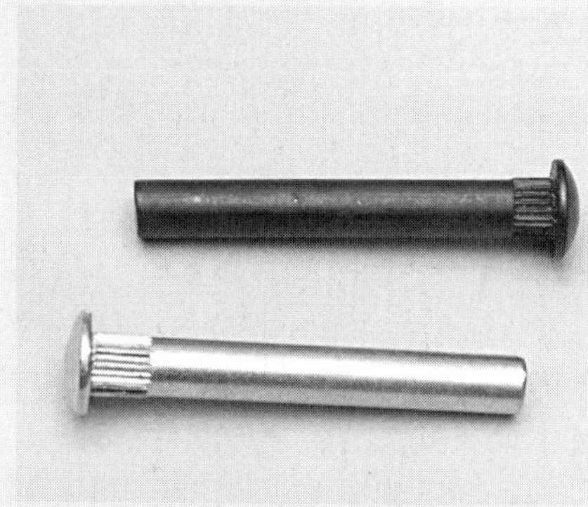

11. Here's another variation with a high-carbon–steel top and cadmium-plated bottom. Cad plating resists corrosion, while the high-carbon steel is more durable. Choose based on what's more important to you.

12. Stamped-steel hinges, first introduced late in the '67 model year, are not as solid as cast-iron hinges. If you own a '69–'78 Mustang, you have to live with them and make improvements. Stamped-steel hinges use shorter bolts with different heads than cast-iron, which is important to remember when replacing cast-iron with stamped-steel.

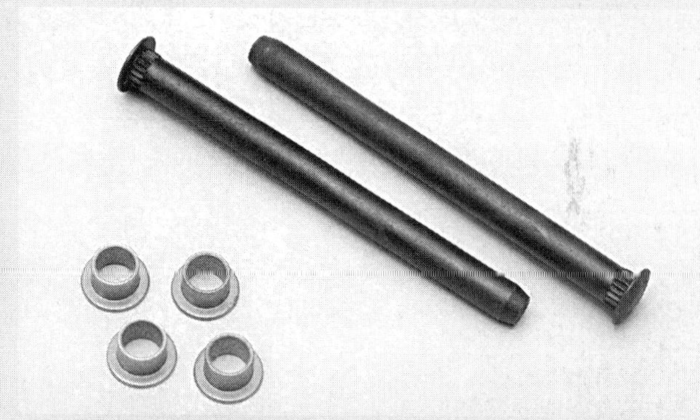

13. Pins and bushings are available for stamped-steel hinges. In fact, we're convinced you can also use these longer bushings in cast-iron hinges if you have access to a good machine shop. This kit is available from Mustangs Etc.

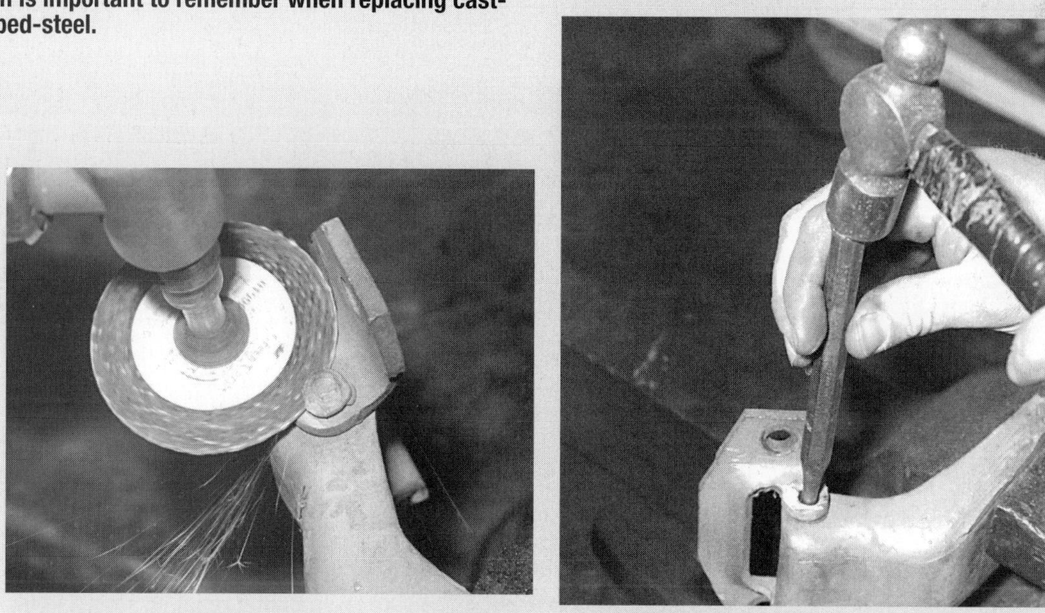

14. To rebuild stamped-steel hinges, grind off the peened tip as shown to remove the pin, and drive the pin out with a punch.

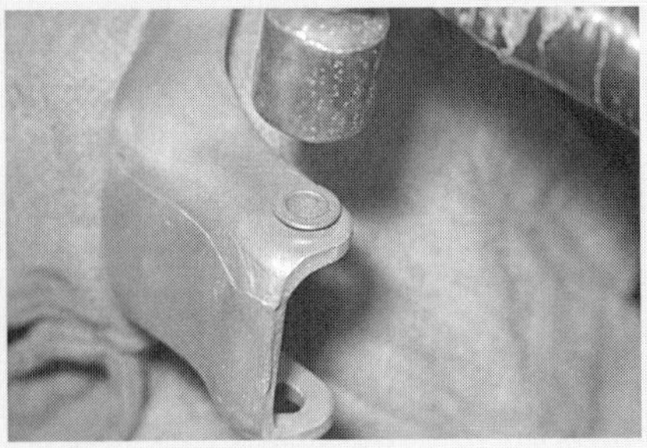

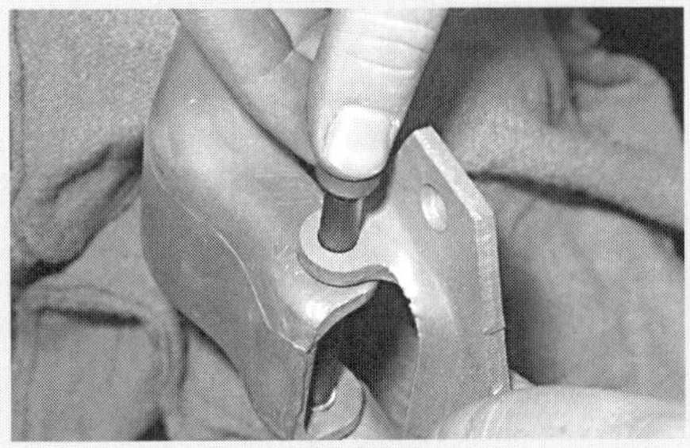

15. Bronze bushings go in like this, just like cast hinges; then drive the pin in. Don't forget white grease. Lubrication is everything to hinge survival.

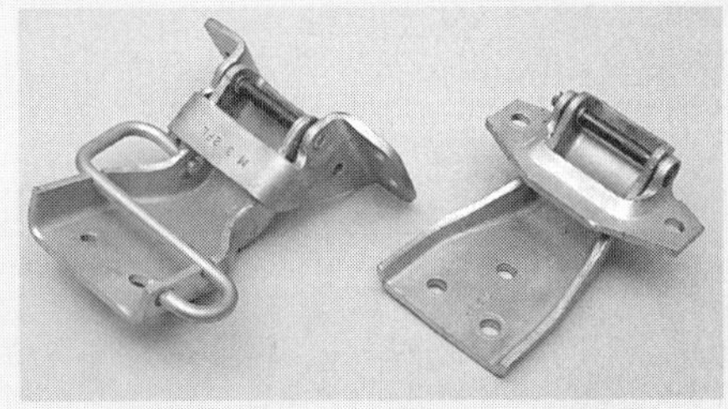

16. Stamped-steel hinges rebuild nicely if you begin with a good set. Because the checks can't be rebuilt, the most challenging part is finding stamped hinges with good checks. Stamped-steel hinges are also available as reproductions.

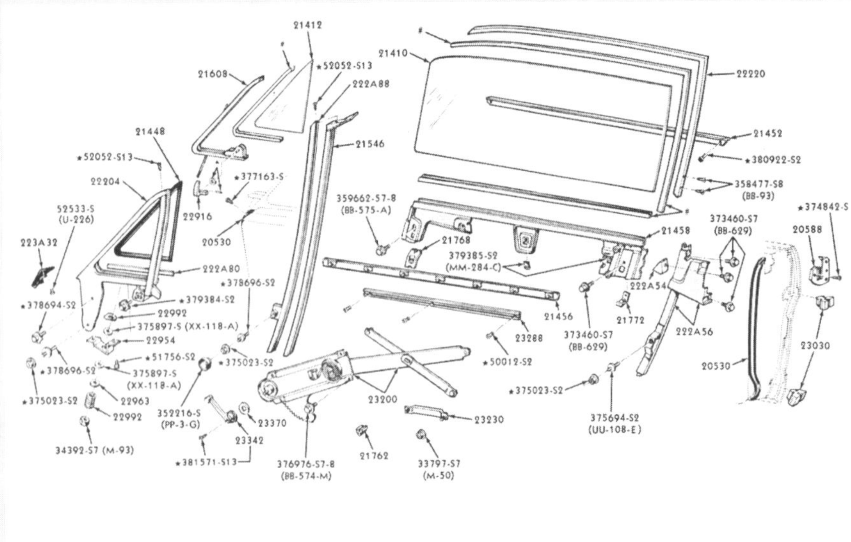

17. Proper window adjustment, which isn't easy to achieve, begins with clean, well-lubricated regulators and tracks. Next is adjustment at three points: two fore and one aft. Window angle is key to smooth travel and seating. Three Allen-screw adjustments control the window angle. Wing window retaining bolts must be loosened before adjustment is possible. Two adjustable window stops fore and aft limit travel at the top.

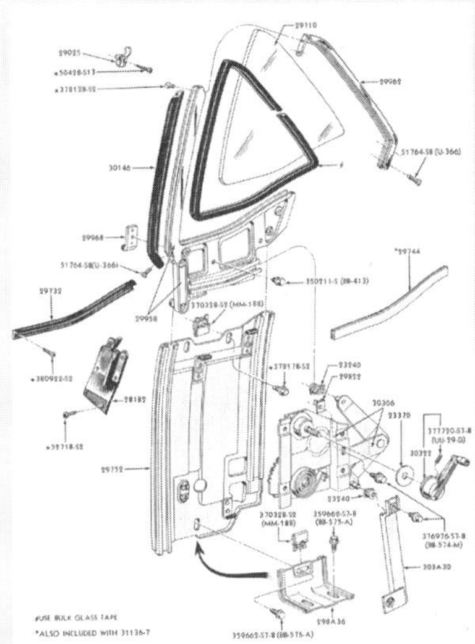

18. Quarter-window adjustment directly affects door-window operation on hardtops and convertibles. Adjust it too far forward, and it will hang up on the door window. The objective is to get the two to seat smoothly without wind noise, yet not hang up when opened.

19. This dusty '67 Mustang has good door gapping and adjustment, but it isn't perfect. First we prioritized at the top in our adjustment, then the top beltline above the quarter-scoops, and finally the bottom. The bottom is off a pinch, but it's less important than those above that we see first.

20. Gaps need to be spot on where the fender meets the door. Note how we prioritize: crown first, then beltlines. Of course, the fender is adjusted to the door to enable perfect gapping of 3⁄16-inch. Because old stampings aren't always accurate, expect challenges here. Ditto for areas with die-cast–metal pieces.

21. Don't forget door strikers in the adjustment process. Move the striker up or down to control door alignment. Move the striker in and out to control door depth. Shim it out to improve door-latch engagement. This is a valuable adjustment tool that requires a No. 3 Phillips screwdriver.

Chapter 32

Replacing a Window Regulator

Text and Photos by Miles Cook

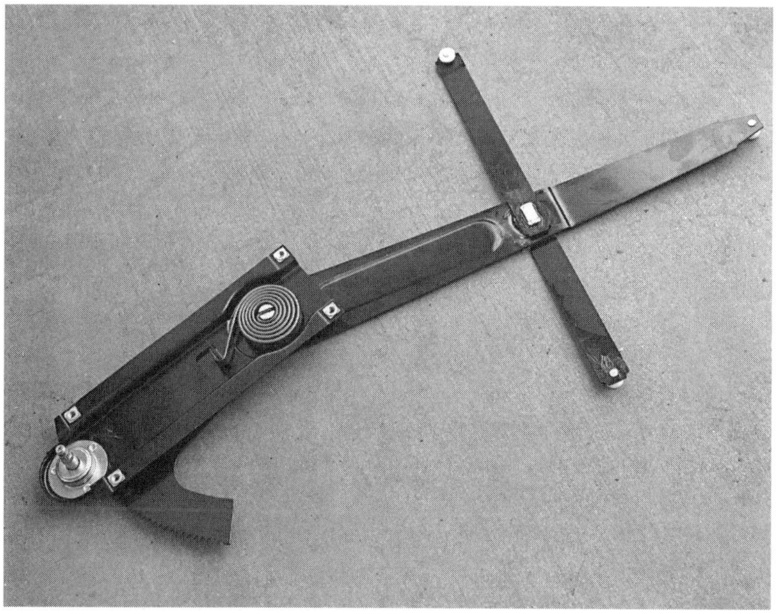

Looking like a pair of big scissors, this is the new regulator we're going to install in the passenger-side door of a '67 Mustang. For those in the Los Angeles area, Mustang Country International is a one-stop shop for parts and labor for detail-oriented jobs like this. It's also an excellent resource for used vintage Mustang parts.

Many vintage Mustangs have windows that don't roll up or down easily. Usually, the solution is replacement of the window regulator, which is the mechanical component that allows the window to roll up and down. While engines, transmissions, brakes, and fancy restomod upgrades are more fun to read and dream about, the basic maintenance and repair items like this keep our cars functioning properly.

The good folks at Mustang Country International supplied the new regulator and installation know-how, while we looked in on the swap being performed on a '67 Mustang fastback recently purchased by our friend, Ed Zoellner. The general course of action shown here is the same for any '65–'68 Mustang, all of which had crank windows and windwing glass at the front of the door. For '69–'73, the only real difference is the lack of windwings and the undesirable glue-in windows in '69 Mustangs. In Chapter 33, we show you how to install '70 door glass in a '69 Mustang.

Source
Mustang Country International
www.mustangcountryintl.com

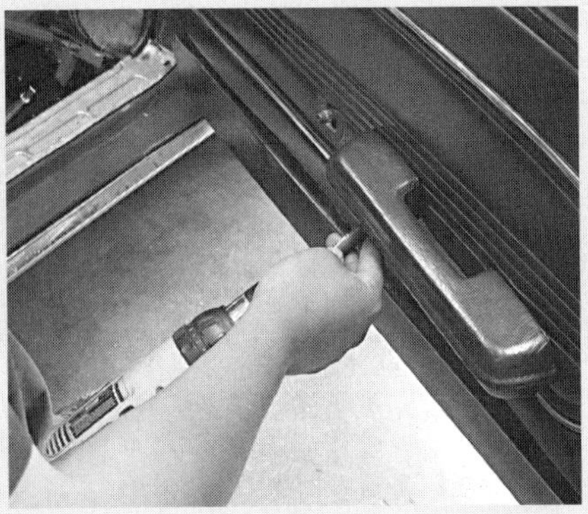

1. The landscape on the interior side of a vintage Mustang door is pretty basic. Removing the window crank, door-release handle, and armrest takes about two minutes with a screwdriver and a socket.

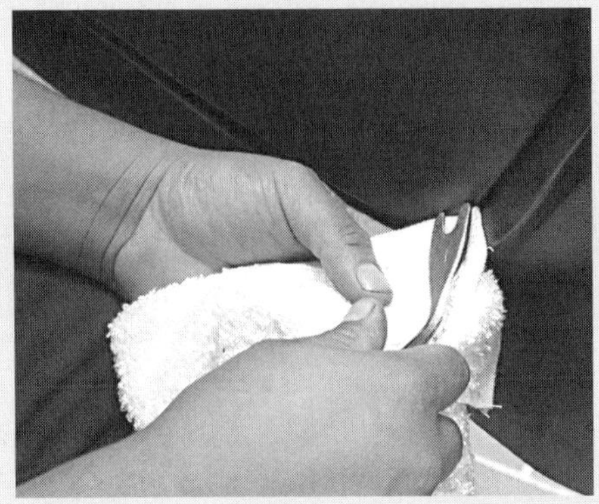

2. Avoid damaging the door panels when removing them. This pickle-fork–shaped prying tool and a soft towel help.

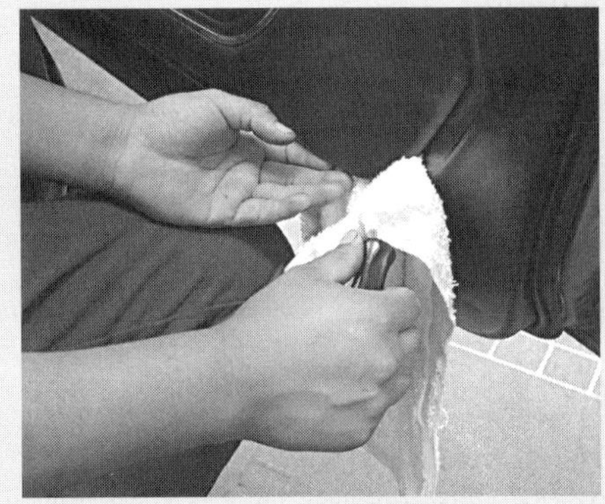

3. The objective is to avoid ripping the cardboard panels while the stronger metal clips stay in the door. The tool needs to be behind the clip, so the clip itself is pried out of the door rather than just the panel. The panel is made of delicate cardboard, so avoid putting any pressure on it.

4. With all the clips carefully pried out of the door, the panel can be removed.

5. Before unbolting the regulator, Mustang Country showed us this trick—a rubber wedge that holds the window in the up position while the regulator is removed. It keeps the window from falling down in the door when the regulator comes out.

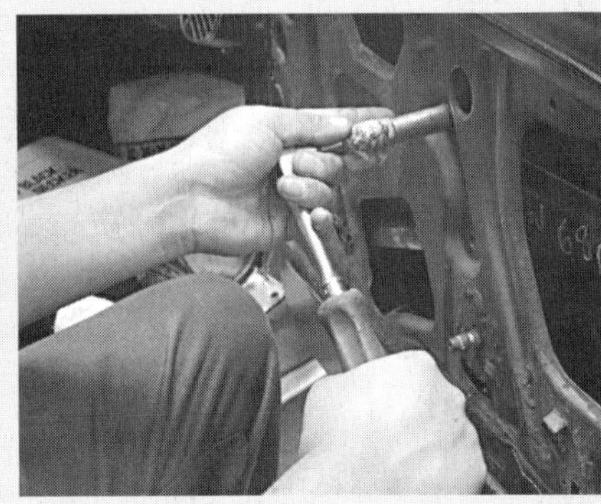

6. Removal of the old regulator is relatively easy. Four bolts toward the front of the door are removed…

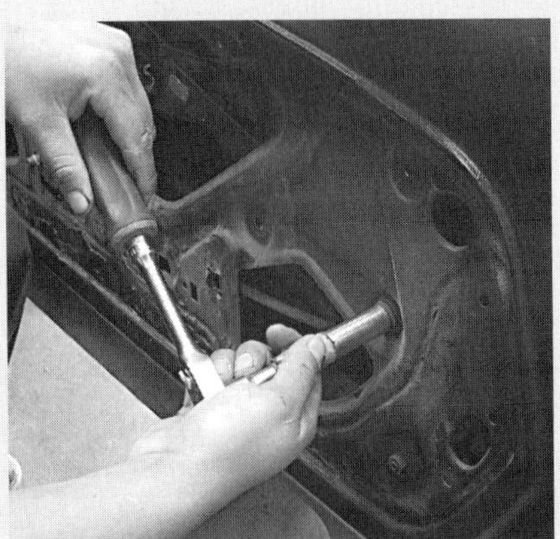

7. …as well as the two nuts near the back of the door. Once these are loose, the regulator is almost ready for removal.

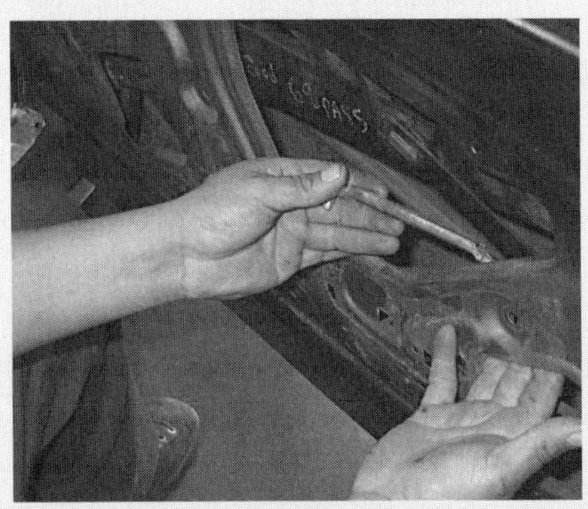

8. With the two rear nuts removed, the rear regulator track (or equalizer-arm bracket, as it's called in the diagram shown in step 18) can be removed. Save this piece, as it will be reused with the new regulator.

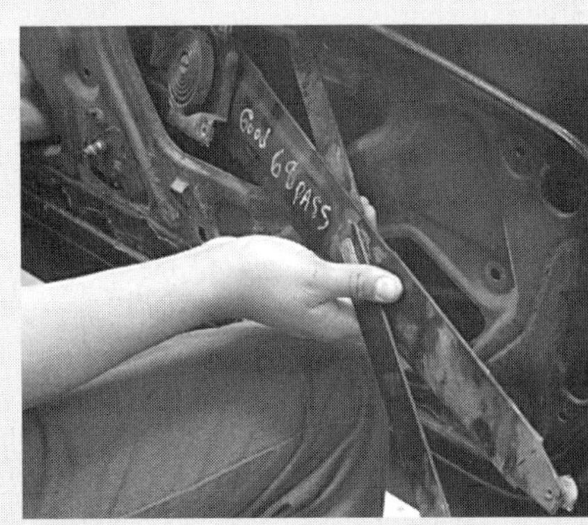

9. After the wheels on the regulator are separated from their tracks, it comes out through the large hole in the center of the door. As you can see from the yellow "Good 68 Pass" marking, this old regulator likely came from a wrecking yard Mustang.

10. When regulators wear out, the up-and-down motion no longer works smoothly. This regulator worked OK, but the shaft splines to which the window handle attached were stripped, thus making it difficult to operate the window. The window hardly operated because the splines on the regulator shaft didn't connect with the window handle.

11. The splines in the window and door-latch handles were also damaged, so they were replaced along with the regulator. Mustang Country stocks window and door-latch handles for all vintage Mustangs.

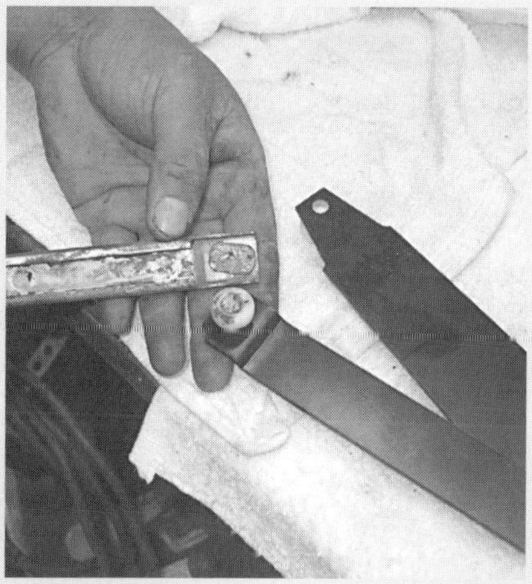

12. According to Mustang Country, it's a good idea to make sure the wheels on the new regulators fit in the tracks because the new ones from various manufacturers might have different-size wheels.

13. Preparing the new regulator for installation includes coating the moving parts with plenty of lubricant, such as the wheel-bearing grease shown here.

14. All regulator pivot points should receive plenty of grease as well.

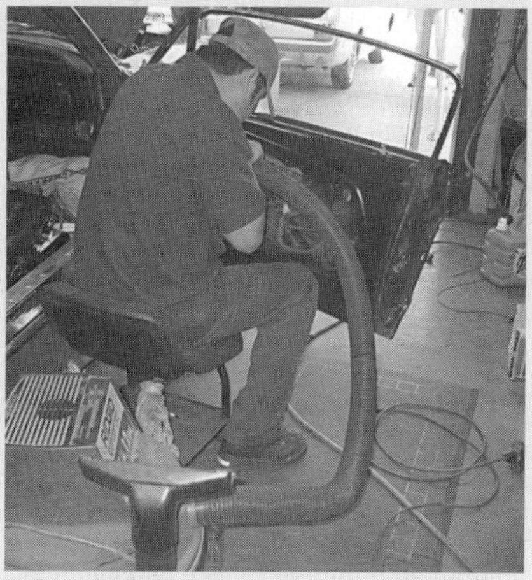

15. This is a good time to vacuum out debris from inside the door before installing the new regulator.

16. We also applied wheel-bearing grease to the main regulator track in the center of the door.

17. With everything checked and lubricated, the new regulator is ready to be installed in the door, using the same hardware that held the old regulator in place. Be sure the three wheels are in place on their tracks before bolting the regulator to the door.

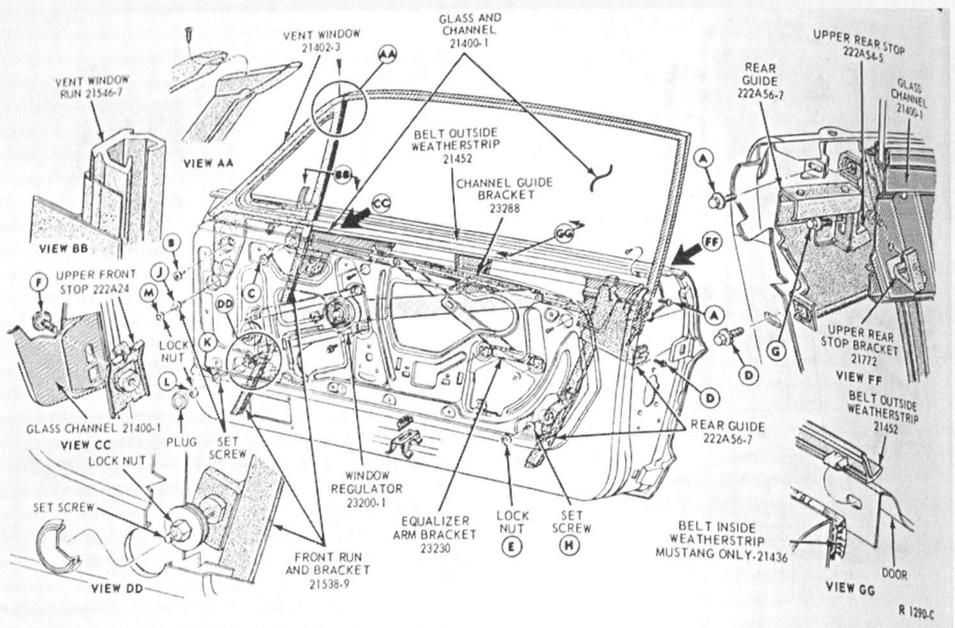

18. This page from a '67 Mustang shop manual shows how the regulator is placed in the door. For any project, a shop manual for your car is useful. Mustang Country stocks them for all '65–'73 Mustangs.

19. With the regulator partially in place, reinstall the rear regulator track using the original hardware. Be sure to place the wheel in the track before securing it to the door.

20. With everything correctly installed, the regulator can be tightened into place. Temporarily install the window handle to test operation of the window before reinstalling the door panel.

21. Reinstallation of the door panel, window and door-latch handles, and armrest completes the remove-and-replace procedure. Carefully push or lightly punch the door-panel clips back into the door as shown to reinstall the door panel. This afternoon project can be accomplished with basic handtools on one door in about two hours.

Chapter 33

Installing '70 Door Glass in a '69

Text and Photos by Miles Cook

Above is what we mean by loose glass in the doors of a '69 Mustang. The best way to solve the problem is to do exactly what Ford did in 1970: replace the glue-in glass with '70-type bolt-in glass. We had to improvise along the way, but once we had all the parts, the conversion on both doors took about a day.

We try to cover a wide spectrum of the vintage and late-model Mustang hobby in our how-to articles. In this case, we have a special treat for those of you who own a classic '69 Mustang.

The '69s (and a few early '70 cars) are the only years of Mustang that Ford glued the door glass into the mounting plates attached to the window regulator. While the '69 Mustang is unique in many appealing ways (for one thing, it's the only vintage Mustang with four headlights), it's also unique in this not-so-great way. Our experience with a '69 SportsRoof was irritating. We tried gluing in the glass, only to have it come out again after a few months. Furthermore, having the glass flopping around in both doors quickly became annoying.

The problem was solved by the folks at Muscle City Glass with reproduction door glass for '70 Mustangs. However, the glass is only part of the picture. You also need '70 plates to attach the glass and regulators together, which we procured from Mustangs Etc.

Finally, we called on Scott Drake Mustang Parts for a set of '70 window regulators. Although they look nearly identical to the '69 versions, there is a slight difference between the two; but you may be able to make the '69s work.

There are many other details you need to know to complete this swap, and we've done much of the R&D work for you. The accompanying photos and captions explain the particulars.

Sources

Muscle City Glass
www.musclecityglass.com

Mustangs Etc.
www.mustangsetc.com

Scott Drake Mustang Parts
www.scottdrake.com

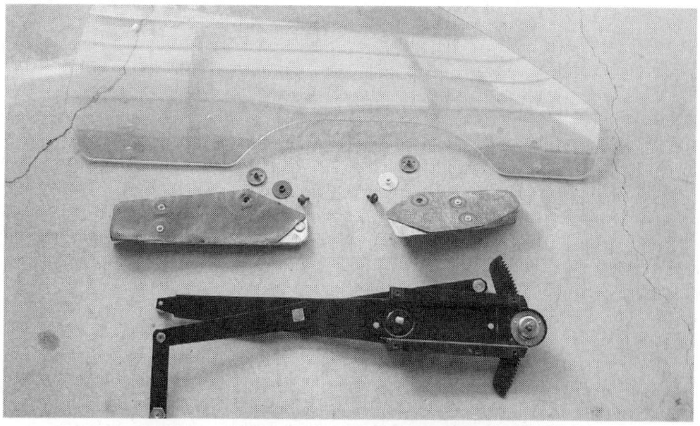

1. This is what you need for the passenger-side conversion. We did our research by performing the driver-side conversion first to determine what was needed. Then we did the photo sequence on the passenger side. The glass is available from Muscle City Glass; the regulators are available from any parts house that stocks Scott Drake Mustang parts. We don't know of anyone currently reproducing the '70 plates between the glass and regulator. Look for them at a shop that carries used parts such as Mustangs Etc.

2. To begin, the door panels need to come off. Remove the window crank and armrest/door pull as shown.

3. If you have nice panels on your car, be careful while removing them because the metal clips are stronger than the cardboard backing on the panel. Pulling on the panel will ruin it, and the clips will simply stay in the door. As shown here, try to get the screwdriver behind the clip and pry the clip out of the door to avoid putting any pressure on the cardboard. We had pretty good luck. Out of 20 clips (10 on each door), we ripped only one mounting hole during removal.

4. With all the clips loose, the door panel comes off the door.

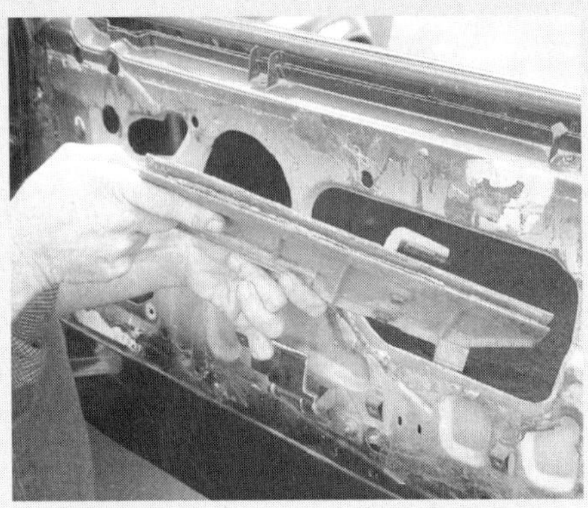

5. Once you have access inside the door, remove the glue-in window plates and regulator. Since the glass was already removed (because the glue wasn't holding), the plates were sitting in the bottom of the door. The rear plate comes out through the largest hole in the door. Here you can see where the glass glues into the groove at the top of the '69 plate.

6. Next, unbolt the arm of the regulator from the front plate.

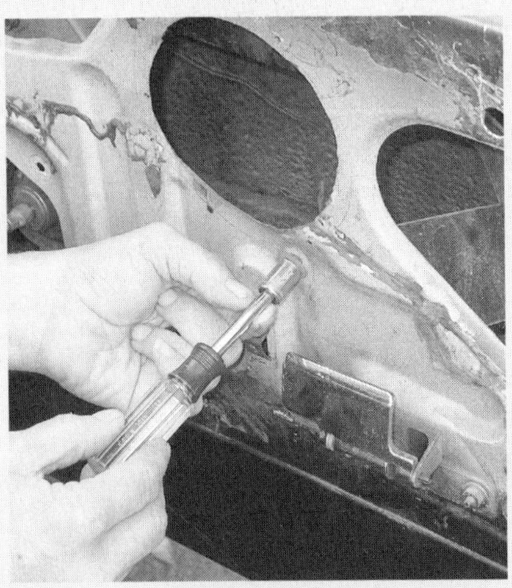

7. There are five bolts that attach the regulator to the door: four toward the front of the door and one near the back that attaches to one of the regulator arms. Remove these five bolts and...

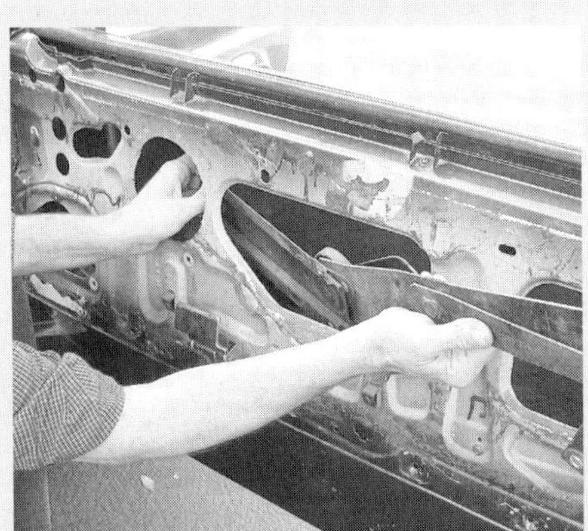

8. ...the regulator can be removed through the largest hole in the door.

9. Shown for comparison are the plates for '69 (top) and '70 Mustangs. As you can see, they are completely different. The '69 plates are glued to the glass, while the '70 plates are secured by three bolts each. Both attach to the window regulator in a similar fashion.

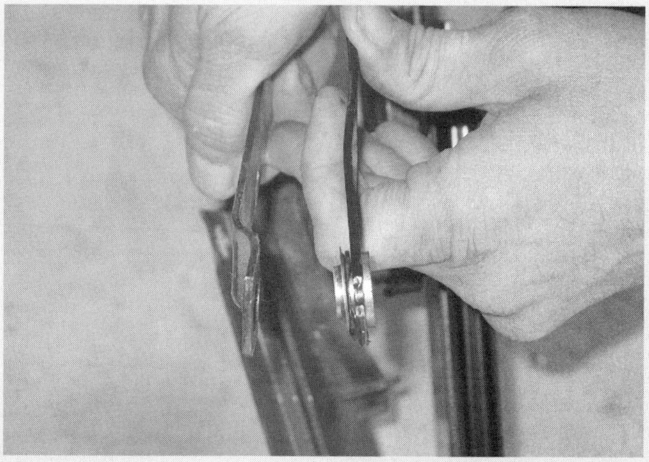

10. Shown here is the main difference between the '69 and '70 regulators. The arm on the '70 regulator on the right is straight, while the '69 version has a small bend in the arm that attaches to the front plate. Although it's a close call, the need for new regulators should be determined on a car-by-car basis because if a car's existing '69 regulators are in good condition they might work with the '70 glass and plates. You can attach the assembled glass and plates and an existing regulator in the door fairly easily to determine if your existing regulators will work. If there isn't enough clearance in the door for the arms to work on the '69 regulators, installing the '70 regulators by themselves is straightforward. Ours were worn out, so we replaced them with the new Scott Drake units.

11. When converting a '69 Mustang to the '70 door-glass mechanicals, the locator brackets attached to the '69 plates can be removed and reused with the '70 plates and glass. If you get locator brackets and rubber insulation with the '70 plates, you won't have to do the following few steps. We did, however, because although Mustangs Etc. had another set of brackets in stock, we knew we would be able to use the locator brackets from the '69 plates, so we removed them to use them with the '70 plates.

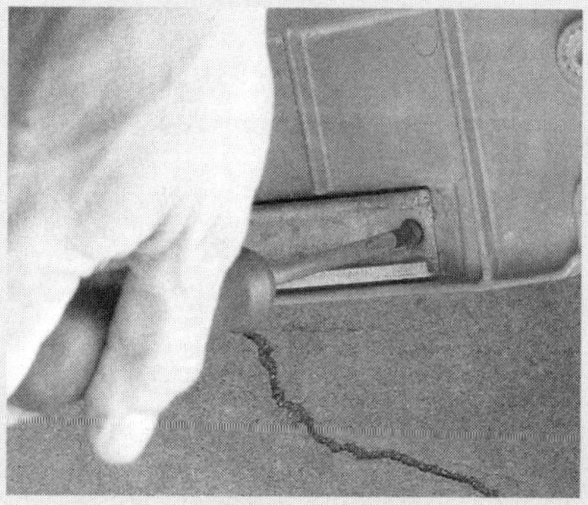

12. We also used the regulator tracks from the '69 plates. Unscrew them as shown.

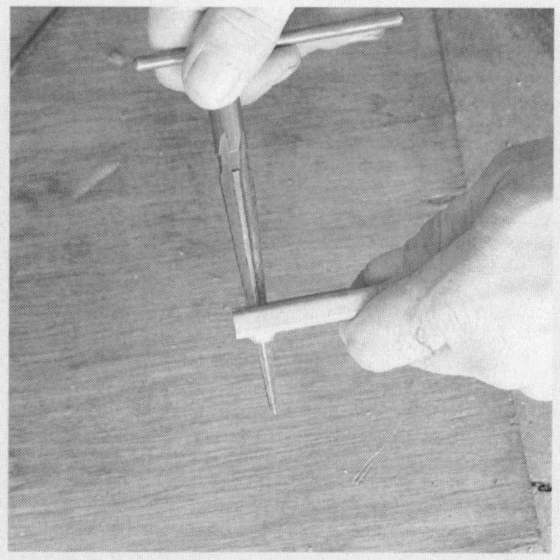

13. The holes in the track need to be enlarged slightly with a hand reamer so they'll fit over the studs on the '70 plates.

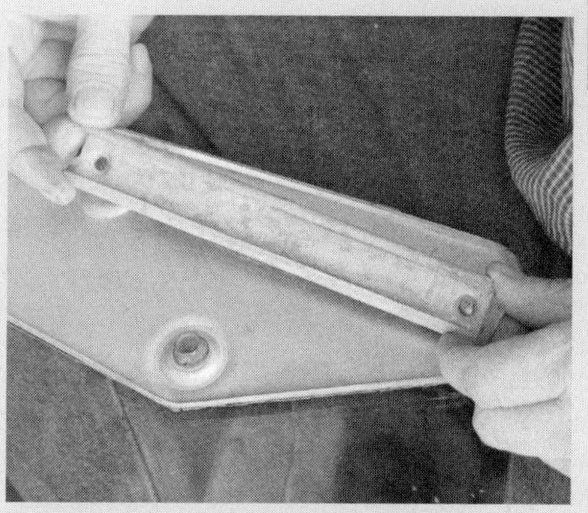

14. Once the holes in the track are large enough, the track drops into place on the '70 plate. Our "magic bolt box" had the correct-size nuts to hold the track in place. A hardware or auto parts store should have what you need.

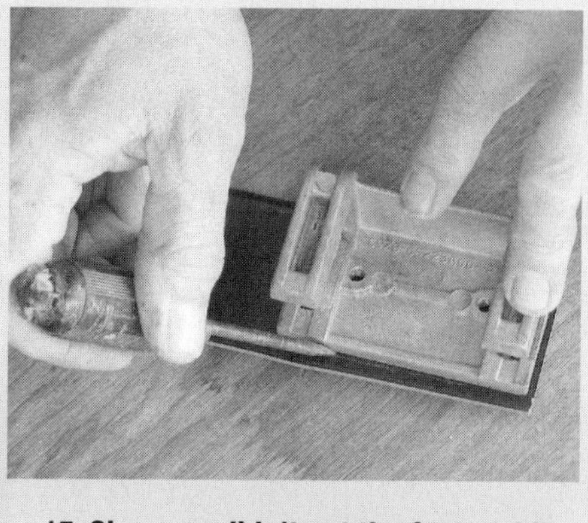

15. Since we didn't get the factory bits from a '70 Mustang, we had to make our own rubber insulation pieces to go between the locator brackets and glass. The metal from any bracket, bolt, nut, or washer should not touch any part of the door glass. It needs to be completely insulated so the glass doesn't break when the car goes over a bump or a rough road. A piece of thin-rubber baseboard molding from a home-supply store works perfectly for this purpose. Use the locator bracket to trace the pattern for the molding, as shown here. A marker or a punch to scratch a line in the rubber works well.

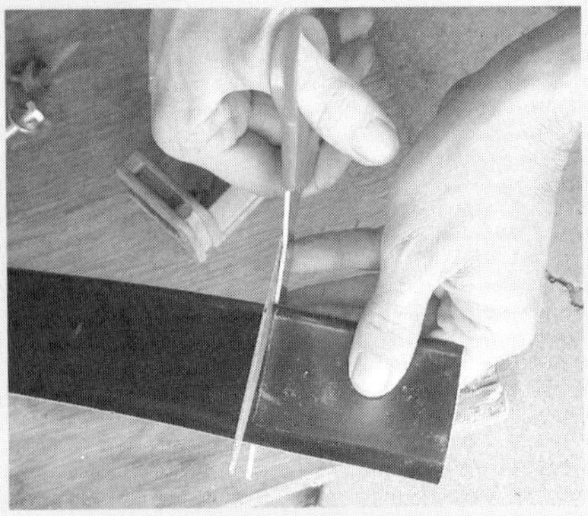

16. With a pair of scissors, cut the piece to the marked length.

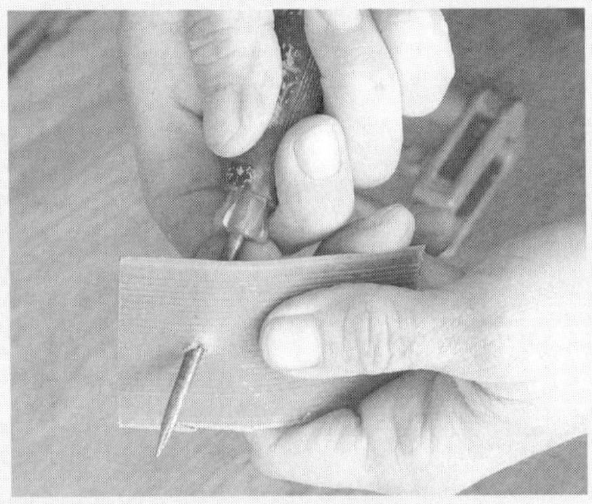

17. Again using the locator bracket as a template to mark the holes, punch the holes in the rubber insulation using a punch tool.

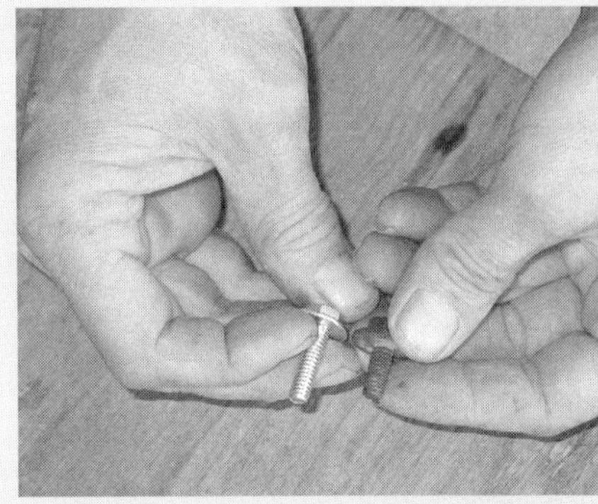

18. On the right are the factory bolts used to hold the locator bracket on the '69 plates. To attach the bracket to the '70 plates and glass, you'll need slightly longer bolts since they pass through the glass in the '70 application, instead of bolting to the window plate. We got these at Home Depot, where we also got the rubber baseboard molding.

19. With all the parts ready, the components go together as shown. The plates go on the inside of the glass, and the new, longer bolts pass through the plates, glass, insulation, and into the locator brackets mounted on the outside of the glass.

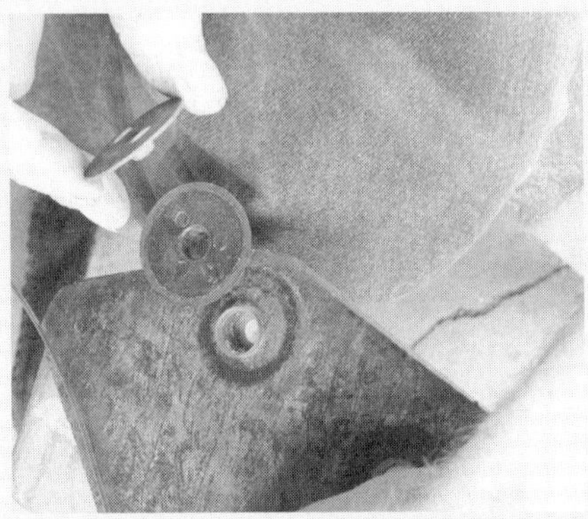

20. Each plate has a separate nut and bolt that attach it to the glass, in addition to the two that attach the locator bracket. The bolt goes into the plate, passes through the glass, through a plastic washer, and into a special nut that fits over the washer. We got these special nuts and bolts from Mustangs Etc. along with the plates. Although you can make do using the locator brackets from the '69 plates, get four of these specialized bolts, nuts, and washers when you locate a set of '70 plates.

21. To provide a clear picture of what the assembled glass, plates, and regulator look like, we assembled the passenger side outside the door. We took it all apart to begin assembly inside the door. The glass, plates, and regulator have to be assembled inside the door because the glass won't fit into the door with the plates attached to it. Once the glass is placed inside the door, the rest of the pieces can be bolted together as follows.

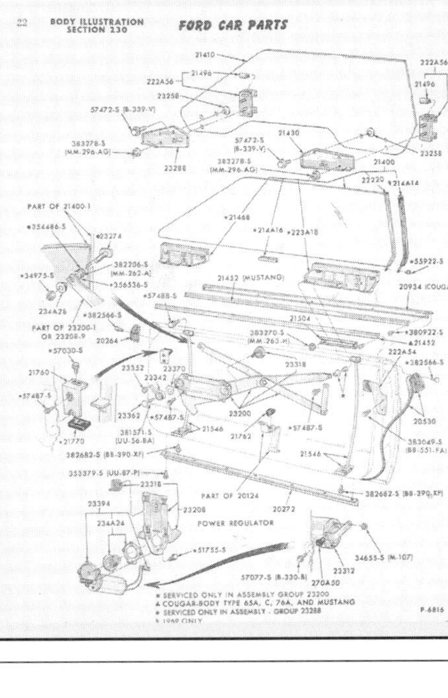

22. We used a diagram from a Ford shop manual that showed an exploded view of the '69 and '70 door assemblies. We noted the different shapes of the lower edges of the glass.

23. The regulator is the first piece to be installed in the door. It's carefully placed in the door and left sitting at the bottom while the plates are attached to the glass.

24. Place the glass into the door through the slot at the top of the door.

25. The plates are the next components to be installed into the door.

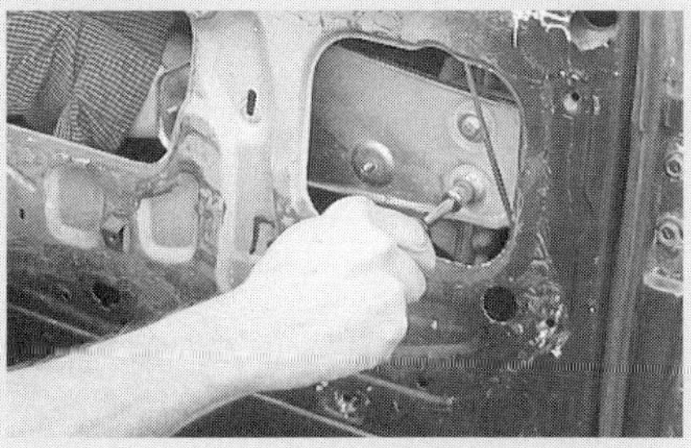

26. Assembly of the glass and plates is easier with two people. Have one person hold the glass in place while another attaches the plates to the glass. Here, the rear plate is tightened to the glass, while the front plate has already been installed. Once the plates are attached to the glass, place them over the tube-shape tracks inside the door and locate the glass.

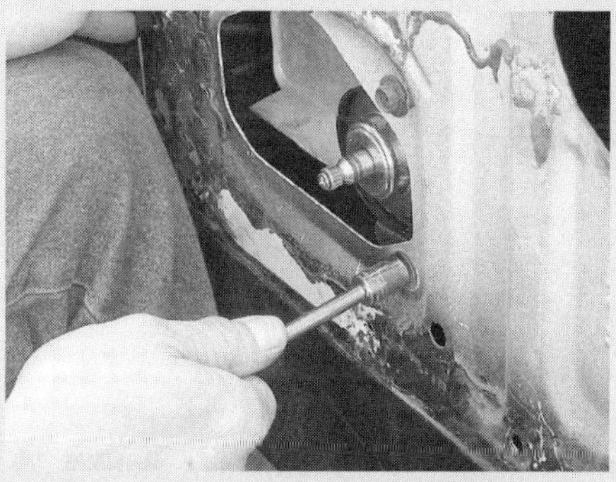

27. Next, reach down in the door to retrieve the regulator and reinstall it into the door using the five original bolts.

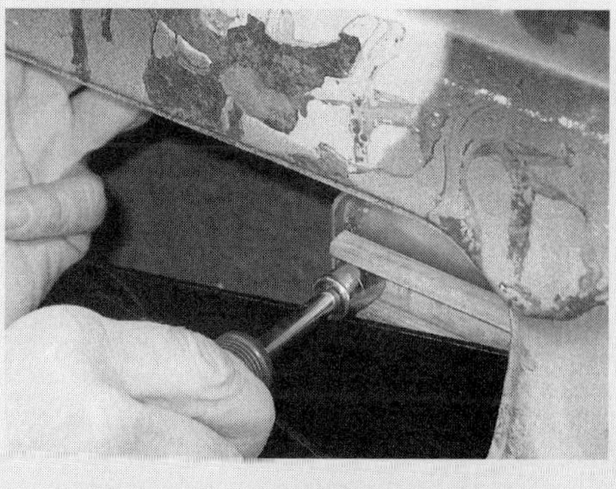

28. Using nuts sized to fit the track studs in the '70 plates, secure the track to the plate. First be sure the wheel on the regulator arm is in place on the track. If you find '70 plates with tracks already in place, you probably won't have to worry about this. But since we used the modified '69 tracks (see step 13), we had to keep it in mind.

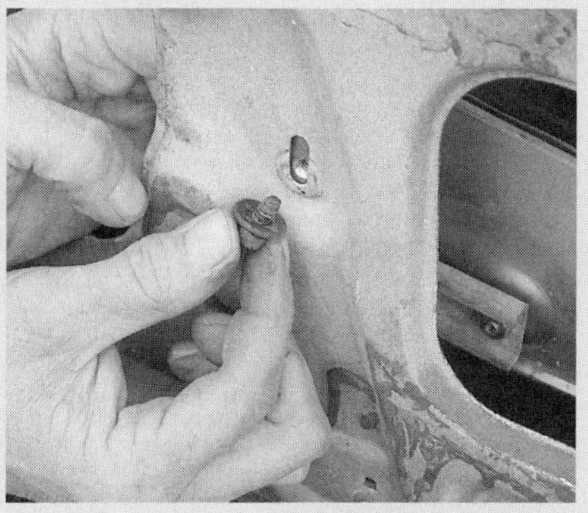

29. Attach the short rear arm of the regulator to the door using the original bolt. Be sure to attach the long front arm of the regulator to the front plate, again using the original bolt as shown in step 6.

30. With everything in place, test the travel of the glass going up and down after temporarily installing the window crank. One of the arms on the new regulator got caught on the regulator spring, so we adjusted the glass out slightly at the bottom of the door by loosening the bolts underneath the door that hold the tracks in place. Once adjusted and the bolts tightened, the window rolled up and down perfectly.

31. The chrome weatherstrip ends on each piece of '69 glass were taken off and reinstalled on the new glass. Basically a press-fit, the ends should fit into place on the '70 glass exactly as they did on the '69 glass. Once you're satisfied with how the windows roll up and down, reinstall the door panel to complete the installation.

Chapter 34

Windshield & Rear Glass Replacement

Text and Photos by Jim Smart

Vintage Mustang windshields are notorious leakers. Here's how to replace yours with a slightly different approach than the OEM manual.

One of the more frustrating aspects of vintage Mustang ownership and restoration is leaking windshields. It's frustrating because no one wants to see their carpet—or feet—soaking wet. Classic Mustangs used rubber gaskets for windshields and rear glass through 1968. Beginning in 1969, Ford started gluing in the windshields, yet stayed with rubber-gasket backlights through 1973. Glued-in windshields have been a blessing because, unless you're really careless, they will not leak. Rubber-gasket installations take a lot of practice to keep them from leaking. Rare is the rubber-gasket windshield that doesn't. If your toes and carpet are wet, you're in good company because a lot of us wrestle with the same problem.

We're going to explain how classic Mustang windshields and backlights are installed, and how to prevent them from leaking. Because this is an inexact science, we offer no guarantees. Sometimes, you can painstakingly install a windshield, thinking you've covered all the bases, and still wind up with water on your floor or in your trunk. We hope this useful information will lead you to a dry interior.

An Unconventional Approach

We invited Antique Auto Glass to visit our shop and install a windshield. These guys use a different approach than the Ford shop manual instructions, which tell us to run sealer around the inside pocket of the gasket as well as the outside pocket, then install the gasket on the windshield. Antique Auto Glass follows the Ford shop manual's instructions to apply sealer to both pockets before installation. Then, they install the rubber gasket on the body first. Once the gasket is installed on the body, they follow with the windshield. Instead of using the rope trick we've all been taught, Antique Auto Glass' Frank Doha lays the glass on the rubber and works the rubber over the glass with a tool.

Sometimes, It Isn't the Windshield

Mustang body leaks aren't always rooted in the windshield and backlight. Cowl vents are the reason for leakage more times than windshields. Check your Mustang's cowl-vent dams for rust-through. Cracks in the lead seams at the roofline and A-pillars can cause a water leak. The same can be said for lead seams at the sail pillars and trunk area. Examine your Mustang's body thoroughly for leak sources. There are many, including rubber grommets around wiring harnesses, accelerator pedals, brake lines, and antenna leads. The rubber seal between the trunk lid and body will leak if exposed to substantial amounts of water. Quarter-panel end caps, taillights, side markers, and even trim pieces can be leak sources. Check them out.

Source
Antique Auto Glass
818/362-1901

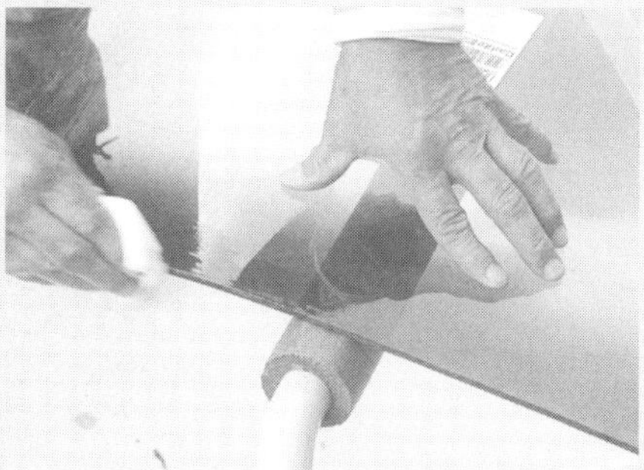

1. Windshield and backlight installations must always begin with clean surfaces. Both the glass and body must be pristine and clean. When cleaning the goop and poop off the body, take extra care not to damage the paint. Scratching the paint down to the primer and bare steel leads to corrosion and rust-out.

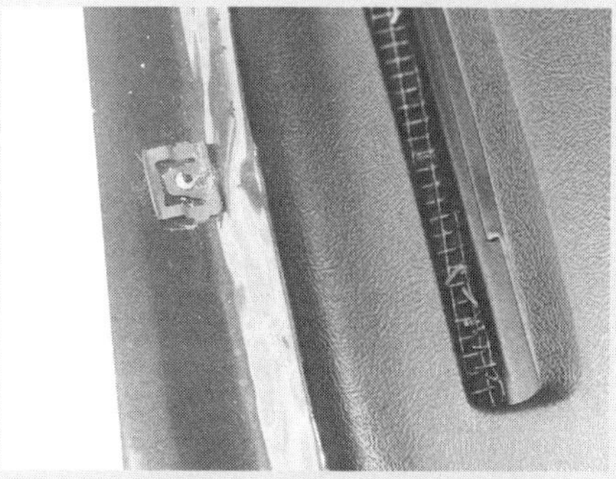

2. Windshield installation should begin with clean surfaces and new molding clips. Always apply windshield sealer between the clip and body to ensure proper sealing. Water tends to sneak around the clips because they provide an air gap. By using a sealer between the clip and body, the air gap is eliminated.

3. 1Installation begins with the rubber gasket. Sealer is applied to the inner and outer pockets as well as the body to ensure proper sealing all around. The gasket is installed next.

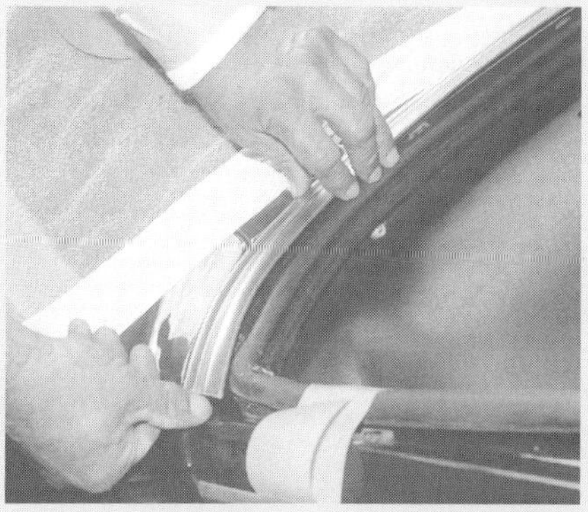

4. The windshield's base molding is installed to provide a solid foundation for glass installation.

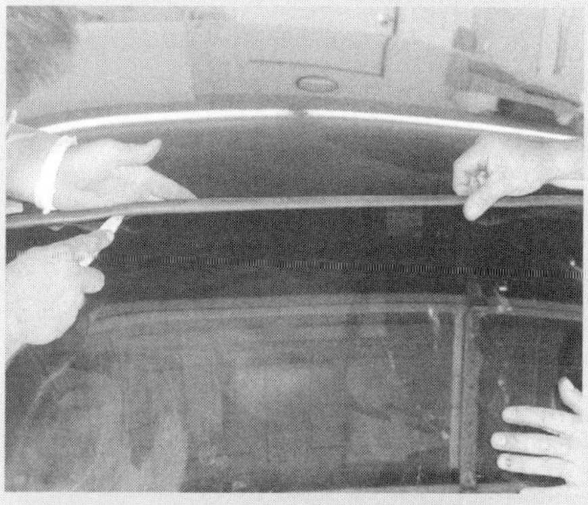

5. Antique Auto Glass begins glass installation at the bottom by setting the glass in the gasket pocket. The glass is laid back against the rubber gasket.

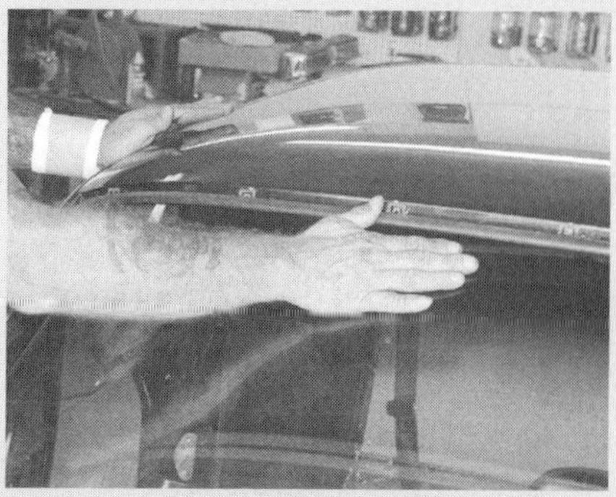

6. Using a soft plastic homemade tool, Frank Dohs pops the rubber gasket lip over the windshield, beginning at the bottom. Instead of using a rope to pop the rubber lip inside the body, Frank pops the outer lip instead, using the plastic tool to get there. The inner lip is already inside.

7. Frank firmly presses the windshield in place, which seats the glass and squeezes sealer around the edges. He suggests firm pressure, but not too much.

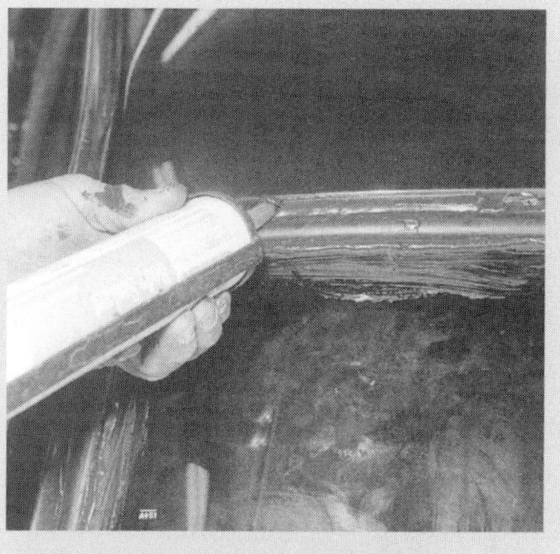

8. With the glass firmly pressed in place, Frank runs a bead of sealer around the perimeter between the glass and rubber gasket. Lacquer thinner and a razor blade are used to clean off the excess sealer.

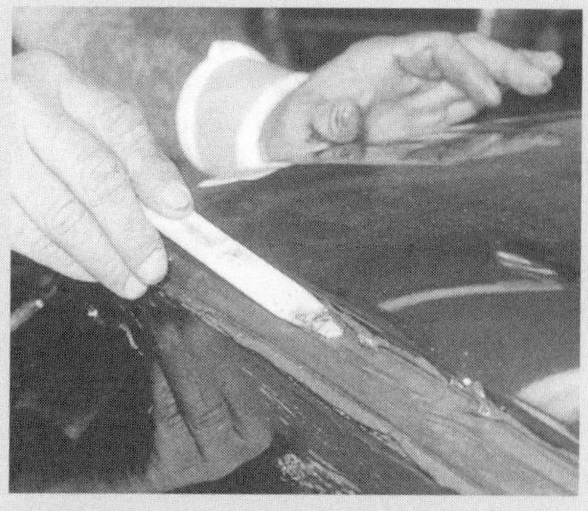

9. Sealer is applied between the rubber gasket and body as shown. This ensures water won't leak between the rubber and body. This is the vintage Mustang's most common leak area.

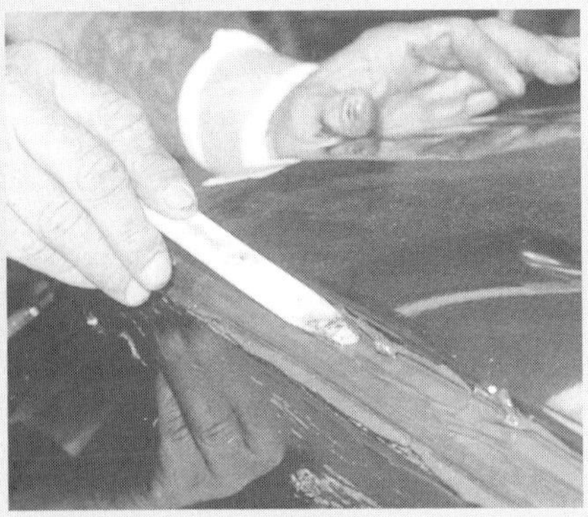

10. The sealer between the rubber gasket and body is spread to fill out the space and keep water out. Sometimes, the molding will breach this seal, causing leakage. If this happens, you have to remove the molding and seal again.

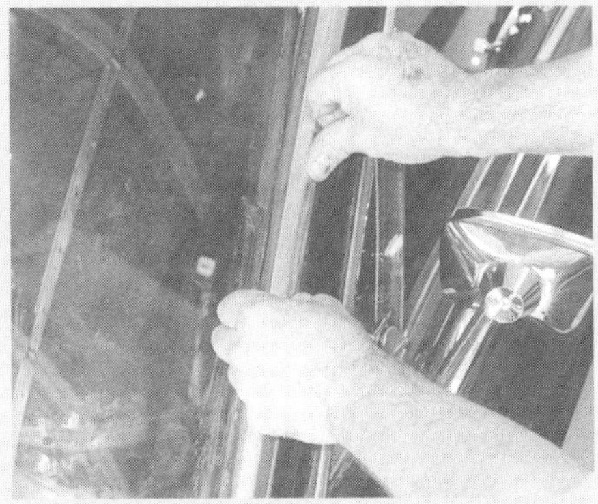

11. We often see mistakes made in windshield-molding installation. Windshield shops sometimes use mallets to install the stainless molding, a procedure that can chip paint and dent moldings. Frank suggests firm pressure on the molding using the heels of your hands. The objective is to pop the molding under the clip. You don't need a mallet, just firm pressure with your hand.

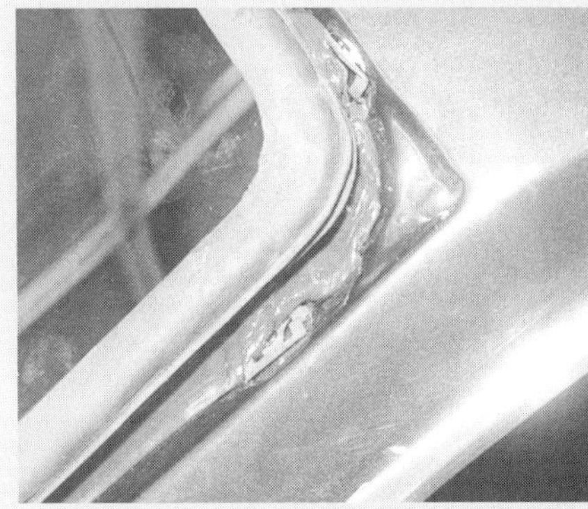

12. Mustang windshields leak mostly in the upper corners because installers frequently fail to fill the huge voids with sealer. Water tends to collect there, finding its way down the A-pillar to the dashboard and kick-panel area. This area must be filled with sealer and properly spread in all directions before the molding is installed. Water needs to be channeled away from this area, down the outside of the A-pillar to the bottom where it safely passes away from the interior. When it does, your feet stay dry.

REAR WINDOW

Classic Mustang rear windows install the same way as windshields. Probably the most unforgiving rear window, also known as the backlight, is in the fastback. It leaks because it lies so flat, which makes it easy for moisture to collect around the perimeter and leak inside. That's why lots of sealer is needed around the fastback's rear window, especially around the outside between the rubber and body. Molding installation follows the same protocol: with a firm, but careful hand.

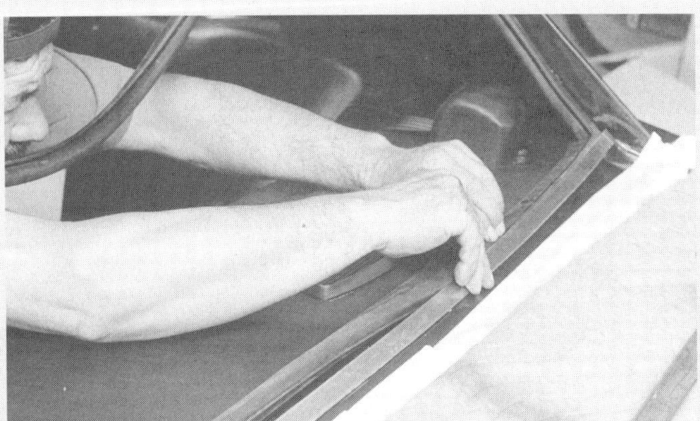

1. Antique Auto Glass installs backlights the same way as windshields, with the rubber gasket going in first. Plenty of sealer is used between the gasket and body, and between the glass and gasket. First, the base molding is installed, then the glass is seated in the base of the gasket.

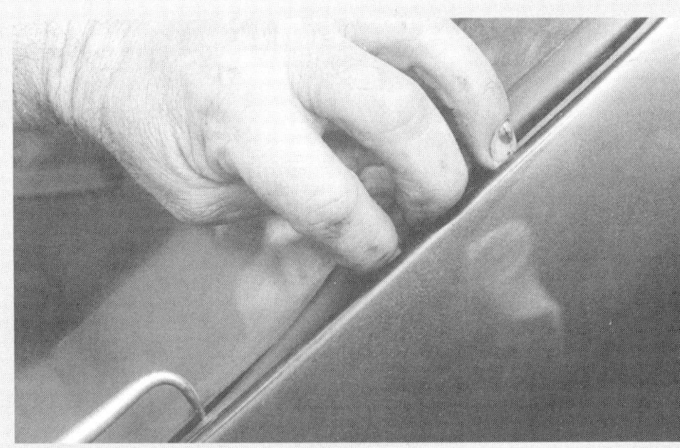

2. The rubber gasket's lip is popped over the outside of the glass as shown. This is a cleaner approach to glass installation. When rope is used, as Ford suggests, windshield sealer gets all over the place.

3. The rubber-gasket lip is popped around the glass perimeter as shown.

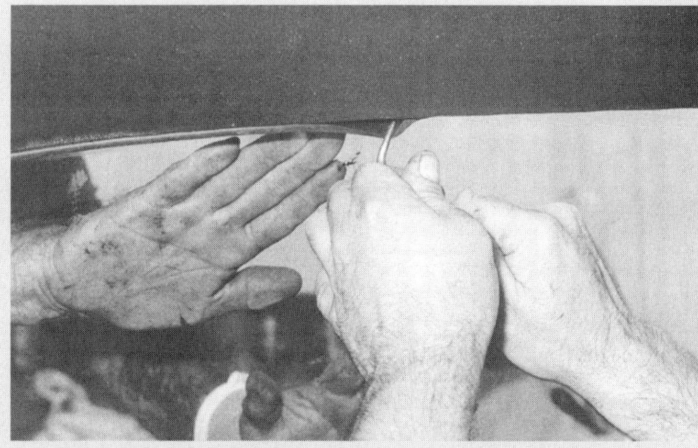

4. Inside, the rubber gasket needs help. We pop the lip around the edges while someone presses from the outside. This way, everything seats into place.

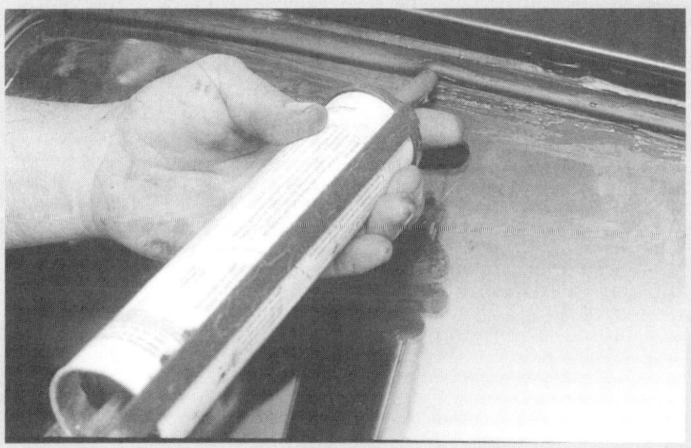

5. Sealer is added between the glass and rubber gasket once the glass is seated.

6. Sealer is then applied between the rubber gasket and body to keep water away from the interior.

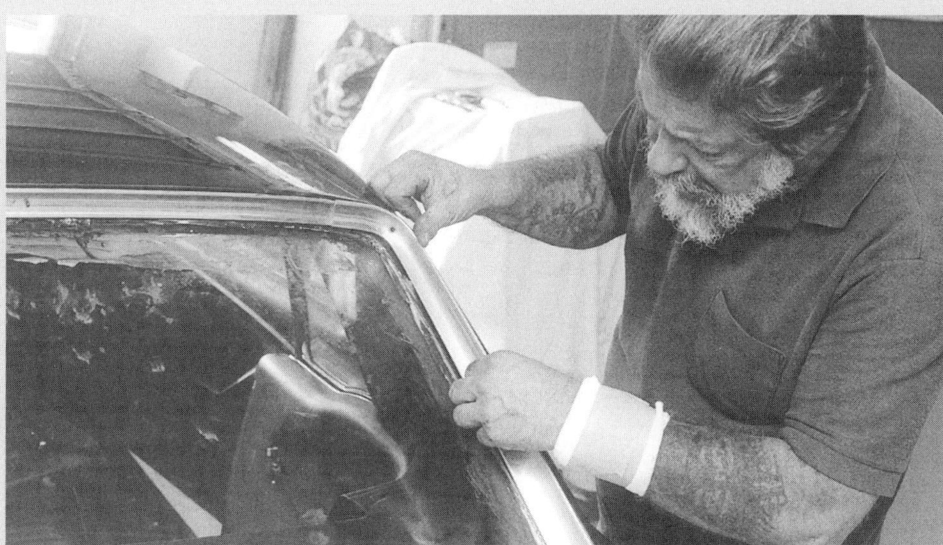

7. Rear-window-molding installation mandates the same gentle method we use with windshields. Popping the molding hard with a mallet promises dents and paint damage.

Chapter 35

How to Install Power Windows

Text and Photos by Jim Smart

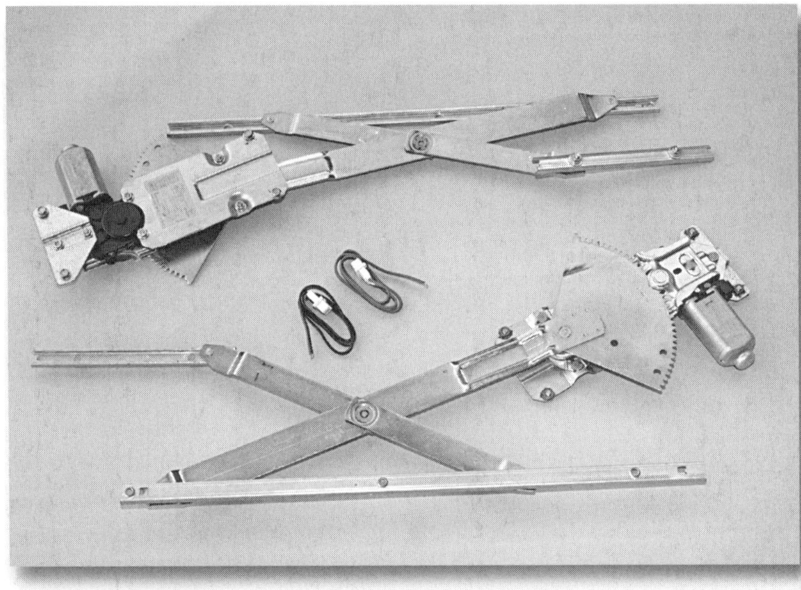

Electric-Life power windows are high-torque, short-duration motor drives that move old windows quickly and effectively. These units bolt in, replacing the factory window regulators with a minimum of fuss. Each motor drive is equipped with moisture-tight connectors and leads.

Imagine motoring up to your favorite cruising spot and effortlessly rolling down your windows. Power windows are an easy classic-Mustang upgrade you can install in a day, taking the grunt work out of cranking those old manual windows. Electric-Life power windows from National Parts Depot are designed to bolt in to replace your Mustang's manual roll-up windows.

To install power windows, you need to prepare. Consider replacing door weatherstripping, window channels, and other soft parts while you're at it. Hit your local AutoZone or Pep Boys for tools of the electrical trade: butt connectors, bullet plugs, wire crimpers, a soldering gun, solder, heat-shrink tubing, flexible rubber or plastic conduit, attachment hardware that isn't included in the kit, sheetmetal cutting tools, a die grinder, 3M's strip caulk or other flexible sealer, water shields, and more.

So let's get started. During this exercise, we learned the hard way what we shouldn't take apart. For example, we undid the aft window track only to discover it wasn't necessary. One downside to the Electric-Life window kit is the instructions aren't detailed enough for those of us with two left hands. Outside of that, the kit is easy to install and operate.

Wiring can be confusing and intimidating. Power windows get power from one source that's energized when the ignition switch is on or in accessory mode. Power can come from the ignition switch or an accessory female pigtail lead that's energized when the ignition is on. We discourage using a power source that is live all the time.

Sources

Electric-Life
www.electric-life.com

Marlo's Frame & Alignment
www.fly-ford.com

National Parts Depot
www.npdlink.com

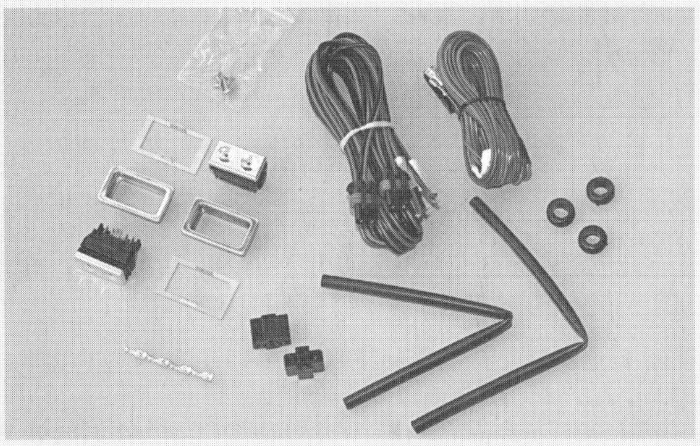

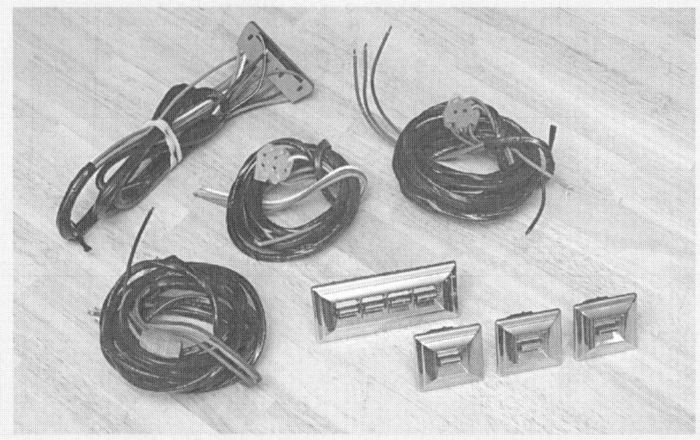

1. There are a variety of power window switches available for classic Mustangs. Electric-Life sent us two types billet (left) and GM chrome (right). The billet switches are easier to install than the GM switches. However, we chose the GM switch for its period look—ideal for those of you interested in a more conservative appearance. The downside to GM switches is fit and reliability. They aren't reliable and must be cut to mount on the doors. Billet switches are less invasive and more reliable.

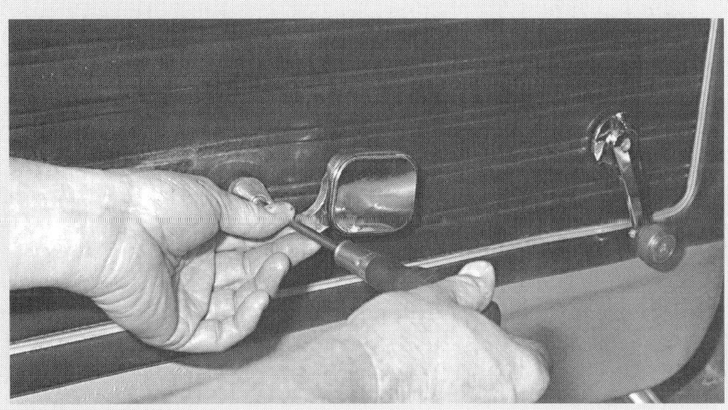

2. To install power windows, first remove the manual regulators; then remove the door panels and water shields. Take off the armrests using a 3⁄8-inch-deep well socket; then remove the door handles using a Phillips screwdriver.

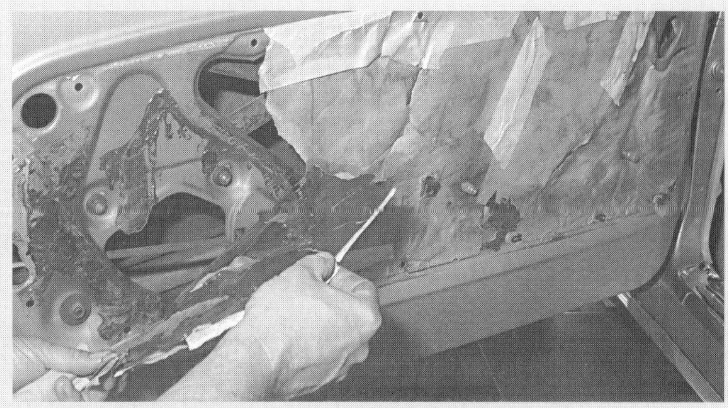

3. Door panels are retained with spring clips. Wrap the screwdriver shaft with cloth, and pop the clips out of the holes in the door sheetmetal. Don't be careless or hurry or you'll damage the door panel and door.

4. This is the manual window regulator. Ironically, it's the original regulator in this San Jose–built '68 Mustang hardtop. A simple scissor mechanism driven by a hand-cranked gear system, the window crank turns a small gear that drives a large 1⁄4 gear.

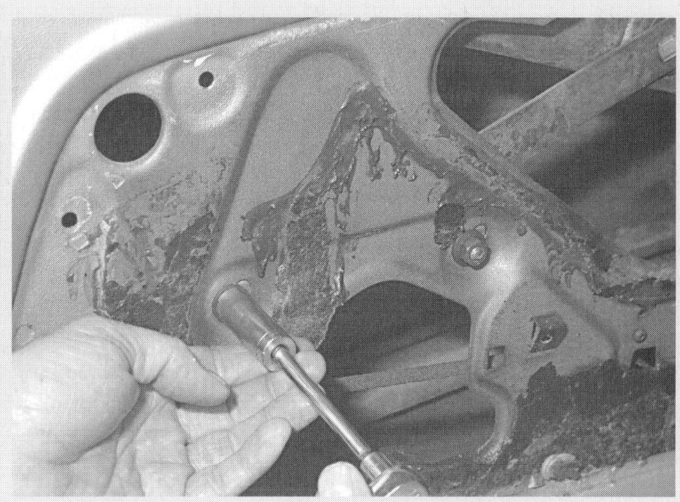

5. Support the window glass in the UP position but not completely topped out. Remove the two idler track bolts and four window regulator bolts.

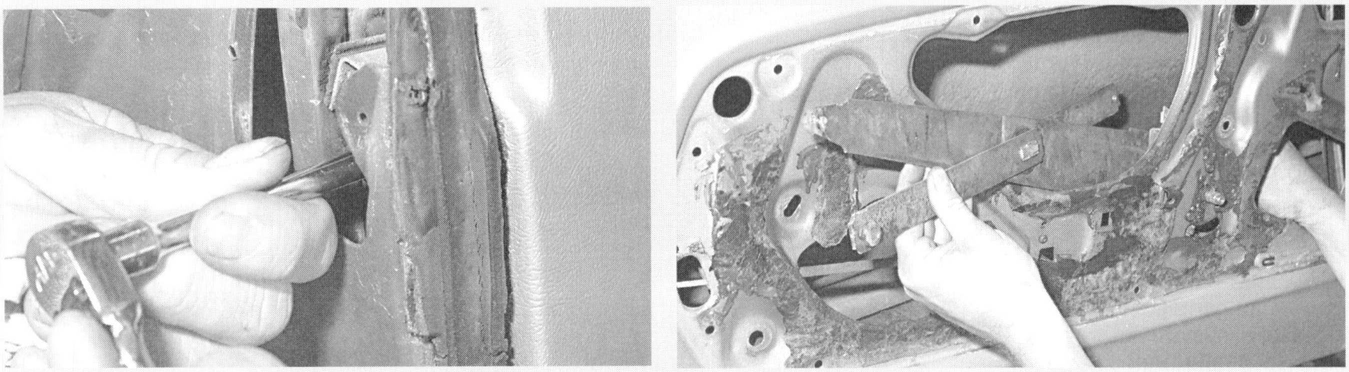

6. You can take one of two approaches to this installation. The aft track can be disconnected from the window, which frees the regulator. This takes a little time, but it's not hard to accomplish.

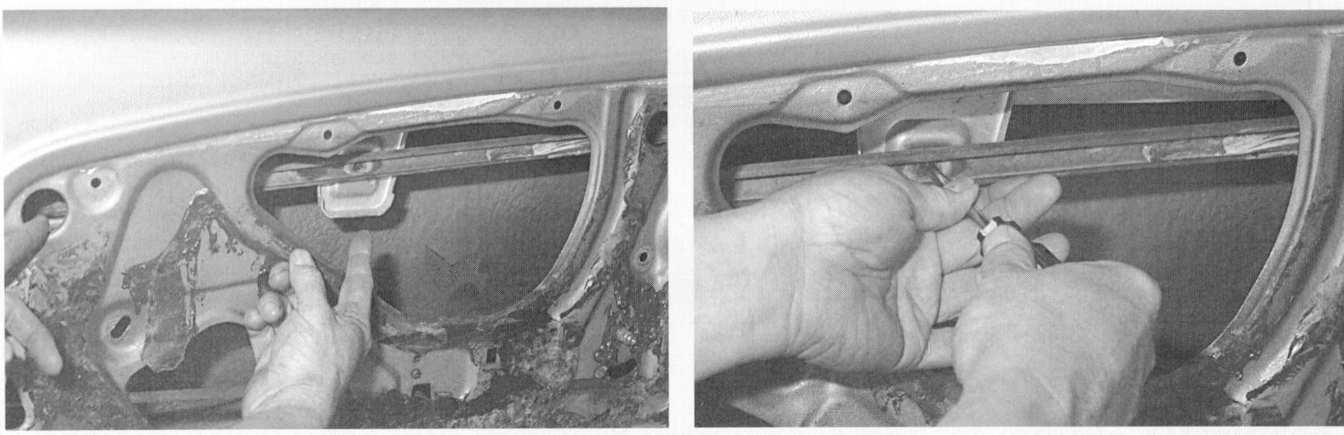

7. Another approach, which you will have to do either way, involves removing three window track screws. Lower the window glass so that all three screws are accessible. However, these fine-thread, countersunk Phillips screws were originally installed with an impact tool and are nearly impossible to remove, stripping out easily. Do it correctly or you'll damage the screw head. If you strip out the screw head, you are, well, "screwed" and will have to drill it out.

8. Our screw head was damaged from a previous removal attempt. We drilled the head out with a 5⁄16-inch bit and removed the track.

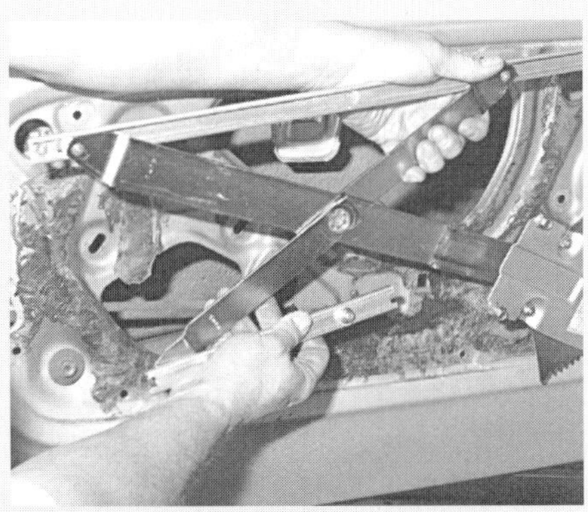

9. Electric-Life's power assembly fits in place of the factory regulator. The top track attaches to the window with three fine-thread, countersunk machine screws. These Grade 8, super-hard machine screws can be found at Lowe's or Home Depot.

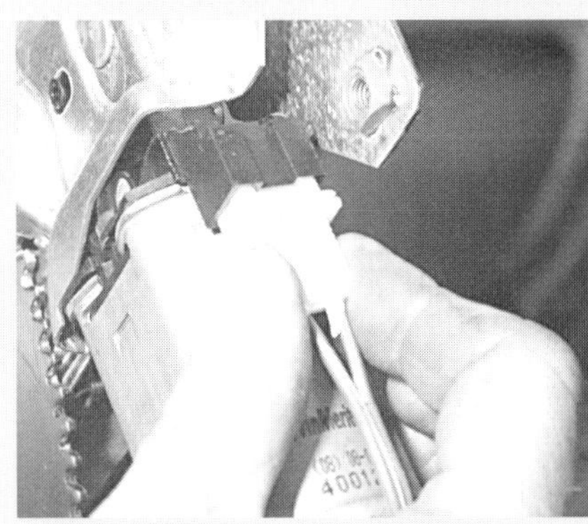

10. Before installation, insert the weather-tight plug. This plug is difficult to install once inside the door. Power comes from the Mustang's accessory circuit through any one of the female accessory plugs near the fuse box under the dash. Power windows must be protected with a 15-amp circuit breaker as fuses tend to blow due to the short-term amp draw of power windows.

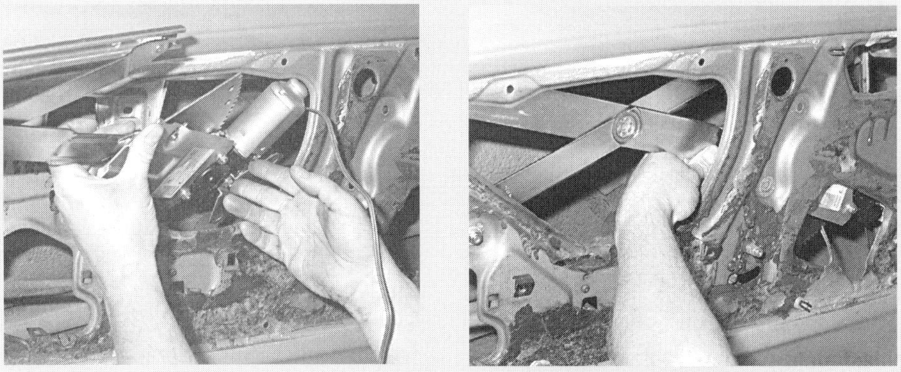

11. The power unit installs as shown. Make sure the window is in the up position for ease of installation. Four 10mm bolts retain the power unit.

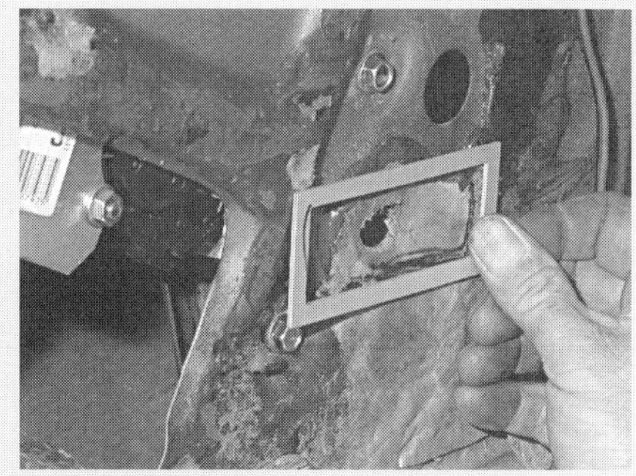

12. To install GM switches, cut both the door and door panel. Because we're going with a gang-switch setup on the driver's door, we cut an opening sized to the switch retainer plate. Based on our experiences, we don't recommend the GM switch unless you intend to build a custom console, and you'll still have to cut sheetmetal somewhere.

13. Getting power to the motor calls for drilling in two locations: at the door-hinge boss and the A-pillar. Find the Ford dimple at the door-hinge boss and A-pillar, and drill a 5/8-inch hole for the wire harness passage. The size depends on the size conduit you use. With door courtesy-light models, run conduit where Ford did.

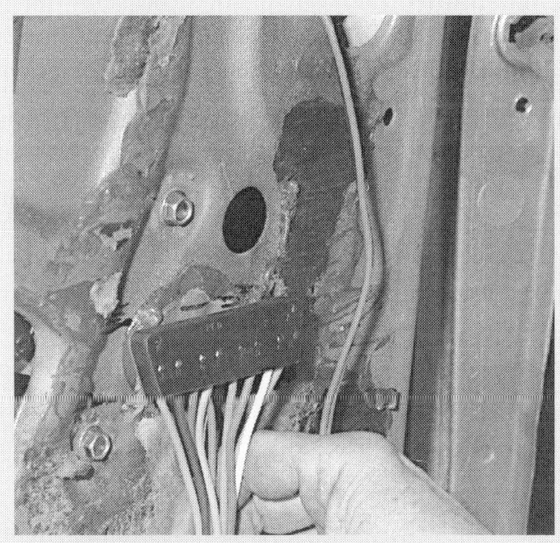

14. Gang switches tie into a main harness that feeds all four window locations.

15. After cutting the opening with a nibbler, clean the surfaces with a die grinder.

16. Our GM gang switch is checked for fit. The retainer clips must fit smoothly, but don't make it so that the switch is impossible to replace.

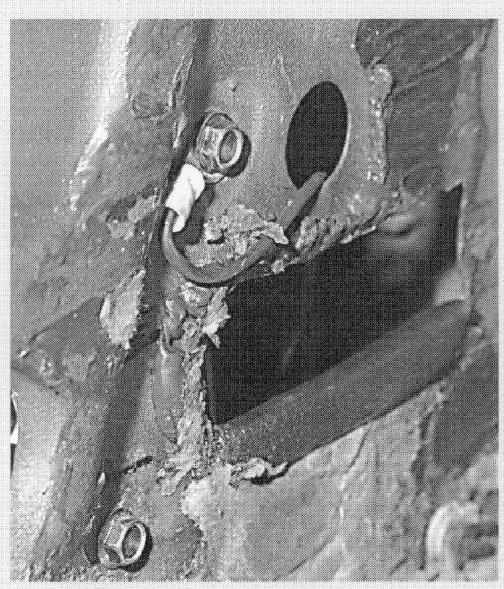

17. Wind lace is installed along the bottom to protect wires from chaffing.

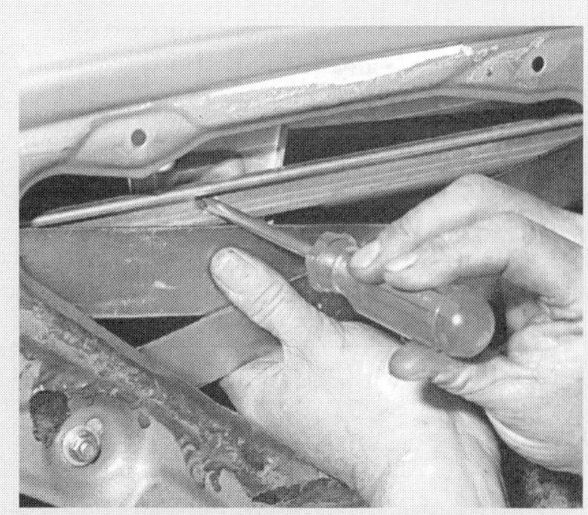

18. The Electric-Life power-window track is tied to the window using fine-thread machine screws. They must be countersunk to clear the rollers. Use a thread locker on these screws. Another important issue is wire clearances. Secure the wiring inside flexible conduit, and tie it ahead of the wing-window assembly and away from the window channel. This keeps the wiring out of harm's way.

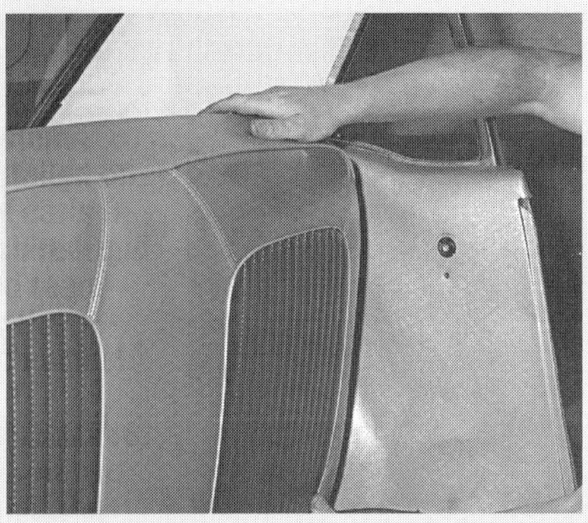

19. Remove the rear-seat and quarter-trim panels for access to the rear quarter-windows.

20. Installing power rear quarter-windows is a cakewalk. Remove three bolts to remove the regulator.

21. Electric Life's rear quarter-window power assembly installs the same as the front. Get the roller into the window track and secure the regulator. As with the front windows, secure the wiring harness, moving it out of the window's path.

22. Take out two more bolts to remove the idler track down below. Slide the regulator out of the tracks; it comes out through the bottom opening.

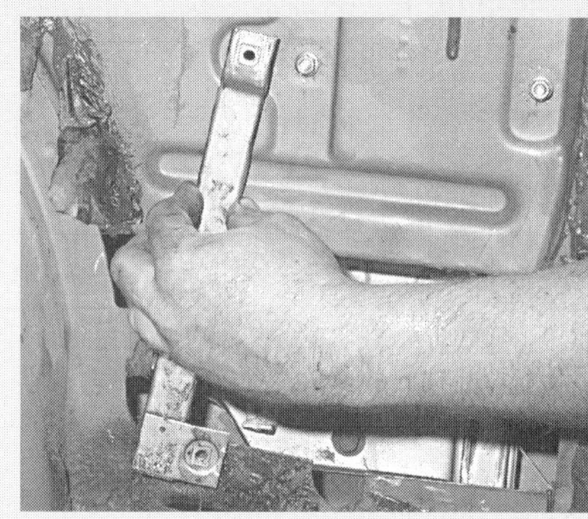

23. The idler track installs like this inside the window cavity, dogleg down and fore as shown. Lubricate all tracks with white grease; wrap the wiring inside flexible conduit.

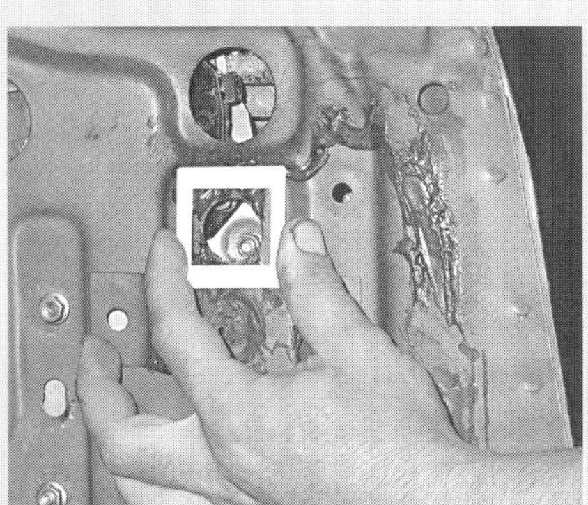

24. If you're going to operate the rear quarter-windows from the rear seat, cut both car and quarter-trim panels. If you'd rather not do that, maintain control from the driver-seat gang switches. Another option is a custom console with rear-seat switches.

25. Rear-seat quarter-window harnesses look like this with GM switches. Electric Life billet switches are simpler with less complicated wiring. In any case, make sure all wiring clears the windows.

Orlando Mustang's Pete and Peter Geisler offer pointers for obtaining the perfect fit.

Chapter 36

How to Fit & Adjust Hoods

Text and Photos by Donald Farr

At first glance, vintage Mustang hoods appear fairly simple. Attached at the two hinges with a mechanical latch up front, hoods shouldn't be a problem. But all it takes is one look around a show field or a Mustang cruise night and you can't help but notice hoods with inconsistent fender gaps, uneven height at the cowl, and a ragged fit with the headlight buckets. Worst of all, the hood and all its faults are right up front where there's no place to hide.

To try to get a handle on vintage-Mustang hood fitment and adjustment, we queried Pete and Peter Geisler at Orlando Mustang. Pete formed a collision services business in 1975, then created Orlando Mustang in 1980 to offer Mustang restoration, appraisals, and parts remanufacturing. Pete Jr. worked alongside his dad for years before joining Orlando Mustang fulltime in 1998, specializing in concours body restorations and paint. Both have had extensive experience in fitting vintage Mustang body panels.

It's easy to assume that hood fitment is a matter of adjusting the hinges, latch, and stops. Both Pete and Peter were quick to point out that vintage Mustangs are more than 40 years old and, in most cases, a lot has occurred over four-plus decades. Accident damage and sheetmetal replacement can tweak a Mustang's body just enough to make perfect hood fitment difficult if not impossible. In fact, Peter points out that proper hood fitment begins all the way back at the quarter-panels.

"You have to identify if you have original quarters and good doors," Peter says. "Match the doors to the quarter-panels, then work your way forward. I always like to fit the doors first, then the hood to the cowl to get the cowl gap,

Source
Orlando Mustang
www.orlandomustang.com

Proper hood fitment begins at the rear quarter-panels and doors. Incorrect quarter-panel repair or replacement affects the way the fenders and hood fit.

Cowl panels on '65–'68 Mustangs are welded in, so there is no adjustment. If a cowl has been repaired or replaced and not reinstalled properly, hood fit will suffer.

Cowls on '69–'70 Mustangs are bolt-in, so some adjustment is available when fitting the hood. Shims can even raise these cowls.

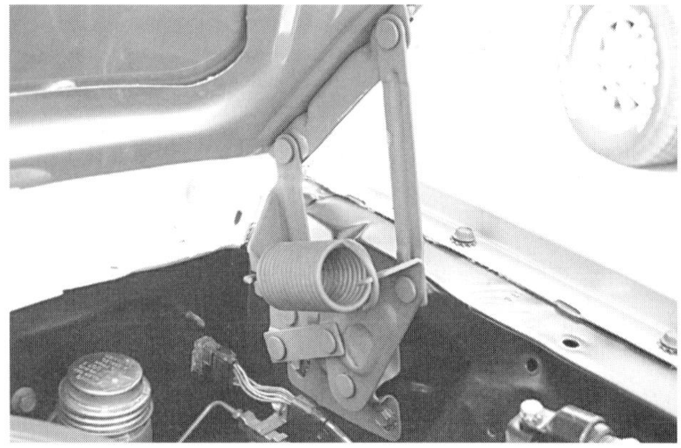

Without properly working hood hinges, it's difficult to get a good hood fit. Most common are worn rivets and elongated rivet holes. Bent hinges, as well as hinges that have been repaired, should be replaced.

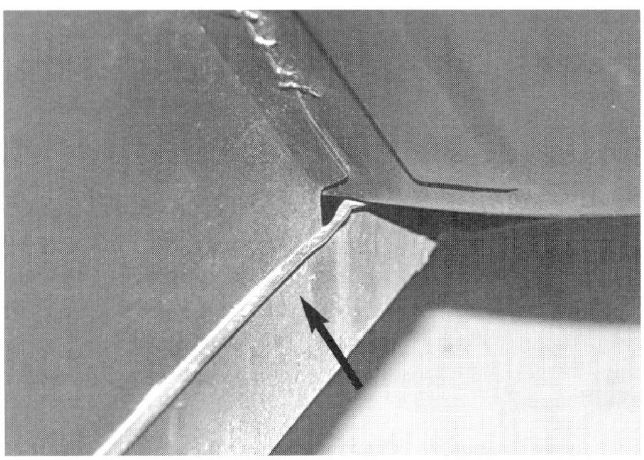

In most cases, old accident damage can be hidden. But even this small front unibody kink (arrow) can throw off body-panel alignment, including the hood.

This hinge is swinging in toward the engine compartment as the hood is lifted. It should stay parallel to the inner fender.

then the fenders last. Headlight buckets are important too. If the buckets don't fit the fenders properly, like you see on a lot of Boss 302s, then the hood fit won't be correct."

Other factors can affect hood fitment as well. For example, vintage Mustangs have always had a problem with rusty cowl panels. If the cowl has been off the car for repair or replacement and was not reinstalled properly, the incorrect placement will affect the hood placement.

More common is accident damage. On the outside, a Mustang may look straight with nice sheetmetal and paint. But if the frame was damaged or bent and not repaired, the car won't be perfectly square all around, affecting panel fit. Pete explains, "Mustangs, being unibodies, can be tapped in the rear—and it can be a light hit and won't look that bad on the car; maybe some bumper damage and a little damage in the rear body panel. But when you look over the rear wheelwells, you can see a kink in the frame, and if no one has properly pulled the car straight on a frame machine, that kink puts pressure on the quarter-panel and pushes it forward. It appears to be a small problem and it's hard to identify, but if you know where to look you can find it. We see cars that are hit in the rear quarter or in the front, and there are actually wrinkles in the floors, under the seats. The cars are completely driveable and safe, and they'll align properly and won't dogleg down the street, but they're squished up a little bit."

Peter adds, "As the quarter-panel buckles in, it closes up the door gap against the quarter. So now you push the doors forward, then move the fenders forward, which moves the hood out. So everything gets pushed forward 1/8 to 1/4 inch."

So, in many cases, there is no way to get a perfect hood fit without practically starting all over with a frame straightening and precise panel replacement.

The hood hinges mount to the car with three bolts and nut inserts in the inner fender. In extreme cases, the nut inserts are often ripped out of the inner fender. Removing the fender to weld in new inserts is the best repair, but standard bolts and nuts can be used.

But what can you do to at least improve the fit on your Mustang?

The hood hinges are a good place to start, say the Geislers. First, make sure the hinges are in good condition. "You've got to start with good hood hinges," Peter says. "You see a lot of them leaning over or they bend in as you open the hood. We see hinges that have been kinked at one time, then hammered and straightened out. Once hood hinges get tweaked or bent, we've found they aren't much good after that."

The hood-hinge rivets are also prone to wear, Pete adds. "As you lift up the hood, the whole system pivots and moves. The hood moves up and down, then it pivots on the axis points. But the average eye can't see the wear in the rivets. You'll have to look for play in the hood when you lift it up and down." The only solution is hood-hinge replacement, either with original hinges or reproductions that are available from Mustang parts suppliers.

But even good hinges have problems, says Peter, because owners don't lubricate them. "Every time someone pulls in to have me look at an oil leak or something, I have to fight to get the hood open. Everyone who comes in here needs to oil the hinges. They're just so tight, which causes the hood to sit up in the back. Do you see guys push the back of their hoods down? Well, if they would just oil the hinges, the hood would settle down by itself."

Pete recommends oiling or greasing all of the car's lubricating points, including the hood hinges and latch, whenever you change your engine oil.

The hinges bolt to the inner fender with three bolts and Nutserts, which are welded to the inner

With a fender removed, you can see how the hood-hinge nut inserts are welded to the inner fender.

The hinges bolt to the hood via a pair of nut inserts inside the hood. On '65–'66 Mustangs, the inserts are welded into the hood; for '67–'70, they mount with clips, making replacement easier. The hood can be moved forward and rearward by loosening these bolts.

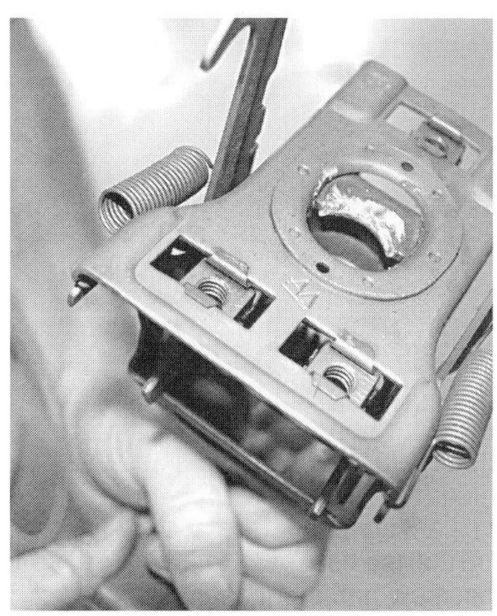

When fitting a hood, some hood-latch adjustment is available with these sliding cage nuts.

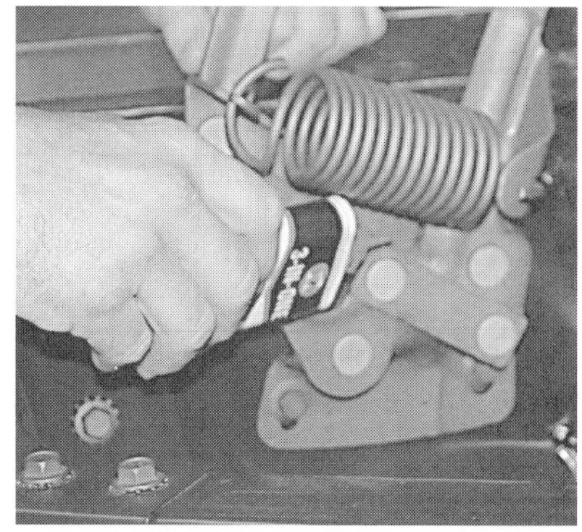

The Geislers recommend lubricating the hinges with 3-In-One oil for smooth, bind-free operation. White grease can be used on the hood-latch mechanism and hood pin.

When fitting or adjusting a hood, the Geislers recommend removing the hood latch so it doesn't affect how the hood opens and closes.

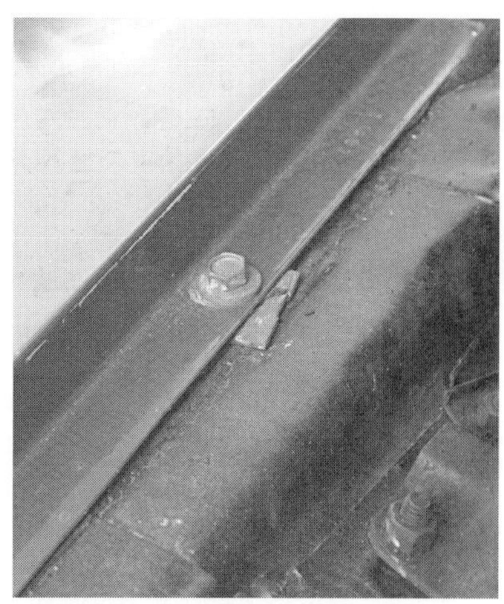

Shims can be used to raise the fender to match the hood. Fenders can also be moved in or out to adjust the gap with the hood.

The hood bumpers are adjustable up and down for setting the hood's front height.

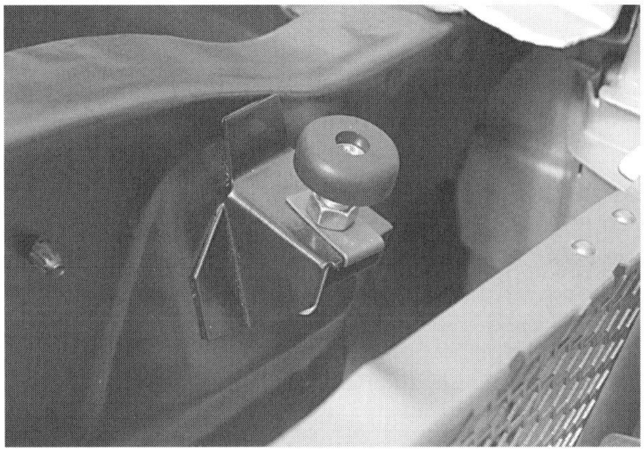

Shelby fiberglass hoods commonly bow in the middle. Orlando Mustang straightens them out by placing weight in the middle for up to six weeks.

Getting the front hood line to match the headlight buckets is tough on '69–'70 Mustangs. With the hood off the car and the front of the hood supported by a piece of wood, the front of the hood can be bent down. Otherwise, it's a combination of hood, fender, headlight-bucket, and hood-bumper adjustments.

fender. Pete says it's common for the Nutserts to rip out of the sheetmetal after years of use. "Damaged hinge-mounting areas have to be fixed first," Pete explains. "One way is to find original inserts and weld them in, which is tedious. The second method is to use a nut with a built-in washer and weld it in from the backside. With it welded in place, one person can adjust the hood."

Peter also suggests, "If you don't have a means for welding, you can use just a bolt with a washer and nut on the backside."

Pete has noticed what appears to have been a problem with '67–'68 Mustangs. "The hood latch failed and the secondary latch didn't hold the hood, so the hood would fly up and go into the windshield and roof. It always ripped the hood hinges right out of the inner fenders."

Problems can also occur inside the hood where the hinge attaches with two bolts. "A lot of cars come in and the hood inserts are damaged or missing, bolts stripped, and such," says Peter. "On '65–'66 hoods, the Nutserts are welded into the hood. The '67–'70s have clips that slide into the hood, so you can change them, much like a fender clip. That's a better design because if you have a problem with stripped threads, you can actually repair them. If the threads are bad in a '65–'66 hood, you have to weld in new nuts."

Once you know the hinge and its attachment are in good working order, you can proceed with adjustments. One trick used at Orlando Mustang is removing the latch mechanism. Pete explains, "That allows the hood to open and close freely, and the latch isn't trying to move the hood forward or side to side. It's like hanging a door in your house. You leave the door knob off to get the door to open and close properly; the same with the hood on a car—you don't want the latch to force it to fit."

Peter offers this tip for making adjustments at the hinges. "When the back of the hood is too high, most people lower the hinge. The proper thing to do is open the hood all the way, loosen the two front hinge bolts, leaving the back bolt snug, then pick up on the front of the hood. That tilts the hinges into a different position, which in effect lowers the back, but you still maintain the same adjustments in front. The front height is completely dictated by the hood bumpers.

"Occasionally you'll see adjustments on the hood where everything fits fine at the cowl and at the front of the hood, but the fenders may be high or low," Peter continues. "There are some slight adjustments most people don't think about. You can put shims under the fender to raise it up. Fenders are actually flexible and easy to move. Or, if the fenders are too high, you can take them off and bend them down."

If you're replacing or installing a hood, Peter says, "I would start by putting the hinge bolts in the center of the slots in the hinges as a starting point, symmetrically on both sides. Slowly lower the hood and make a note where the hood fits in. If it's too high on one side, adjust one side at a time—don't try doing both sides at once. Make small adjustments. Once the rear is done, start on the front. Lower the hood and watch where the hood pin enters the hood latch. The latch has slotted cage nuts, so you can move the latch left or right. There is no forward-to-back adjustment. That's why it's critical to get the hood lined up. If the radiator support has been moved in, you're going to have to push it back out. Get the pin lined up so the tip is just catching on the latch. When you get a nice, solid popping sound, you know you've got it just right."

Shelbys, with fiberglass hoods, offers its own set of problems. Most common are '67–'70 hoods that are bowed up in the middle. Orlando Mustang has discovered that placing weight on the hood can straighten it out. "To correct the bow," Peter says, "place the hood on sawhorses, one at the front and one at the rear, then pad the hood for protection and put some weight on top. We usually use tires and rims, like 100 pounds. Let it sit for about six weeks and it will slowly flatten the hood without damaging it. Then use a lighter-weight spring; they're the same dimensionally but have less tension. They came from the factory on the '67–'68 Shelby but they'll work on a '69–'70. They'll still keep the hood up, but will help prevent the bow from coming back."

The fiberglass can also be inconsistent. In some cases, it's too thick where the hinges mount, so the back of the hood sits too high. Peter slots the fiberglass to get the height correct.

Hood fitment and adjustment can be tedious, but getting the gaps right can be well worth the time and effort.

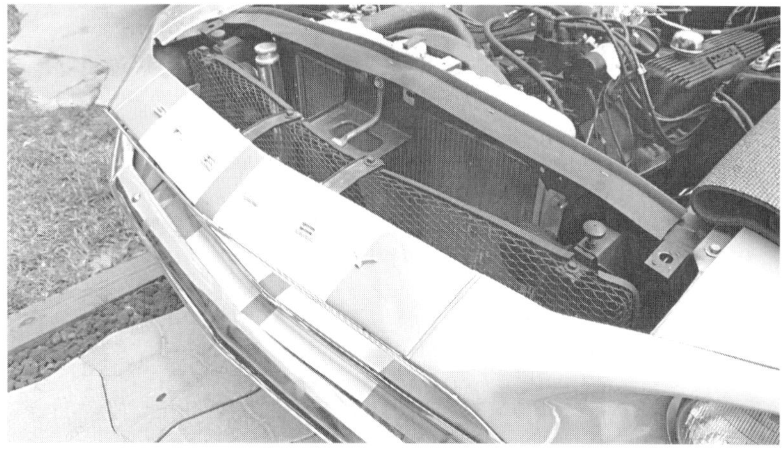

Thanks to the front header panel, '68 Shelbys are especially difficult when fitting a hood. Not only do you have to get the gaps right at the back and side, the front gap with the header panel is also critical.

Lighter-tension hinge springs were used on '67–'68 Shelbys. They can also be used on the '69–'70 to lessen the chance of bowing.

Ah, sweet perfection. This is the type of consistent gaps you're looking for when fitting or adjusting a hood.

Chapter 37

Installing '69–'70 Front & Rear Spoilers

Text and Photos by Miles Cook

National Parts Depot's reproduction spoiler kit for '69–'70 Mustangs comes with the same flat-black finish as the original versions. The spoiler looks great right out of the box as shown here, or it can be easily scuffed for paint if you're going for a custom look, such as painting it body color. In general, we're reluctant to add superfluous knick-knacks to our '69 fastback project car because we don't want to upset its bare-bones, road-racy vibe. But a factory spoiler, especially one that was used on Trans-Am Boss 302 Mustangs, won't detract from that look.

Rear Spoiler

The optional trunk-mounted rear spoiler was one of the many unique features found on '69 and '70 Mustang fastbacks, including Mach 1s, Boss 302s, and Boss 429s. The spoiler added a muscle car image, and for that reason, it remains a popular add-on. In the past, the opportunity to buy a new spoiler was something of a hit-or-miss proposition, with either a used one or an N.O.S. unit as the most likely options. Not anymore, as National Parts Depot now offers a nice reproduction that's also approved by the Mustang Club of America for concours show duty.

Features of the NPD spoiler include lightweight ABS plastic construction, which makes it light enough to use without the need for a prop rod, which was used to hold the trunk lid open with the original solid wings. However, for fully restored cars, NPD offers a prop rod. Additionally, just like an original, the angle on these reproduction spoilers can be adjusted for appearance purposes as well as functionality should you use your car for high-speed open-track activity or running down the dragstrip.

The spoiler comes with die-cast pedestals and stamped brackets that result in a completely correct installation. NPD's thorough installation instructions, along with the pair of support brackets that mount on the underside of the trunk lid, are other features worth noting. Not only do the brackets support the weight of the spoiler, they also serve as a template for drilling the holes in the lid for accurate placement of the spoiler. We had the spoiler installed on our '69 fastback in about three hours.

Source
National Parts Depot
www.npdlink.com

NPD's reproduction front spoilers are built to match the original piece. Thermoformed to the original contours, they're made of cross-linked, high-density plastic. Unlike fiberglass, this material flexes and rebounds after hitting curbs, low driveways, speed bumps, and other road hazards. For '69 and '70 Mustangs, NPD offers two different spoilers for each year, and they mount differently. For '69s, order NPD PN 001A74-1A, and for '70s, the spoiler carries PN 001A06-1A. NPD also offers front spoilers for '65-'68 and '71–'73 Mustangs. Of course, the '65–'68 versions are non-original.

Front Spoiler

Now let's move to the front of the car for the installation of an original-style chin spoiler that you often see on '69 and '70 Mustangs. Standard equipment on Boss 302s and optional on others, it's a functional piece that provides a notable amount of down force and allows less air to get underneath the car, which causes lift.

While some upgrades require a lot of work and might not provide much of a visual or performance improvement, a front spoiler is quite the opposite. Installing a front spoiler, such as this very nice reproduction piece from National Parts Depot, is easy. It takes only 2-3 bolts, 14 1/8-inch holes in the spoiler and the car's front valance panel, and installing sheetmetal screws.

With the front of the car up in the air for better access (such as with the ramps we used), you can install one of these truly trick-looking front spoilers in well under an hour. Even with photographing the procedure, we had the installation complete in about 45 minutes.

Rear Spoiler Install

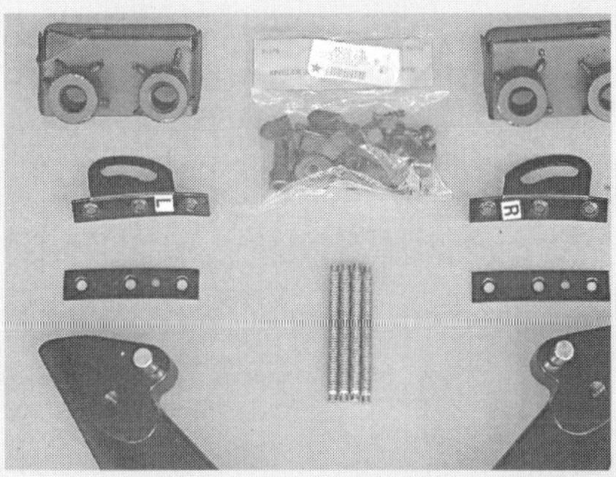

1. Besides the wing itself, the quality of the kit is also evident in the installation hardware. Shown here are the stanchions, stanchion gaskets, retainer plates, studs, and AMK Products fastener kit. The reinforcement plates are sold separately.

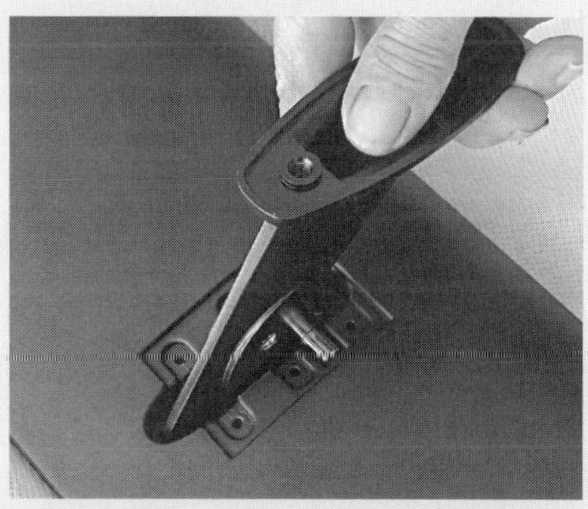

2. Assembling the spoiler is easy. Begin by putting a stanchion in place in the spoiler slot. The stanchions aren't side specific, so they fit either side.

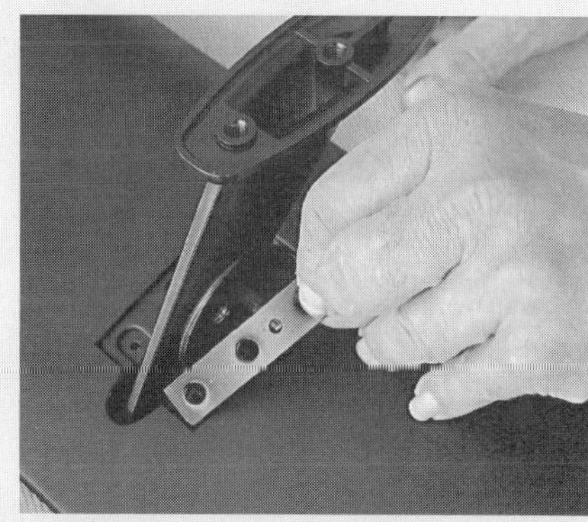

3. Next, install a flat retainer plate (again these parts aren't side specific) on the outside of the stanchion …

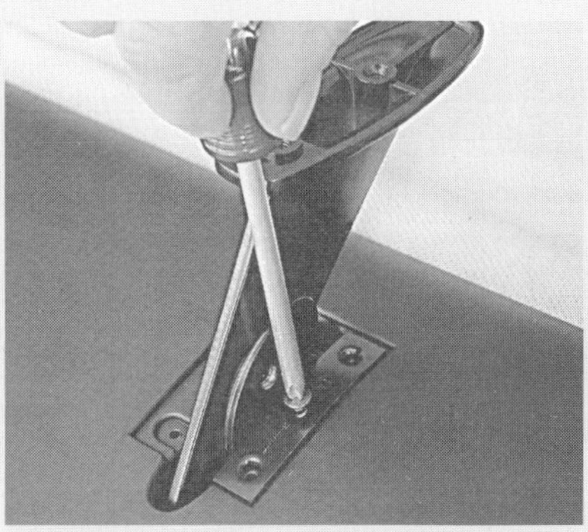

4 … and secure it in place with the supplied Phillips-head sheetmetal screws. You'll be creating new threads in the ABS plastic spoiler with the screws, so they'll be somewhat harder to turn than usual.

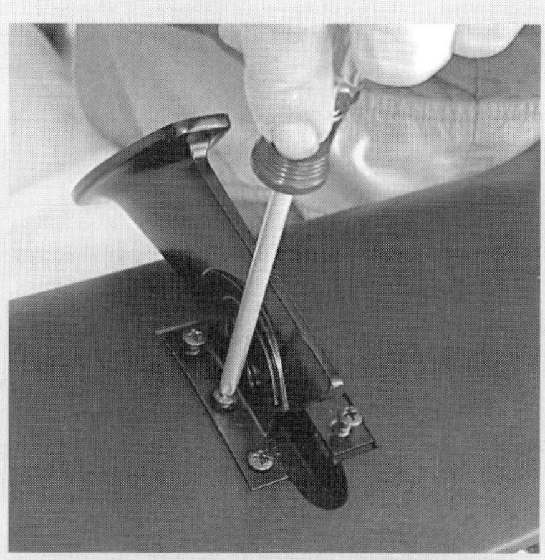

5. The pivot plates are side-specific (and marked as such) and install inside the stanchions using the same supplied screws. Tighten the screws securely after you have all six started for each stanchion.

6. With a supplied washer, install one of the supplied, large Allen-head bolts through the pivot plate and into the stanchion's threaded hole.

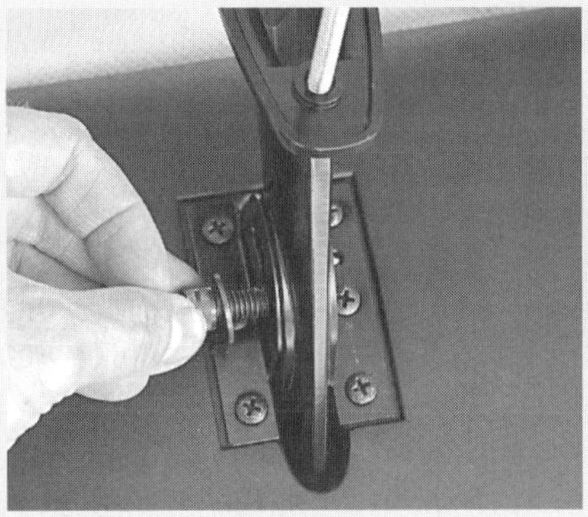

7. Tighten the Allen-bolt to secure the stanchion; this is how the spoiler is adjusted. Once installed, its rake can be altered by loosening the Allen-head bolts and moving the spoiler to the desired position, then retightening the bolts. It's a little tight with the spoiler close to the trunk lid, but it still only takes a minute or two to adjust the spoiler with an Allen wrench.

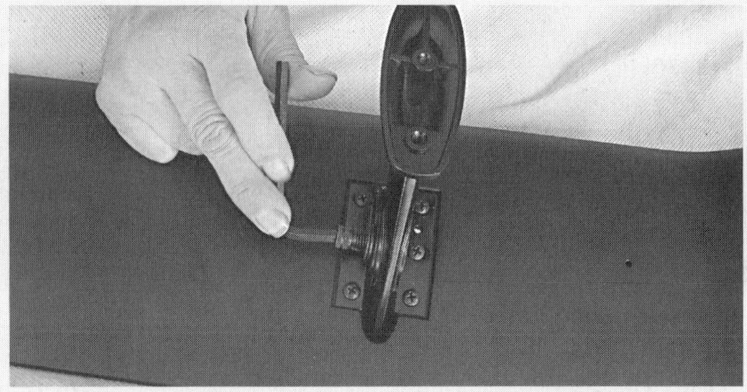

8. Thread the studs into the bottom of the stanchion. You can also carefully tighten them with a pair of locking pliers by gently locking the pliers on a part of the threads that won't be used, but be careful not to damage them.

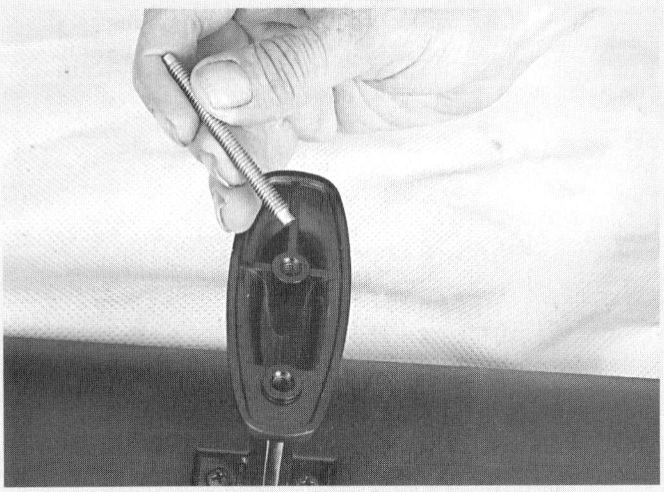

9. The completed installation of the stanchion, retaining plates and studs on the driver-side of the spoiler looks like this. Repeat the process for the other side of the spoiler.

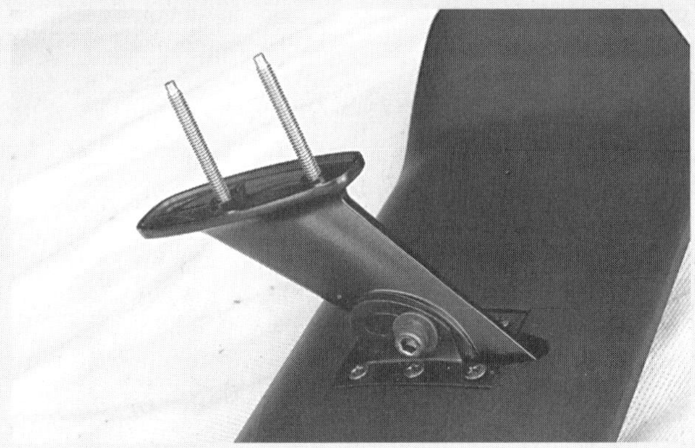

10. With the spoiler assembled, installing it is accomplished more easily by removing the trunk lid. Mark its position before removing the bolts on the hinge brackets to ease alignment on reinstallation. Since our car was painted with the bolts in place, it was easy to relocate the lid using the different paint color underneath the bolts.

11. With the four bolts removed, the trunk lid easily slides up from the hinges and off the car.

12. Working on a soft surface (such as a table with a towel over it to protect the trunk's paint) and with the lid upside down, the reinforcement brackets, by virtue of the fact that they mount with a hinge bolt, serve as templates for drilling the holes in the lid. However, in addition to the trunk-lid surface where the spoiler mounts, there's another level of sheetmetal on the bottom side of the lid that must be cut out to accept the reinforcement brackets, which are side-specific. On cars originally equipped with a spoiler, the lid has an oval cutout for each reinforcement bracket and holes for the spoiler itself. But on lids that weren't equipped with a spoiler (such as ours), an oval cutout must be made to allow the bracket to mate directly to the underside of the sheetmetal. Begin by marking where the bracket needs to be placed. We used a permanent marker and the correct bracket for the side we were working on. The hole in the side of the bracket is where a trunk-lid bolt will go through. Eyeball this hole as close as possible with the hole in the side of the lid while the bracket is sitting on the surface that will be cut away. Draw a circle on the surface around each round collar on the bracket to correctly locate where you will drill.

13. Use an electric drill with a 1-inch hole-saw bit to drill out the sheetmetal. Be careful not to drill into the second layer of steel; it's the top of the trunk lid.

14. With one hole drilled out already and one to go, this is where we're at along the way. Remove the round piece of metal inside the trunk before proceeding.

15. With both holes drilled out, there's still some material in between the two holes that also needs to be removed. Use an electric or air-powered tool equipped with a cutoff wheel to carefully remove the metal in between, creating an oval-shaped opening large enough for the spoiler reinforcement bracket to drop into.

16. With the opening created, the bracket fits into place as shown.

17. A neat trick is to use a 1⁄2-inch drill bit to locate the pilot marks for the spoiler holes. The 1⁄2-inch bit fits perfectly in the holes in the bracket. If there are a few burrs, rotate the bit until it moves freely inside the bracket.

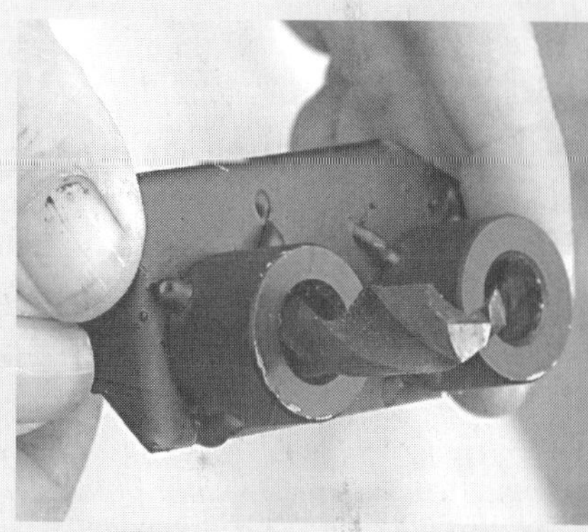

18. With the bracket in place in the oval opening (and the side hole lined up with the hole in the lid for the hinge bolt), twist the bit by hand to mark the center point for the hole, which will be drilled with a smaller 1⁄4-inch bit. It's the ideal way to correctly locate the critical holes that will be drilled in the top of the trunk lid.

19. Using a 1⁄4-inch bit (the same size as the studs in the spoiler stanchions), carefully drill the holes in the trunk lid. Here the support bracket is still in place to ensure we had the hole started properly in the center point. Once started, you can stop for a moment to remove the bracket, and continue drilling the holes all the way through the lid.

20. After drilling the final holes in the trunk lid, we temporarily removed a stanchion from the spoiler, making it easier to test how the studs fit into the holes. We found the studs fit quite well; only some minor elongating, mainly just deburring with a file, was required. Before test fitting the stanchion, we also installed the supplied sealing washers as shown.

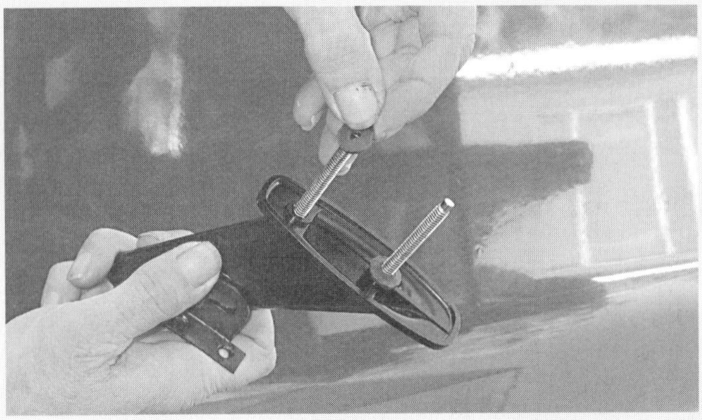

21. During our test fit, we also made sure the stanchion gasket was in place.

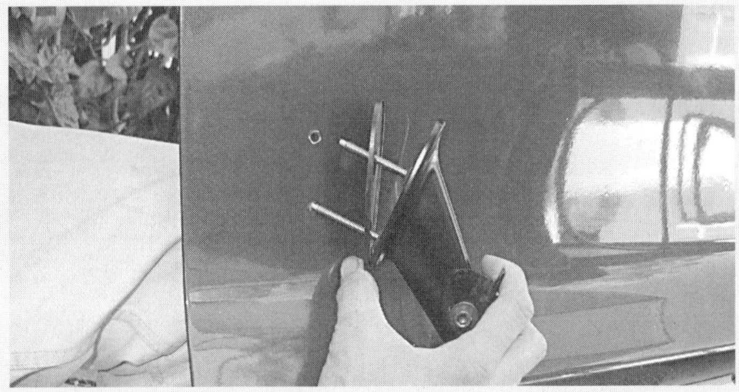

22. With everything squared away and the spoiler reassembled, we moved on to the final installation of the spoiler on the trunk lid. Carefully fit the studs through the holes, again being sure you have the stanchion gaskets in place. In the end, we were content with the four holes drilled into the lid as the spoiler dropped into place with no trouble, and its placement on the trunk lid was right on the money.

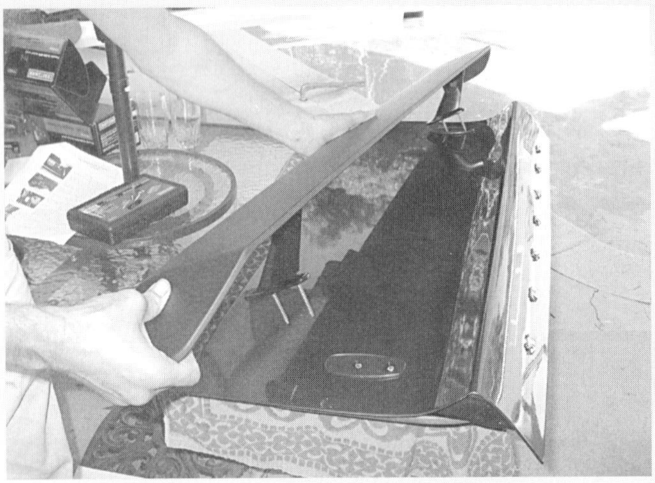

23. With the spoiler installed, carefully lay the spoiler and trunk-lid assembly back on a soft surface to install and tighten the four supplied nuts with spin washers. Be sure the gasket lips are surrounding the stanchions on each side before final tightening.

24. Here's the completed assembly, ready for installation on the car. The flat-black finish of the spoiler contrasts perfectly with the '68 Highland Green finish on our '69 Mustang project car.

25. The trunk lid is notably heavier (around 10–12 pounds) with the spoiler in place. Thus, it's easier to reinstall as a two-person operation. The hinges need to be threaded into the holes of the lid and lined up to reinstall the bolts.

26. With the lid in place, securely tighten the bolts. The two rear bolts go through the reinforcement brackets to further strengthen the spoiler's attachment to the trunk lid.

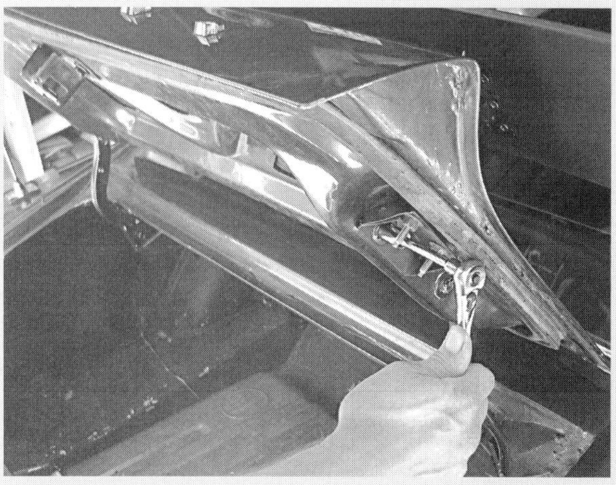

27. Complete the installation by covering the exposed stud ends with the supplied rubber caps.

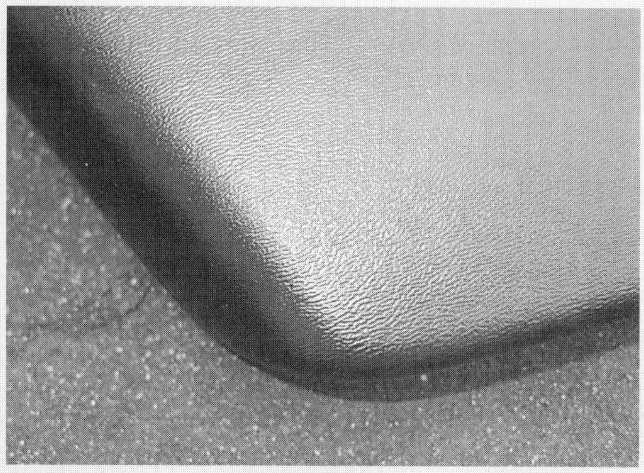

28. Now let's move to the front. This closeup of one of the ends of the front spoiler shows the original-type textured finish of the spoiler surface.

29. Installation hardware is supplied with the spoiler. This includes 3 factory-style nuts and bolts, along with 14 sheetmetal screws.

30. With the front of the car on ramps, remove the two lower bolts that hold the lower edge of the valance panel in place. A 1⁄4-inch drive ratchet and socket is ideal for this task.

31. There are three large holes already in place on the spoiler where the supplied bolts fit.

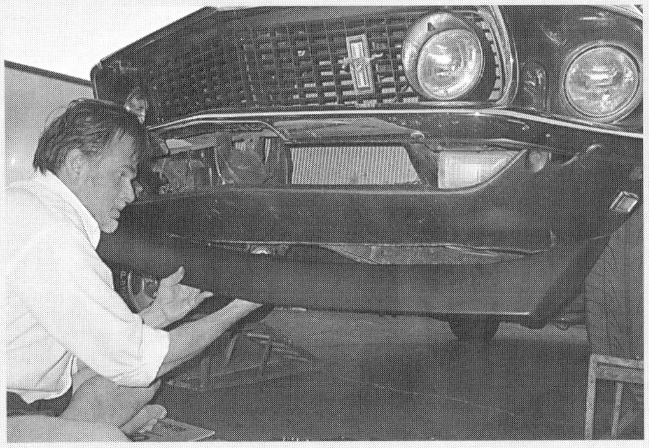

32. The spoiler is lightweight, and getting the center bolt fitted through the valance while holding the spoiler in position is pretty easy. Next, install a nut onto the bolt and tighten it partially.

33. When all three bolts are in place (the two outside ones don't need nuts as the bolts install into the threaded speed nuts that are already on the car), tighten them evenly and securely. With the spoiler essentially installed at this point, all that's left is to drill 141⁄8-inch diameter holes for the sheet metal screws. While the three larger bolts hold the spoiler in place fairly well, use all the screws for a solid installation. As the supplied instruction diagram illustrates, the 14 evenly spaced screw locations are as follows: 4 screws go on the outside of each of the 2 end bolts (8 total); 3 go on either side of the center bolt but inside of the 2 end bolts (6 total). Drill the holes using the spoiler itself as a template. Drilling the holes is easy since the spoiler is soft plastic and the valance is relatively thin steel sheet metal. Drill through the spoiler and the valance at the same time to ensure the holes line up.

34.Complete the job by installing and tightening the sheet metal screws. Tighten the screws relatively firmly, but not so tight that you could strip the holes in the valance, which will then reduce the screw's clamping force.

ELECTRICAL

Chapter 38

Understanding Mustang Electrical Systems

Text and Photos by Jim Smart

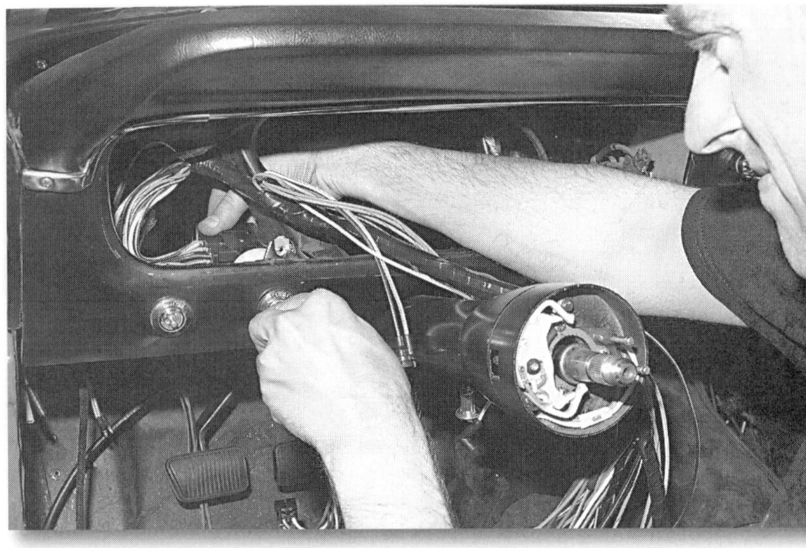

Virginia Classic Mustang has solutions that will help you see the light.

Source
Virginia Classic Mustang
www.vamustang.com

Mustang electrical systems are clever pathways that accomplish a big job whenever we twist the key and hit the road. Yet, they don't always work as they're supposed to because parts wear out, people make mistakes, and connections corrode and fail. What's more, electrical systems sometimes do weird things that we're just not going to understand, even with a degree in electrical engineering. At their best, electrical systems illuminate the lights, vibrate the sound system, ignite the fuel/air mix, and help warm or cool the cabin. Without the electrical system, we would all be walking, and sometimes we do.

Automotive electrical systems work differently from home systems, yet the overall approach for getting electricity to do our work is basically the same: current flow from positive to negative, which explains two-prong outlets in the home and a two-terminal storage battery in our Mustangs. At home, we use 110/115 volts of alternating current. In Mustangs, it's 12–15 volts of direct current.

Power for Mustang electrical systems begins and ends at the battery, which stores 12–15 volts of direct current to keep a Mustang operational. The charging system is designed to keep the battery charged by converting the engine's rotary motion to electrical energy by way of an alternator, or a generator in the case of '64 1/2 models.

Cars utilize a single-wire electrical system, meaning current flows through wires to each accessory and back to the battery's negative terminal through the car's body. In homes, electricity flows through two wires with an extra ground wire on hand to help prevent electrical shock. Automobiles don't need the return wire because the steel or aluminum body is

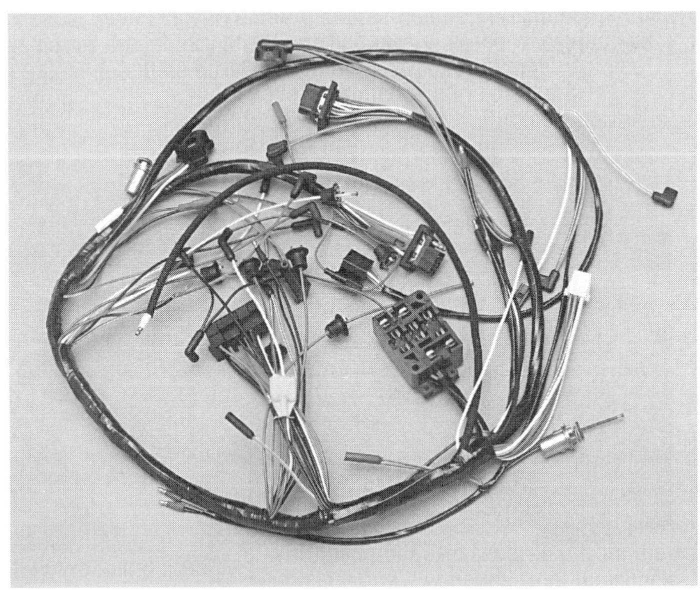

Did you know you can buy a complete new electrical system from Virginia Classic Mustang? This is the main wiring loom for a '66 Mustang. Although the electrical system is intimidating, there's virtually nothing to it. Replacement of the entire electrical system can be performed in a day.

In '65–'66, Ford used firewall bulkhead connectors to get power to and from the engine compartment. The bulkhead connector behind the engine handles the engine's senders and ignition power. The bulkhead connector near the master cylinder carries power from the battery to the electrical system, plus power to the headlights, parking lights, and horns. Beginning in '67, Ford moved these vulnerable connections under the dashboard to protect them from the elements.

used as a conductor to complete the circuit. This is where the term negative ground comes from.

Sometimes a ground wire is needed to carry electricity when the body can't, such as between the engine and firewall. Why? Because the engine rests on rubber mounts that insulate it from the body and chassis. The alternator or generator's ground connection helps ground the engine to some degree, but it's not enough.

Weak and faulty grounds explain many of the electrical gremlins that are difficult to troubleshoot. When electrical components start acting weird, such as dim headlights, a stereo that mysteriously quits, an ignition system that leaves you stranded, turn signals that suddenly blink rapidly or not at all, and other unexplainable surprises, the problem is a weak ground somewhere. This is why there are no unimportant grounds. They all serve a purpose.

There's no real mystery to electricity. It is simply the flow of electrons, tiny subatomic particles of matter smaller than an atom. These energetic guys are tiny, but they do a lot of work when they become a team on a mission. This takes us back to the importance of proper grounding. When electricity can't follow a proper path, it finds alternative paths. If you don't ground the engine, for example, the ignition system will find an alternative path to ground. When it can't, the engine quits. Weak headlight or taillight grounding

The fuse box is the first point of entry for most of the electrical-system power. Fuses handle light-duty electrical loads like the heater fan, instrument lights, some accessories, radio, courtesy lights, and more. Headlights and parking lamps are protected through a circuit breaker in the headlight switch.

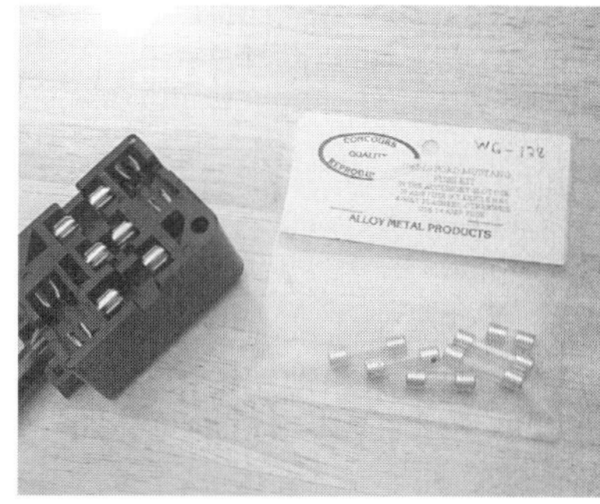

When you order your electrical system from Virginia Classic Mustang, don't forget this fuse kit, which includes every fuse required. It's a good idea to purchase one or two additional kits just in case.

Do your headlights cycle off and on when they have been on a while? If so, the circuit breaker is faulty and the switch needs to be replaced. Virginia Classic Mustang can help with everything you need to replace the switch.

We suggest using the best headlights possible for nighttime safety and driving pleasure. Xenon lamps seem to be the best headlight out there now. Make sure they are properly aimed.

Headlight switches are multi-taskers because they do many things, including dimming the instrument-panel lights through the variable resistor shown here. Turn it counterclockwise to the limit and it illuminates the courtesy lights. As its name implies, the variable resistor impedes the flow of electricity, dimming the lights.

Brake-light switches are found behind the instrument panel in all Mustang applications since August of 1964. Prior to that, the brake-light switch was a pressure switch located on the master cylinder.

Tail and parking-lamp bulbs employ two filaments, a dim one for parking lights and a bright one for turn signals and braking. The shorter, thicker filament is the brighter one because there's more resistance. The taller, thinner filament is dimmer for parking lights. We see a few of these wired backwards, meaning bright for parking lamps, which leads people to believe you are braking.

Grounding terminals exist throughout your Mustang's electrical system. All of them are important to proper electrical-system function. For best results, always ensure each of them is tied to the body/chassis as shown. Too many of us overlook the engine-to-firewall ground lead. Don't forget this one. If you do, expect all kinds of ignition and charging system woes.

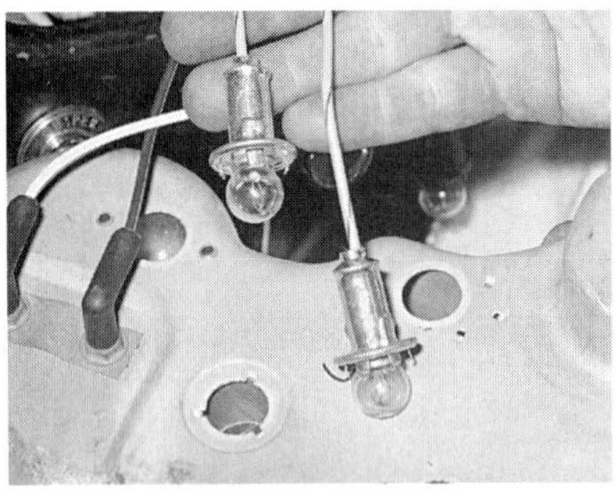

Brighter bulbs are available for poorly lit classic-Mustang instrument panels. However, be cautious about bulb heat. While you're at it, replace the instrument-voltage limiter, also known as a voltage regulator. When instruments don't work or max out, the voltage limiter is faulty and must be replaced.

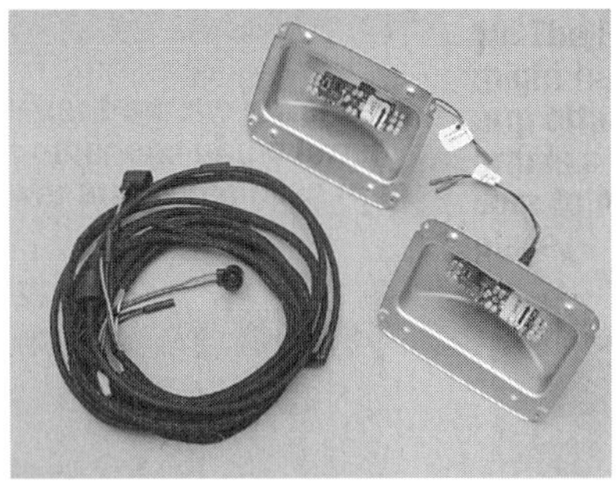

This is a typical Mustang three-speed fan switch. It is logical to assume it is a variable resistor because it offers three positions, but it is actually an electrical traffic cop that routes power to a variable resistor in the heater box.

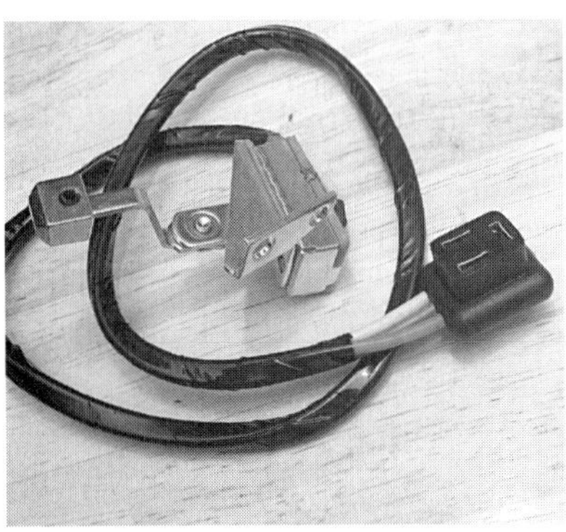

Three wires route power from the fan switch to the resistor package and fan motor. Power passes through the fan switch to the motor, then to the appropriate resistor to ground.

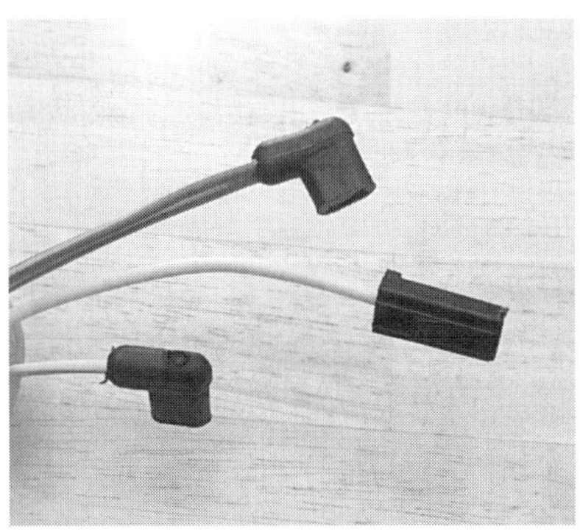

This is the variable resistance package, located on the heater box to aid resistor cooling.

sends electricity searching for alternative paths to ground, sometimes bringing the radio to life when the ignition is turned off. That's what we mean by electrical gremlins that cannot be explained. Most of the time, it's a faulty ground.

Think of your Mustang's electrical system like you would a home plumbing system. Electricity flows through wiring like water flows through pipes. When there is resistance to the flow of water, we get less water. Electricity works the same way. When there's resistance to the flow of electricity, there's less of it. However, when we create resistance to the flow of electricity, we get something we don't get from water: heat. Heat and resistance are what make light bulbs glow. Heat and resistance are also what make cigarette lighters get hot. They can also start fires when resistance comes from a weak connection.

Resistance is created with switches, resistors, resistance wire, filaments, and other controls. We use it to control fan-motor speed, light intensity, sound-system volume, ignition-coil voltage, and more. When we create resistance, we impede the flow of electricity.

A Word or Two About Lighting

Automotive lighting has come a long way since 1965. Today's lighting technology places greater demands on an old Mustang's electrical system. Halogen and xenon headlamps require greater amounts of power, which places quite a load on headlight switches. This means shorter service life and the need for as much alternator as you can muster. The most power from an alternator you can expect with externally regulated systems is 65 amps. Go to a single-wire alternator and you can count on upwards of 150 amps.

Halogen taillight bulbs are much brighter, which is a great safety feature. However, Halogen lamps run much hotter, which mandates special taillight lenses and venting of your taillight dishes. LED taillights run cool as long as you opt for ones that are visible from a distance.

Fan Speed Control: A Lesson in Resistance

There are a lot of misconceptions about how heater-fan controls work. There is but one truth. Heater-fan switches are not variable resisters. Resistance is the same in both (two-speed) or all three (three-speed) positions. Resistance comes from a resistor package installed in the heater plenum. The resister package consists of two lengths of resistor wire, which control resistance not to the fan motor but to ground. When the fan switch is in Low, resistance to ground is high, which impedes the flow of electricity to ground. This lowers fan speed. Move the switch to Medium and resistance to ground becomes less, raising fan speed. Move the fan switch to High and bypass the resistance wires completely, going directly to ground. The fan then runs at full speed.

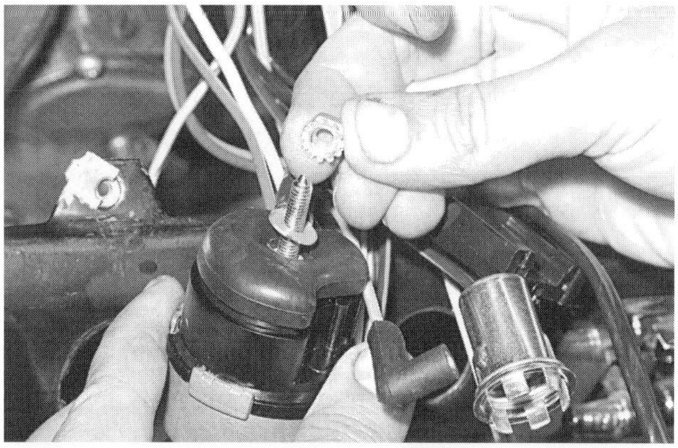

We have seen all kinds of wiring molestation through the years, none of it pretty or necessary. Do you know how many sources for power there are in a classic Mustang? At least four that we can think of, which means you never need to cut a wire to get power. Power is available at the ignition switch from '65–'67. There are also two triplex plugs for accessories and instrument lighting. All you need are bullet connectors to get power.

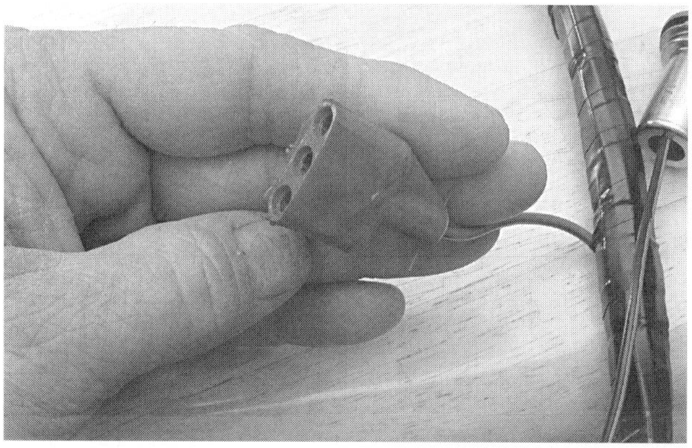

This is the instrument-lighting triplex plug for auxiliary lighting. Look for the yellow plug for auxiliary power to run accessories.

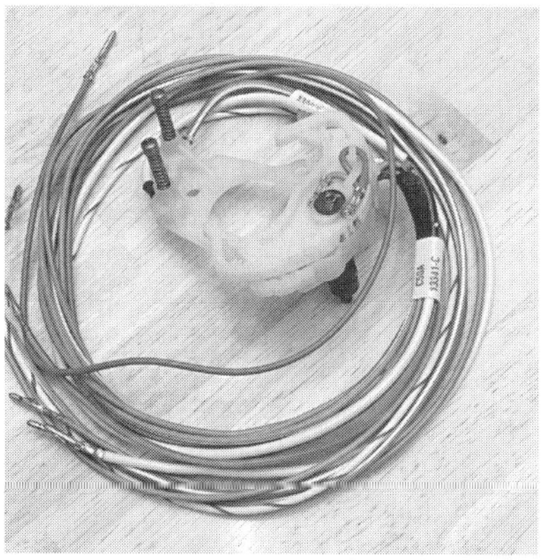

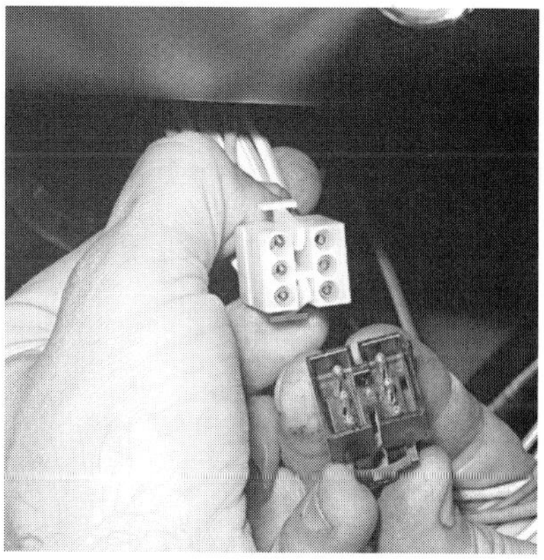

Whenever you replace the turn-signal switch, don't forget to keep the original turn-signal plug on the switch harness. Throw it away and you will have to find a replacement.

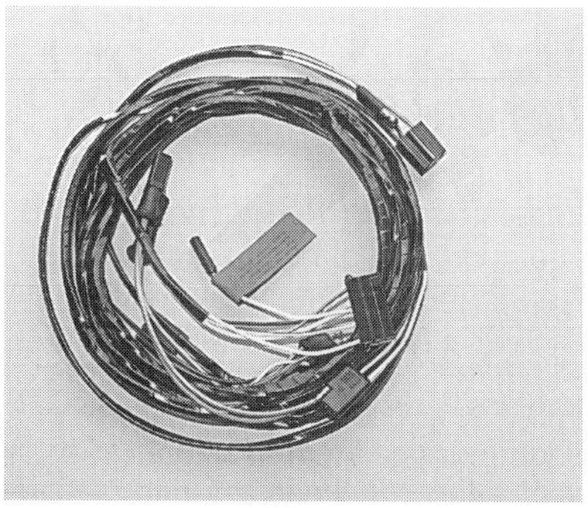

Detailing the engine compartment is the best time to replace all underhood wiring, which is exposed to extremes of heat, cold, moisture, and dryness. Play it safe and refresh all underhood wiring whenever you detail. Don't be afraid to protect it with flexible conduit while you're at it.

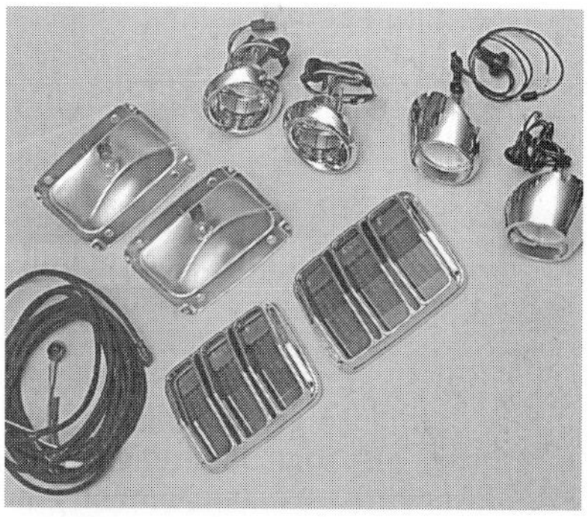

Virginia Classic Mustang has virtually everything you need for exterior lighting. This is the taillight harness and everything that goes with it. New taillight dishes, back-up lamps, and parking lamps make a big difference in safety and the quality of your restoration.

Chapter 39

How to Install New Wiring

Text and Photos by Jim Smart

Replace your Mustang's electrical system with new harnesses from Mustangs Plus.

What can you do about flickering lights, a dead sound system, psychotic instruments, undependable turn signals, and horns that don't always work?

Our classic Mustang electrical systems are notorious for grief and frustration, mainly because most are now reaching four decades of service. After 40 years, insulation becomes brittle, exposed copper corrodes, terminals become pitted, and electrical tape begins to roll off. With all of these elements in play, there's potential for problems, even disaster. Exposed copper wiring can ground out against the body or chassis, smoking the harness and starting a fire. Brittle wiring can break, thrusting you into darkness on the roadside. And imagine a bad wiper connection in a rainstorm.

We visited JMC Motorsports in San Diego, California, while they were installing a new stock electrical system from Mustangs Plus. Replacing a vintage Ford's factory electrical system is easy because everything plugs together one way. Firewall connections are easy to pop out and in, and switches plug in quickly. Almost anyone can do it.

When we arrived at JMC Motorsports, the old electrical system had already been removed. Removal of the old system isn't that difficult. It begins with disconnecting the battery. We suggest shooting photographs of your Ford's old electrical system before removal. Zero in on the instrument wiring and everything behind the dashboard. Because Ford designed this stuff for rapid assembly on an assembly line, plugs mate only one way. Grounds speak for themselves, especially when you photograph them before removal. Firewall plugs mate at the firewall to tie the underdash main wiring loom to the headlamp and engine looms. Plugs that tie the main wiring loom to the taillight harness are also behind the dashboard on the left side. The turn-signal switch plugs into the main wiring loom at the base of the steering column.

Always examine potential problem areas before they can arise. For example, we've seen fuse-box screws accidentally run through wiring, causing a short circuit. We have also witnessed wire chafing, which caused the wiring to ground out. Though reproduction wiring looms are wrapped and protected exactly like the originals, we can make them even safer. Plastic anti-chafe can be used to minimize contact, and you can also wrap the wiring loom with electrical tape for greater protection. Examine how the wiring is routed, which can also be a factor in protection. Sharp metal edges can chew right through wiring, especially when you consider road shock and vibration.

Sources

JMC Motorsports
www.johnsmustangs.com

Mustangs Plus
www.mustangsplus.com

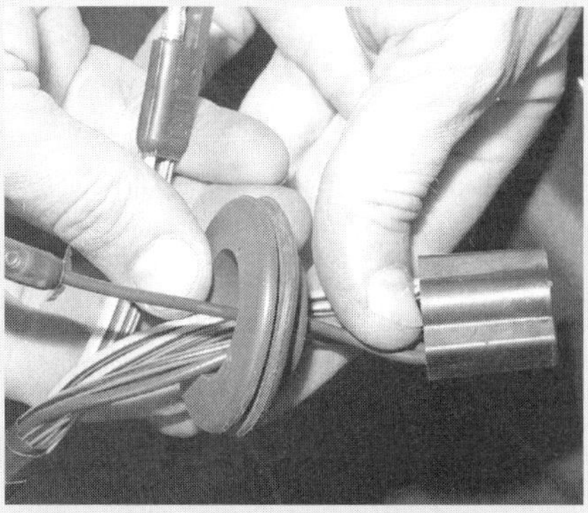

1. You can begin the new wiring harness installation virtually anywhere in your Mustang. Dave Toth of JMC Motorsports begins at the radiator support and works aft. Dave uses an aftermarket rubber grommet where the headlamp/parking lamp wire looms penetrate the left and right inner fender aprons. Once the leads are completely through to the headlamp door, Dave connects the parking lamp and ground. The heavy rubber grommet is not stock, but it protects.

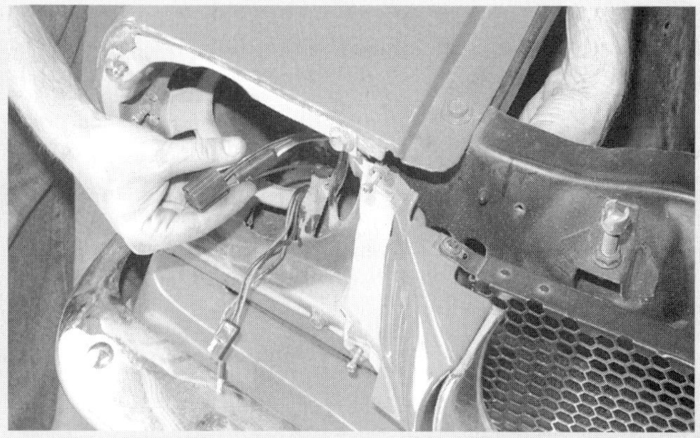

2. Dave does the same thing on the righthand side, connecting the parking lamp and ground.

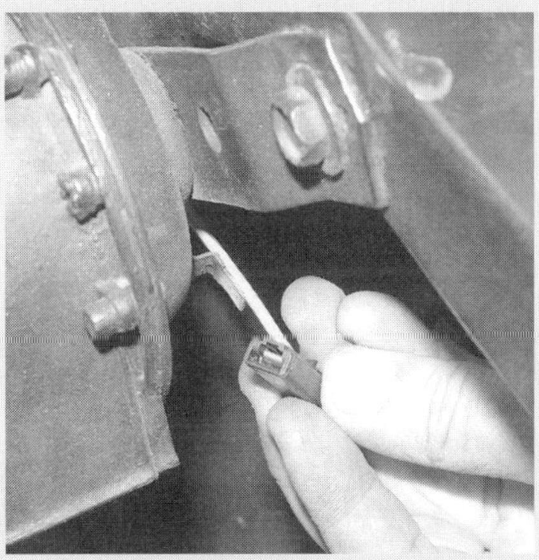

3. Horn leads are routed through the radiator support and connected.

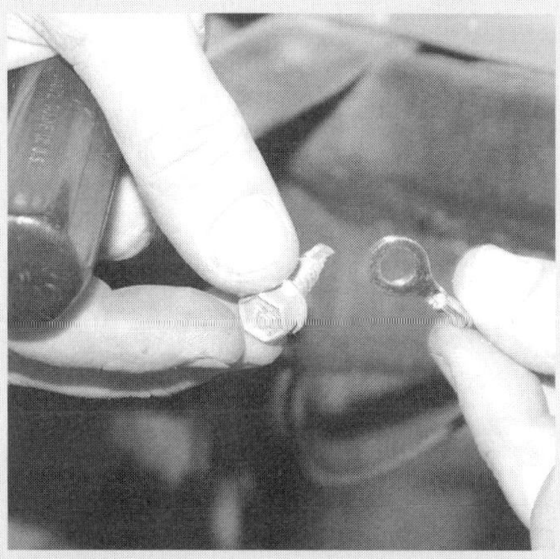

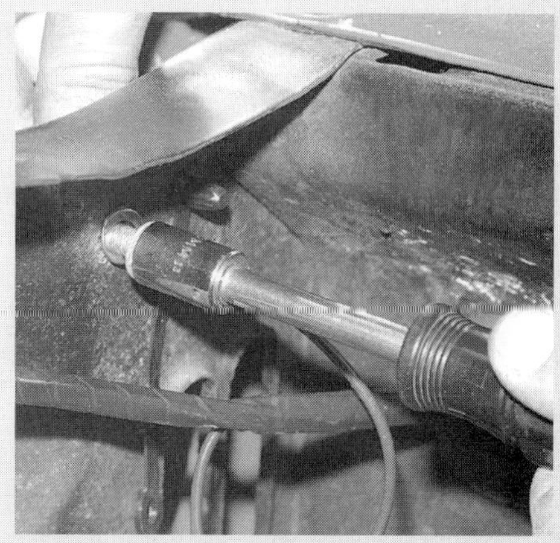

4. At the radiator support, there are two grounds. Connect both, using the proper hex-head sheetmetal screw with a star washer for solid continuity. Bad grounds, often caused by bad connections or no ground at all, create all sorts of electrical-system gremlins.

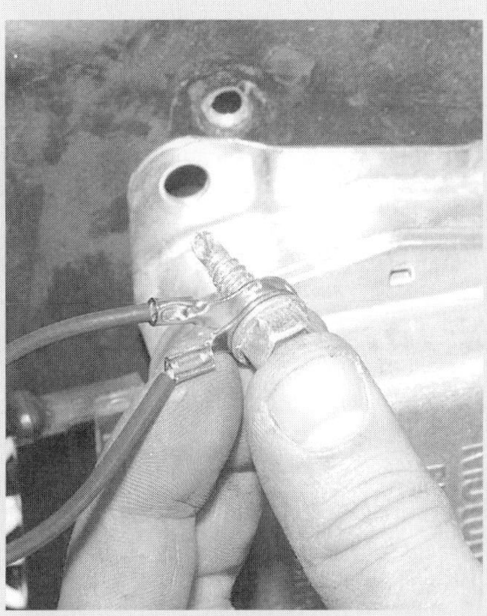

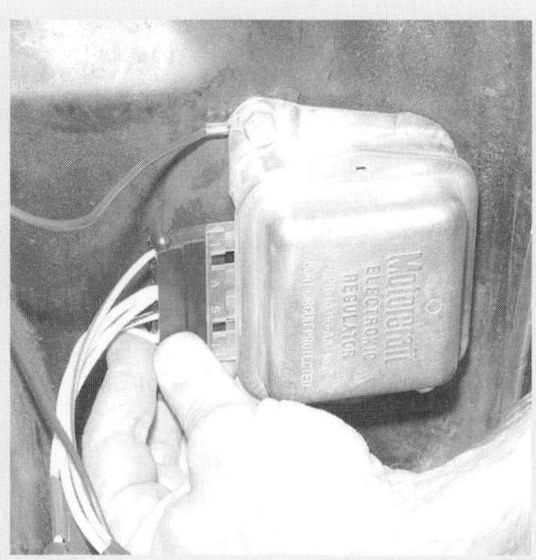

5. The voltage regulator installs with two hex-head sheetmetal screws. There is also a ground, shown here. Ford installed a radio-noise suppressor to minimize static and hum from the charging system. Make sure the ground is solid.

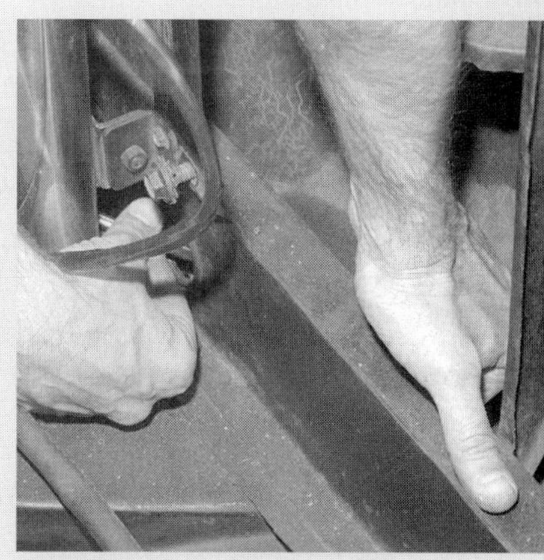

6. The charging system segment of the harness is fed through the unibody's front framerail box. Make sure you have plenty of protection around the wiring where this loom enters and exits the box. This terminates at the starter solenoid and alternator.

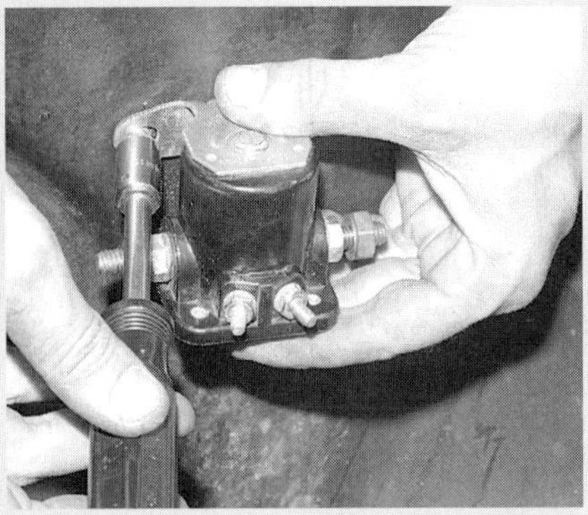

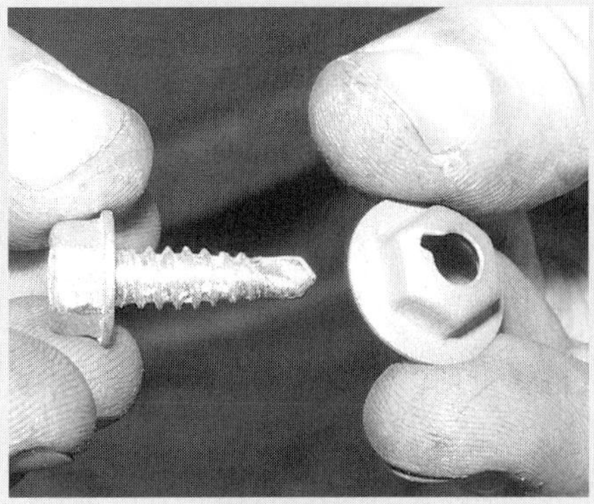

7. Dave installs the new starter solenoid. Because so many suffer from stripped-out screw holes, they need tinnerman's nuts on the back side. With the fender installed, the tinnerman's nut cannot be seen.

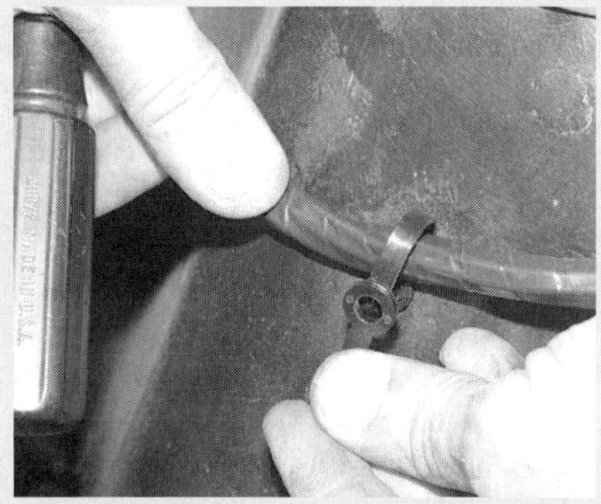

8. Wire-loom loops have a special madness when it comes to proper installation. Dave uses a 5⁄16-inch socket to press them into place. The loop should always hang down, never up. When you position the loop up, it stresses the loop, resulting in failure—especially in a hot engine compartment.

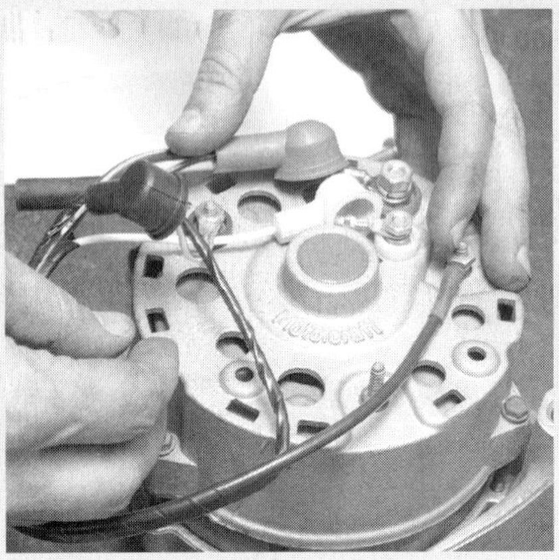

9. The alternator harness is next. Each Mustang engine calls for different harness routing on the back of the alternator, so check your Ford shop manual. The most important thing to remember is solid connections that are protected. Determine proper connections as well before connecting the battery.

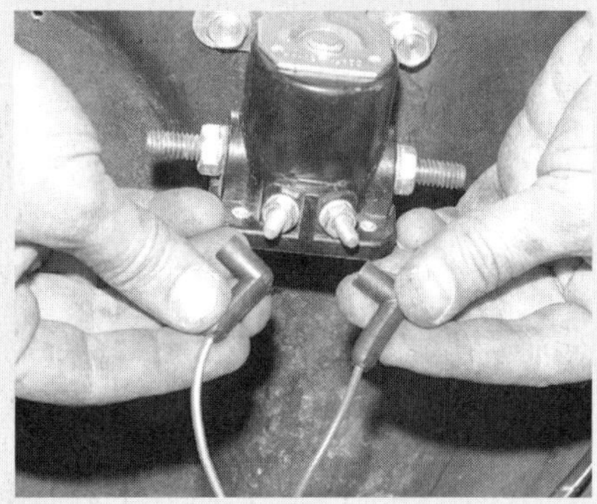

10. The engine gauge feed is next. We call this an engine gauge feed, but it also connects to the starter solenoid, which enables us to activate the starter. When you turn the key to "Start," power flows through one of these two wires to close the solenoid contacts, which channels power directly from the battery to the starter. Power also flows to the ignition coil during cranking via the solenoid. Three engine-gauge feed wires go to the coolant-temperature sender, oil-pressure sender, and ignition coil.

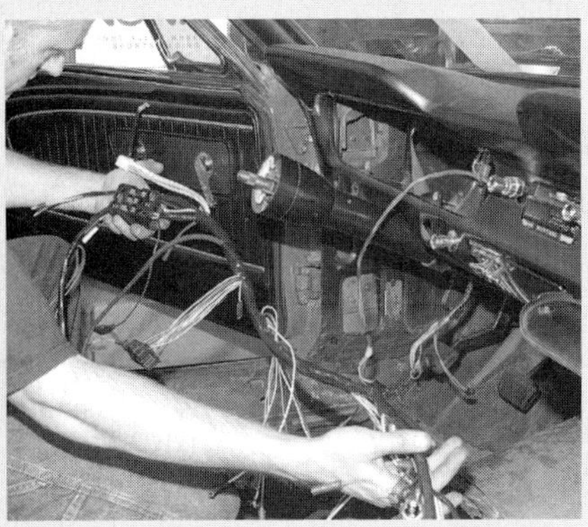

11. Dave Toth lays out the main wiring loom, which goes behind the dashboard.

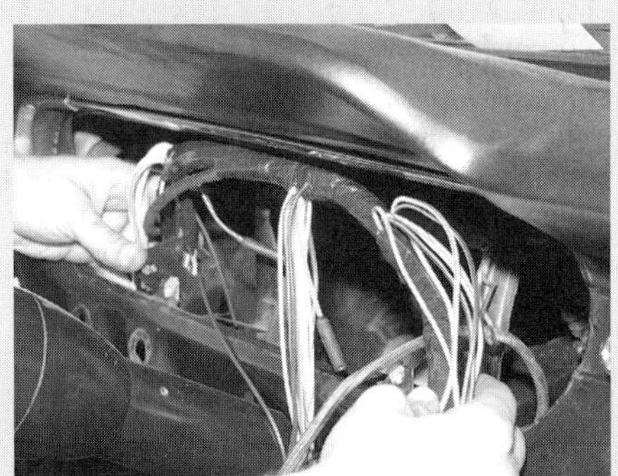

12. The main wiring loom is fed into the dashboard. Instrument cluster connections, naturally, are positioned at this opening. Get a grasp of what connections you are dealing with, and make sure they're where they belong.

13. The fuse box on '65–'66 Mustangs is located on the left-hand side above the kick panel. Dave takes extra care not to pinch any of the wires behind the fuse box. The fuse box's two hex-head sheetmetal screws can penetrate wire insulation, making an ugly short circuit.

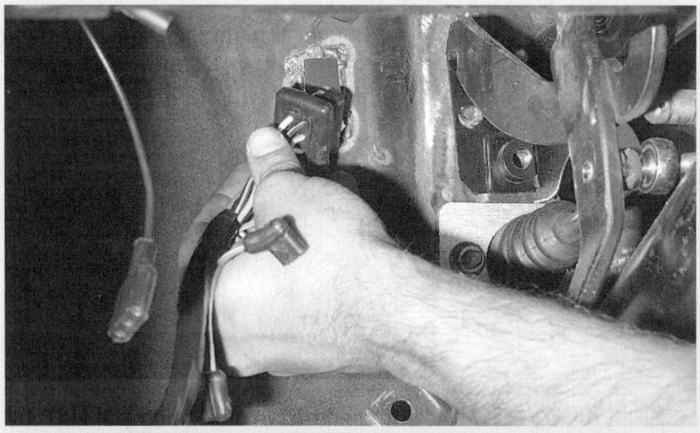

14. The headlight-harness feed begins at the main wiring loom behind the dashboard on the far left-hand side. Push the firewall plug through until the retainer clips pop through the firewall. This is the way the pins should look.

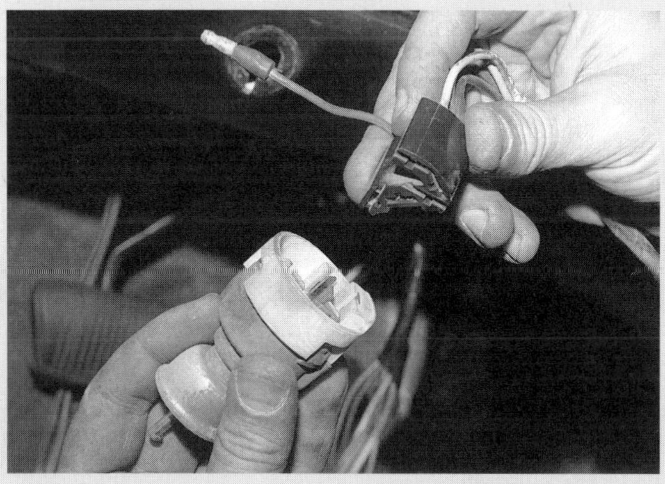

15. The windshield-wiper switch installs like this. JMC Motorsports uses a special tool to tighten the column. You can order this tool, and other switch collar tools, from JMC. Connect the leads and install the switch. The knob is retained with an Allen screw.

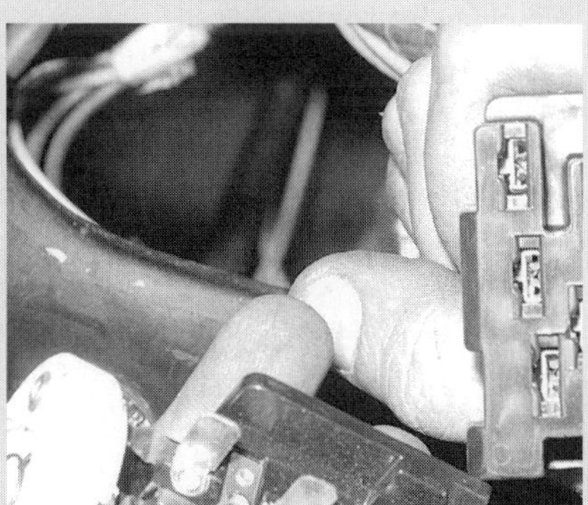

16. The headlight switch is next. A gang plug gets the job done here. Position the switch in place and tighten the collar with a common screwdriver or the special tool mentioned earlier. Install the headlight switch knob and shaft. Push the shaft all the way in and turn until the shaft seats and locks.

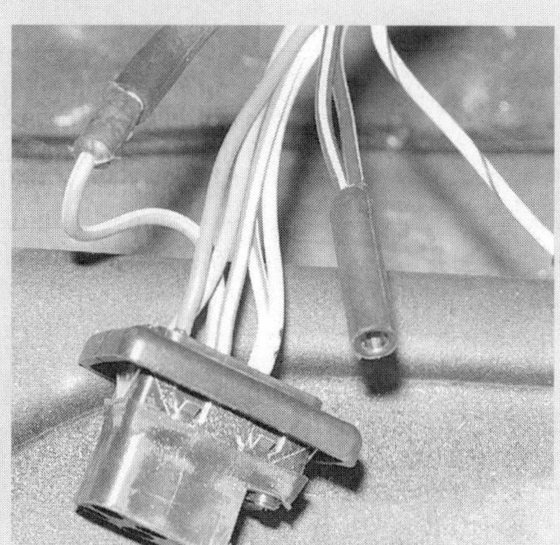

17. The engine gauge feed firewall plug installs mid-firewall. Behind the dashboard, there are pigtail leads for accessories and the automatic transmission indicator light. Push this plug through the firewall until the retainers click. It should look like this installed.

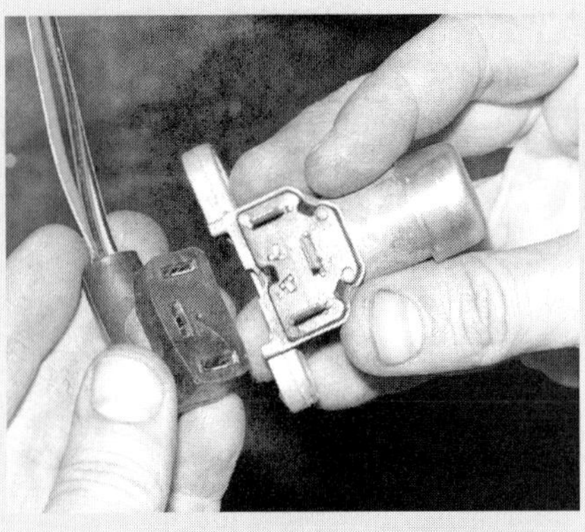

18. The headlamp dimmer switch plugs in like this. The plug should be on the left side of the switch when the switch is installed. Use the plastic switch protector between the switch and floorpan.

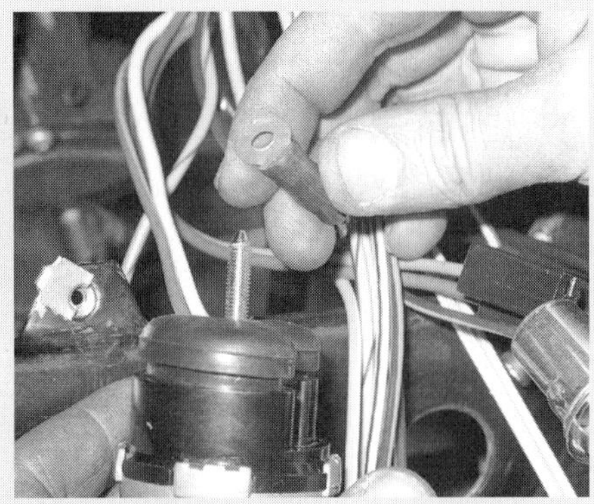

19. The ignition switch is an involved installation process. For '65–'66, there is a gang plug, then a main accessory lead or two from the center post, which is retained with a lock nut. You can add power accessories from this post. These power accessories must be protected with an inline fuse. We also suggest capping this post with an insulator to prevent short circuits.

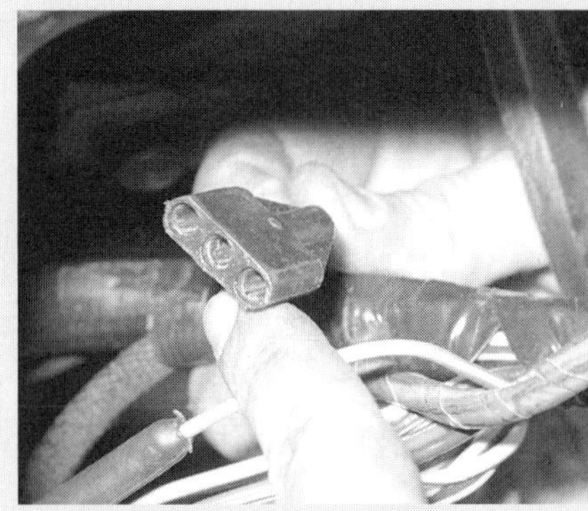

20. When you need power for accessories, lights, and the like, you don't have to cut and splice connections. Ford provided female bullet plugs throughout our Mustang wiring harnesses for instrument lights, sound systems, cruise control, and more. Use these plugs for light-load accessories only. Heavy loads, like high-power sound systems and power windows, need an auxiliary electrical system with its own protection. This is a three-hole plug for instrument lights, including the radio and Rally-Pac. There are other plugs for courtesy lights and accessories. Remember, do not cut the harness when you don't have to.

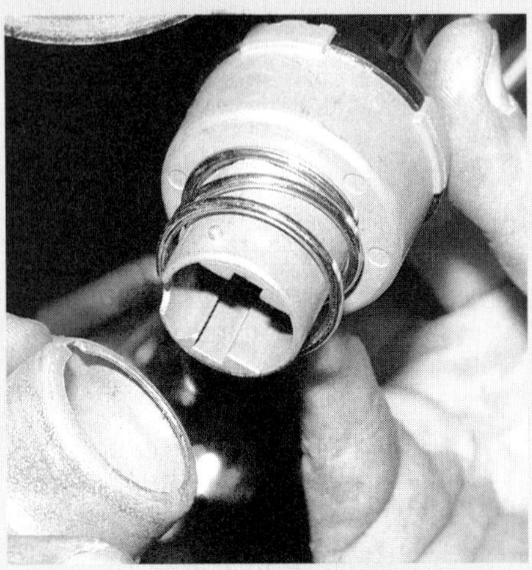

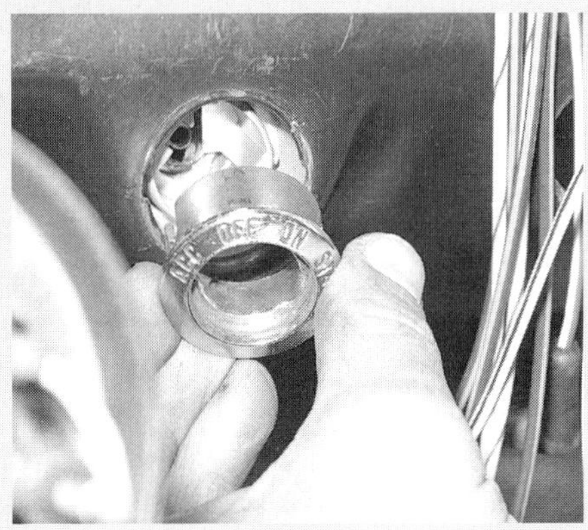

21. Put the ignition switch in place, then slip the escutcheon in place. Push in and turn against the switch, which will lock the switch in place.

22. While you're at it, replace the switches. Headlamp switches deteriorate with age, causing the circuit breaker to cycle your headlights off and on. Contacts in switches become pitted and useless with time. Corroded contacts create increased resistance, causing heat, and possibly starting a fire. It's also a good idea to replace all of the light bulbs, including headlamps.

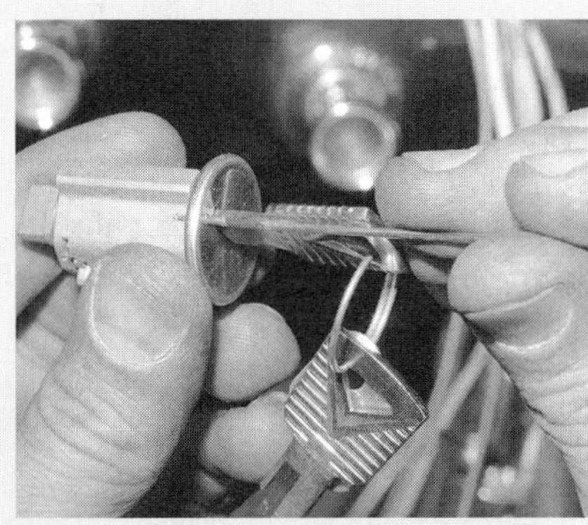

23. Installing the ignition lock cylinder is a project all by itself. Turn the switch to "ACC" with a screwdriver, then insert a paper clip wire into the cylinder to install the lock cylinder. The lock cylinder should click into place. You will probably have to try it a couple of times before the cylinder locks.

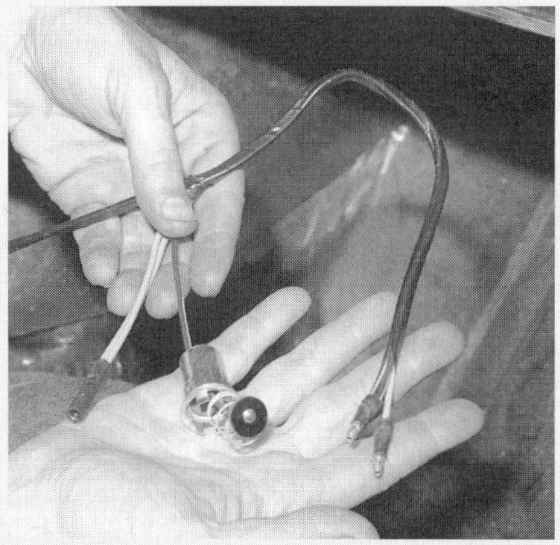

24. For courtesy lamps and switches, each door switch completes the circuit on the power side. Grounding occurs at the light.

25. The brake-light switch looks like this for non-'64 1/2 Mustangs and is located on the brake-pedal linkage at the master cylinder or power-booster rod. For '64 1/2 Mustangs, the brake-light switch is a pressure switch screwed into the master cylinder.

26. The engine-gauge feed and headlamp harnesses are connected to join the main wiring loom to these two harnesses.

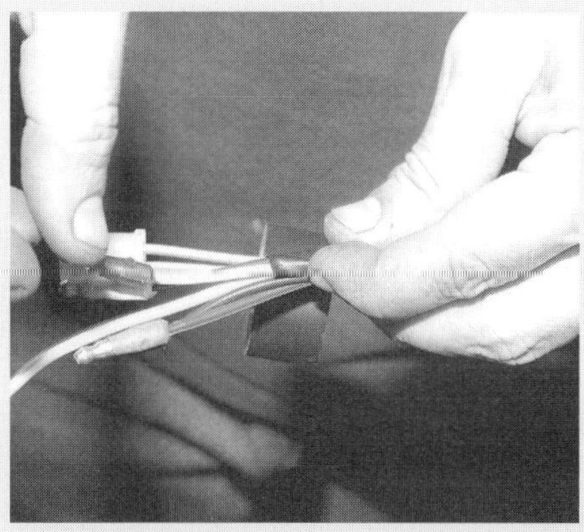
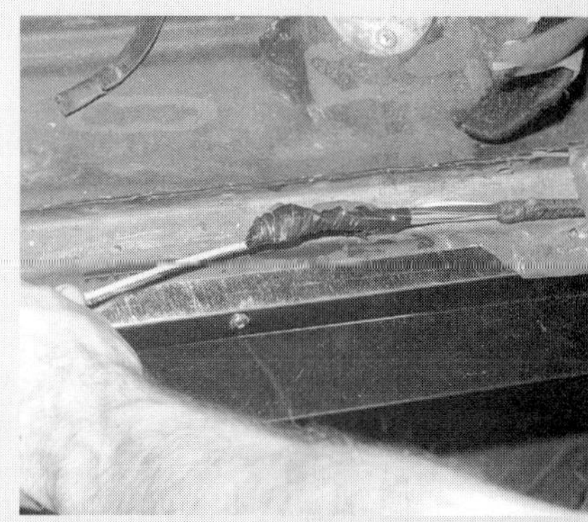

27. The taillight harness is installed by wrapping the small plugs together in a tight bundle, then tying a length of wire to the bundle with electrical tape. Feed the wire through the wheelhouse area, and grab the wire from the quarter-trim-panel area. The wire is fed to the rocker-panel area, then pulled through. The wire bundle comes out at the rocker panel.

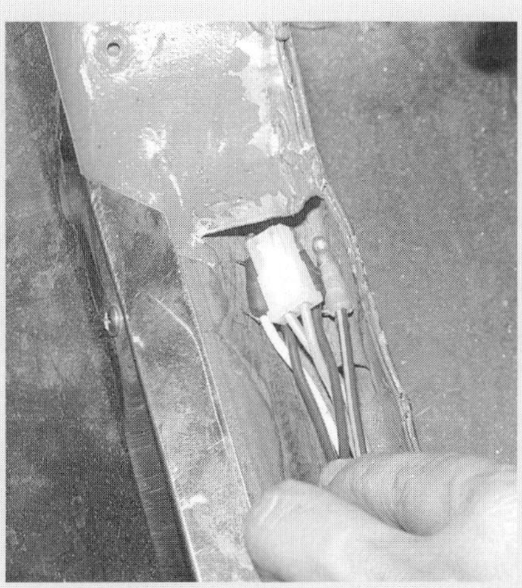
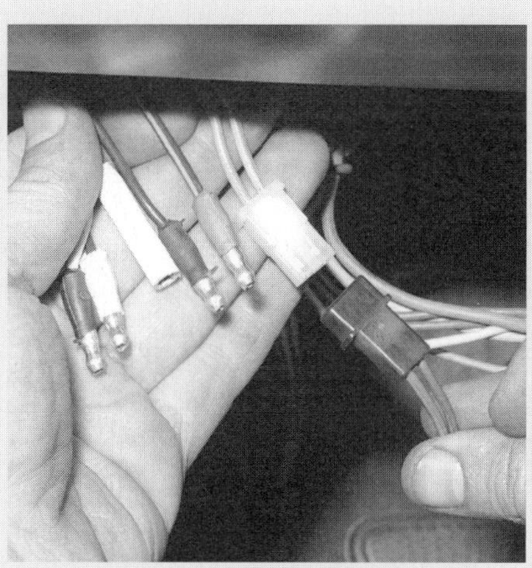

28. The taillight wiring bundle is inserted into the A-pillar as shown, where it comes out behind the instrument panel. Connect the taillight harness to the main wiring loom. All wires and plugs are color-coded.

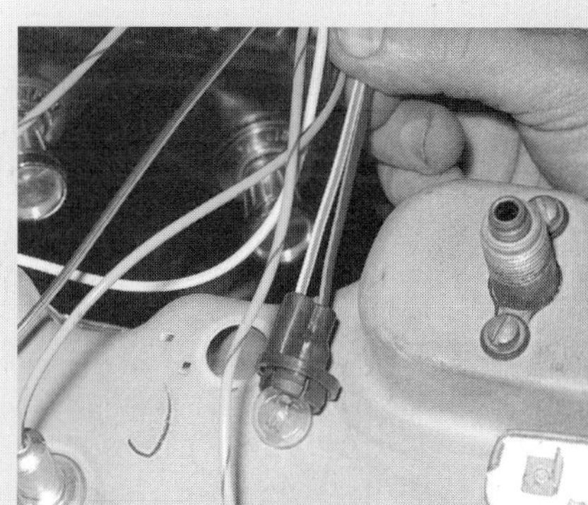

29. Let's wrap this one up at the instrument panel. For '65–'66, the instrument lamps and plugs are part of the main wiring loom. Install a new instrument voltage regulator (also called a voltage limiter) while you are there. Instrument-wiring color codes remain the same with Falcon-style or five-dial clusters. See your Ford shop manual for details.

Chapter 40

How to Install a Single-Wire Alternator

Text and Photos by Jim Smart

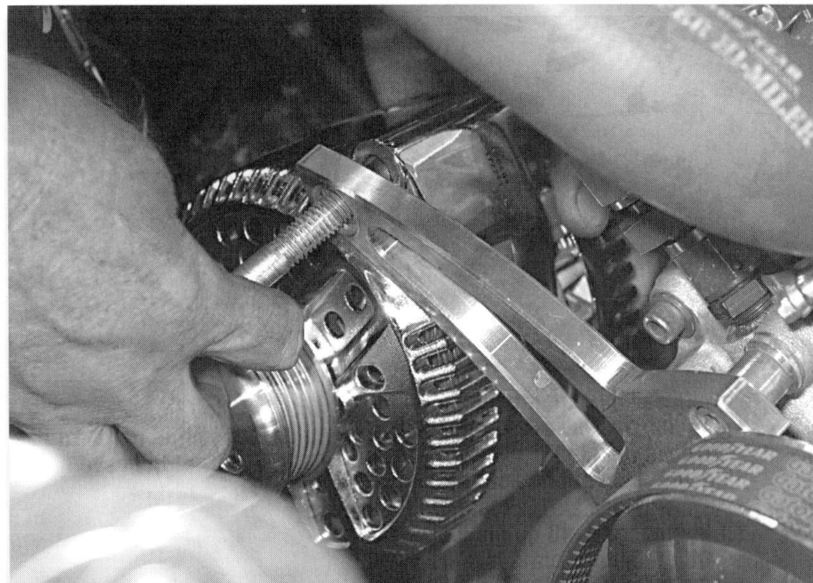

Powermaster Motorsports makes it possible to throw away your old-style alternator and voltage regulator.

Charging systems tend to baffle us because they involve the magic of electricity, a mysterious element we can't see happening unless contact is made or we feel it in an unpleasant way. Unpleasant can range anywhere from getting jolted by a spark-plug wire to checking a Christmas-light socket with your pinky (not recommended).

Classic Mustang charging systems get real unpleasant when you try to start the engine and nothing happens. There's that telltale sign of trouble when the starter solenoid clicks as it struggles to receive direct current from a weak battery. Even worse is dead silence.

First trickle-charge the battery to perform a load test. If the battery fails the test, you need a new one. Some auto parts stores will check the specific gravity of the electrolyte in each cell, which tells only part of the story. A load test reveals the rest.

If you have a healthy battery that keeps going dead or lighting that's relatively dim at night, the blame may lie with your Mustang's battery-charging system: the alternator (or generator for '64 1/2 only) or voltage regulator. Vintage Mustangs were fitted with 35-65-amp Autolite ('65–'71) or Motorcraft ('72-up) alternators with external voltage regulators mounted on the radiator support, inner fender, or other location. On '74–'78 Mustang IIs, they were near the battery tray in the firewall area.

Maintenance is the downside to those classic alternators with external regulators. When the charging system calls in sick, it could be the alternator or it could be the regulator. Most of the time, it's the regulator if you're running one of the old mechanical types. Solid-state regulators are more forgiving of the harsh elements in an engine compartment.

Source
Powermaster Motorsports
www.powermastermotorsports.com

Alternators quit most of the time due to diode failure or worn-out bearings. In any case, there's more to worry about with our old-fashioned alternator charging systems.

With Powermaster's single-wire alternator, it's easier to live with Mustang charging systems. As the name implies, there is one electrical cable between the positive side of the battery and alternator. That's it. The single-wire alternator has a built-in solid-state voltage regulator that's easy to replace if it fails. Brushes inside the alternator can be replaced when you replace the voltage regulator. That's how simple these late-model, high-output alternators are.

High-output means 115 amps at idle and 165 at 6000 rpm. This enables you to run all kinds of accessories, such as a sound system, air conditioning, additional lighting, powerful headlamps, auxiliary driving lights, and more.

With the single-wire alternator, you can remove the external voltage regulator or leave it in place to maintain a stock-looking appearance. When you disconnect the original alternator, you eliminate the voltage-regulator part of the old charging system.

The cable supplied by Powermaster is cut to proper length with a new eyelet and installed on the alternator. This cable goes to the battery side of the starter solenoid.

1. This is the Powermaster 115/165-amp high-output alternator for vintage and late-model Mustangs. When it arrives on your doorstep, it's fitted with a serpentine belt pulley. However, the package includes a V-belt pulley for older Mustangs. Absolutely never use a hammer to remove the pulley because bearing damage will result. Use a pulley puller and be careful. Tolerances inside these high-amp alternators are tight. If you're careless, you can damage the armature inside.

2. This is the solid-state internal voltage regulator, which is easily replaced and serviced. It also contains the brushes, which are easy to replace.

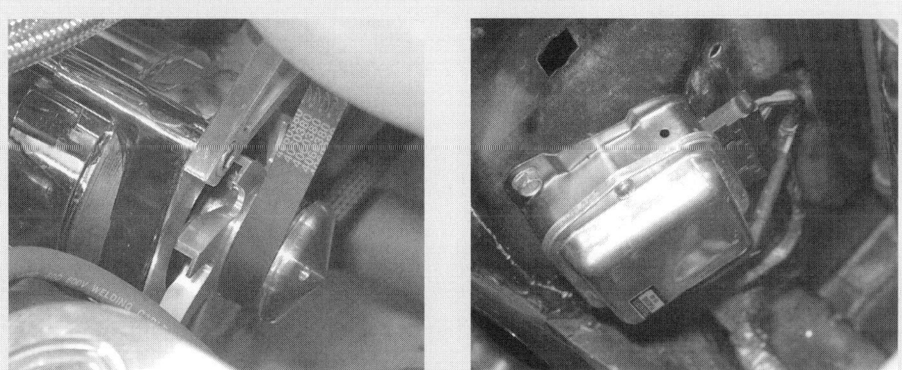

3. This is a classic Mustang's original charging system (left): a 35-65-amp alternator with an external voltage regulator. At right is a solid-state voltage regulator. The Powermaster single-wire alternator eliminates this.

4. We're working with Don Vincent of Southern California, who owns this '68 Mustang hardtop. Don opted for a March serpentine-belt drive when he built his 450-horse 302ci small-block. That makes this conversion different from a stock V-belt application.

5. Don begins removal of the old Motorcraft alternator by first disconnecting the battery, then loosening the adjustment down under at the 6-o'clock position. Don uses an Allen wrench on this March pulley setup. Remove the adjustment bolt completely.

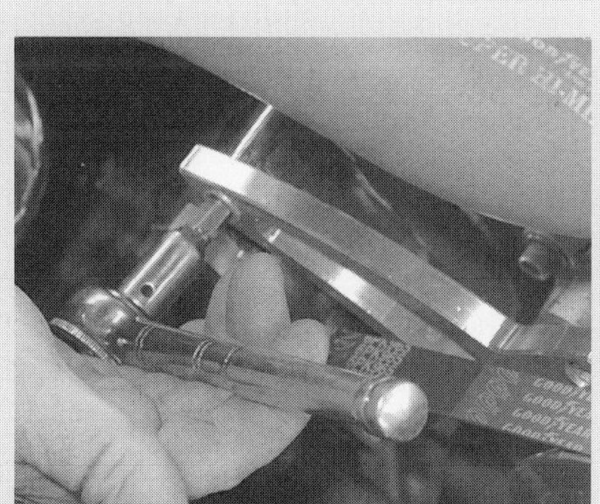

6. Next is the top pivot bolt, which is removed with an Allen wrench in this case. You'll need a 5⁄8-inch socket for your Mustang stocker.

7. Remove the V-belt in a stock application. Here, Don removes his serpentine belt.

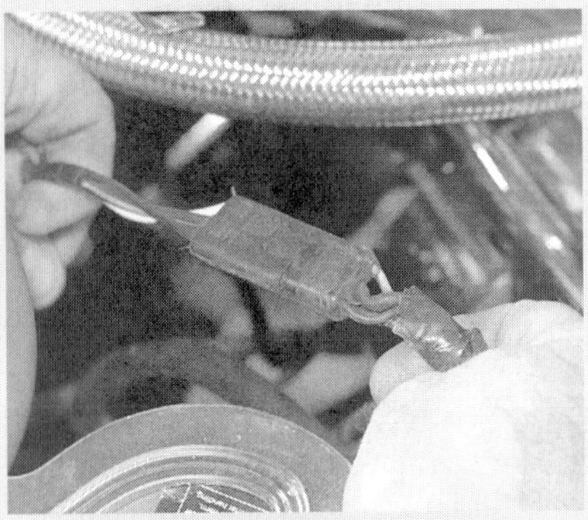

8. The alternator/voltage regulator harness is disconnected next.

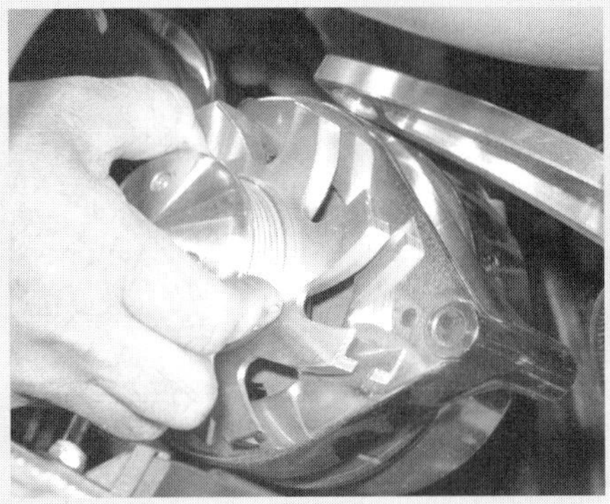

9. The pivot bolt is removed next, then the alternator.

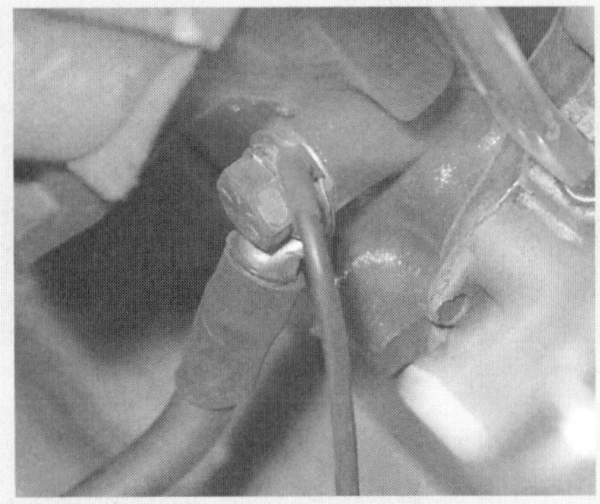

10. Don't forget to disconnect the alternator's ground at the engine block on the righthand side. Reinstall the battery negative/ground cable and bolt.

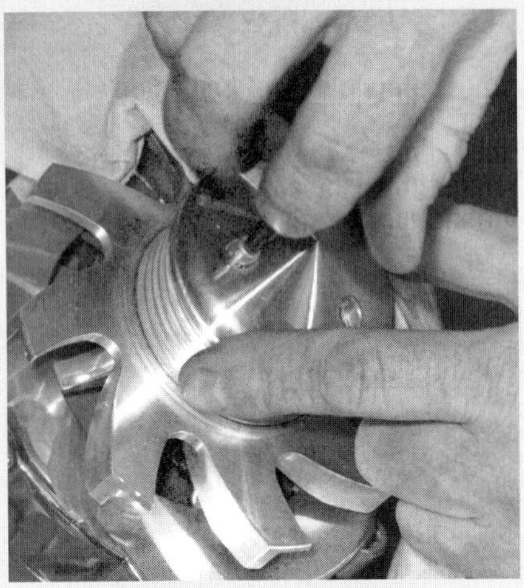

11. Don has to move the March pulley from the Motorcraft alternator to the Powermaster high-amp alternator. The cap is removed along with the pulley.

12. Side by side, the Powermaster (left) and the Motorcraft (right) are similar in size, but not in function and weight. The Powermaster is much heavier due to its abundant copper winding inside. The Motorcraft's March pulley will be moved over to the Powermaster.

13. Don removes the March pulley from the Motorcraft alternator with an air-impact gun.

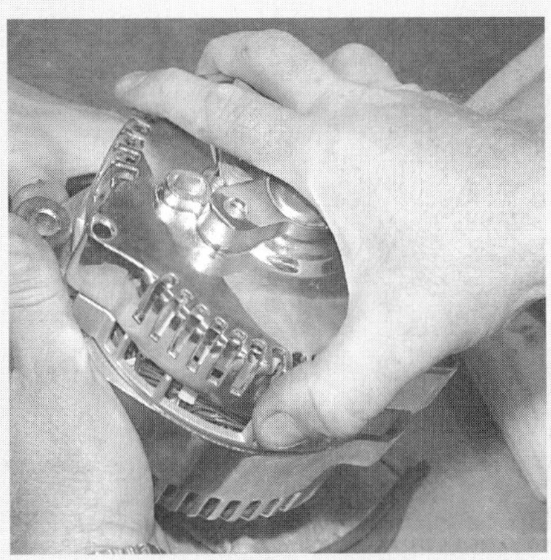

14. As it comes out of the box, the Powermaster locates the voltage regulator and cable right at the cylinder head of a small-block Ford. Three bolts tie the two halves of the Powermaster together. Simply remove the bolts (with the pulley off) and rotate the front of the case so the connection is away from the cylinder head.

15. The cable supplied by Powermaster is cut to proper length with a new eyelet and installed on the alternator. This cable goes to the battery side of the starter solenoid.

16. The March serpentine pulley is installed on the Powermaster alternator. No spacer is required behind the pulley.

17. Don hangs the Powermaster on the same pivot bolt used for the smaller, lighter Motorcraft. It's a perfect fit.

18. This adjustment turnbuckle from March Performance is a nice way to tighten the drive belt.

19. The last thing we have to do is get power to the alternator from the battery side of the starter solenoid. There are no other wires—just this one.

20. This is the Powermaster alternator installed in a classic Mustang. It's a good-looking piece that gives us abundance of amperage at idle. It's also more reliable than the original charging system.

Chapter 41

How to Install GT Foglights

Text and Photos by Donald Farr

A new kit from Virginia Classic Mustang makes it easy to convert from the standard grille to GT foglights.

On April 17, 1965, exactly one year after the Mustang's public introduction, Ford celebrated the first anniversary by introducing a pair of major option packages. One was the Interior Décor Group, better known today as the Pony Interior. The other was the GT Equipment Group, which added foglights, side stripes, trumpet exhaust tips, fast-ratio steering, handling suspension, front disc brakes, and five-dial instrumentation (also part of the Interior Décor Group). In particular, the grille-mounted foglights established a tradition for Mustang GTs: All subsequent Mustangs GTs except '69 have incorporated foglamps, mounted in the grille for '66–'68 and in the lower front valance from '82–'04. For '05, the GT's foglamps returned to a prominent position in the grille.

Adding foglights to non-GT Mustangs has been a popular practice since '65, when even Ford encouraged the GT modification by offering accessory foglight kits through Ford dealerships. Now Virginia Classic Mustang has made it easy to update standard '65–'68 Mustangs with its new Foglight Conversion Kits. The kits include everything needed to install factory-style GT foglights, including lamps, wiring, a switch, and the lightbars for '65–'67 models. For example, '65 kits use the car's existing running-horse corral, and '68s don't require a corral or bars at all.

Sources

Classic Creations of Central Florida
www.classiccreationsfl.com

Virginia Classic Mustang
www.virginiaclassicmustang.com

We followed along with the foglight kit installation on a '66 Mustang at Classic Creations of Central Florida. It's a basic bolt-on other than drilling holes in the radiator support for the wiring and in the instrument panel for the switch. However, you can avoid drilling permanent holes by routing the wiring over or around the radiator support and mounting the switch further under the instrument panel. The foglight wiring connects into the factory headlight-switch harness so the taillights illuminate with the foglights. It's helpful to have a factory wiring diagram handy for making the proper connections.

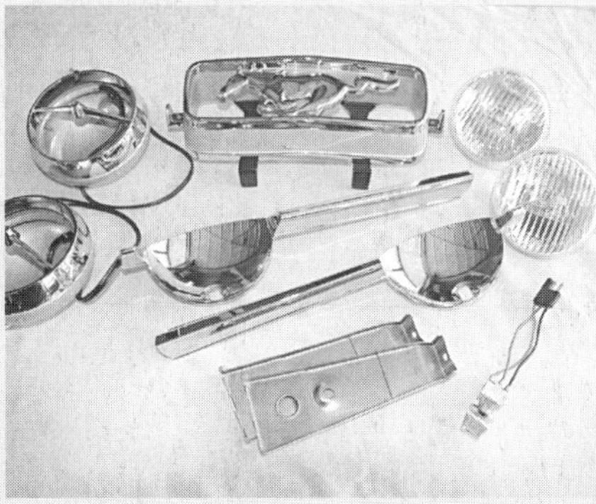

1. The main components in Virginia Classic Mustang's Foglight Conversion Kit for '66 Mustangs are the running-horse corral ('65 Mustangs use the existing corral), foglamps, lamp housings, bars, grille brackets, and the switch.

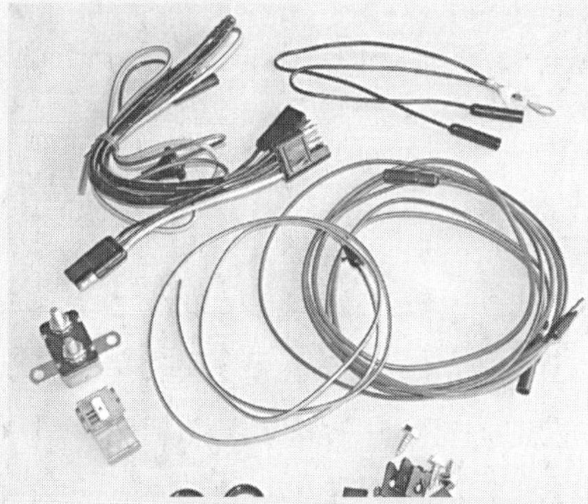

2. Wiring is also included with the kit, which connects the foglights to the underdash switch. Also included are the circuit breaker, the radiator support grommets, and attaching screws.

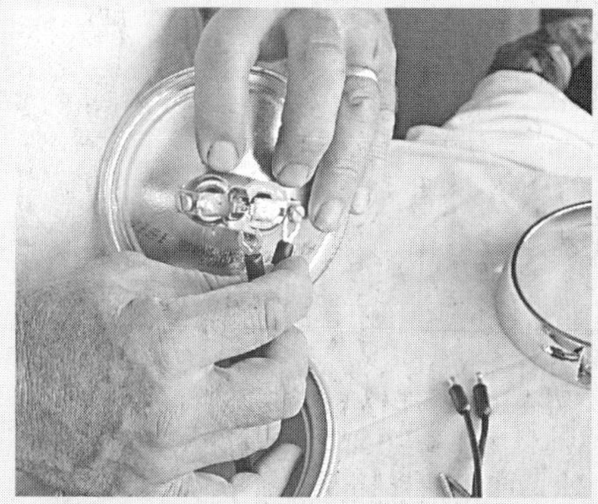

3. Start the conversion by assembling the corral, bars, and lights. First, separate the lamp housings (two screws on each), route the lamp wiring through the rear housings, connect the wiring to the lamps, and secure with the screws.

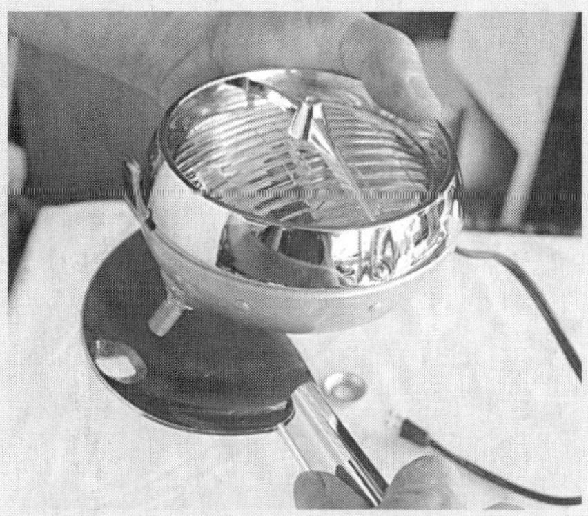

4. Reassemble the foglight housings with the lamps installed, and assemble them to the bars using the studs on the rear housings.

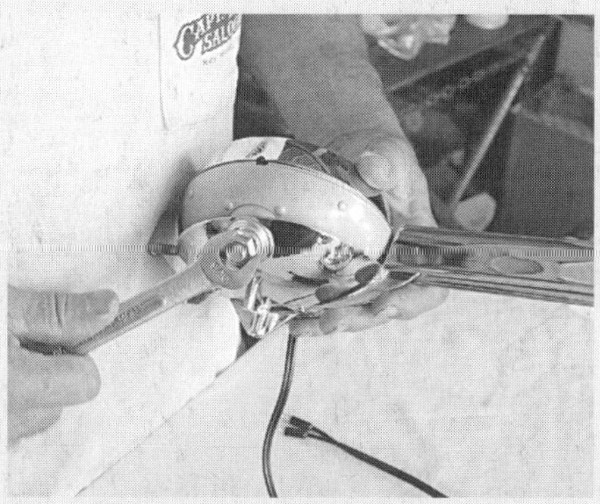

5. Secure the bars to the housings with the nuts. Be sure to use the special washer between the housing and nut.

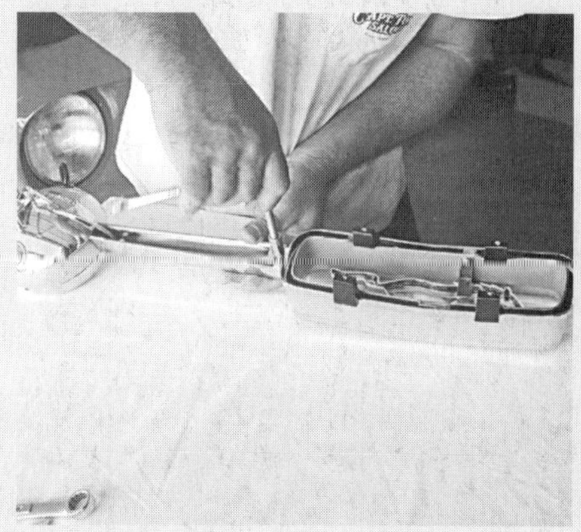

6. Attach the assembled bars to the running-horse corral. A new corral is supplied with the kits for '66 Mustangs; the existing corral is used for '65 Mustangs.

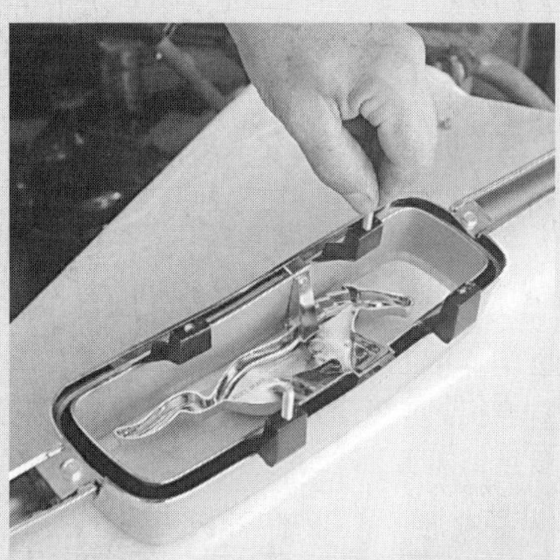

7. Install the mounting studs, four in the corral and one in each of the lightbars.

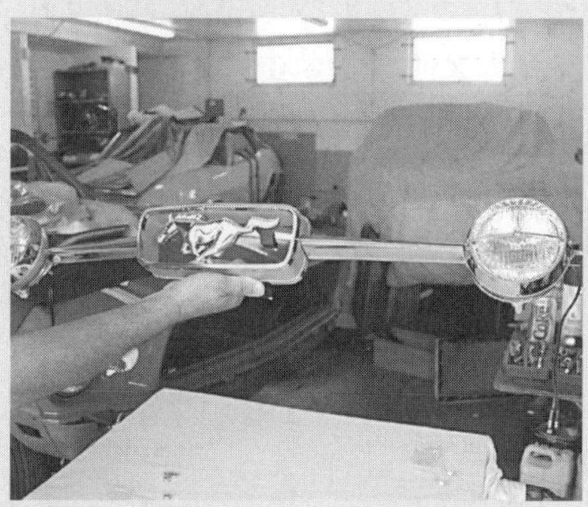

8. Merv Rego from Classic Creations of Central Florida shows the assembled foglights, which are ready to mount on the '66 grille.

9. Unbolt the grille and remove it from the car.

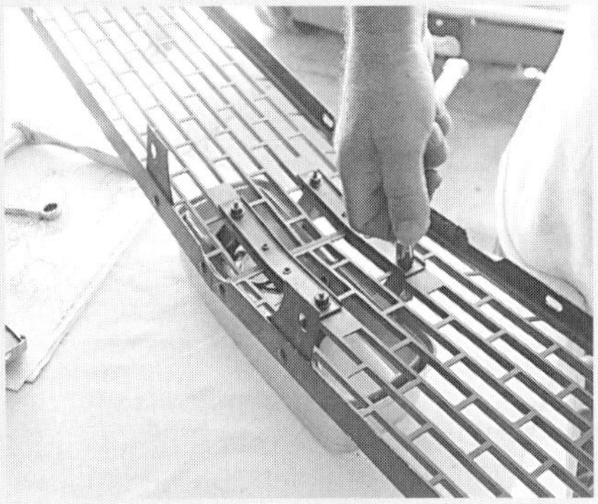

10. With the grille on the workbench, remove the four nuts that retain the non-GT running-horse corral and remove the corral. On '66 Mustangs, the foglight corral is different from the standard corral due to the ears for mounting the lightbars.

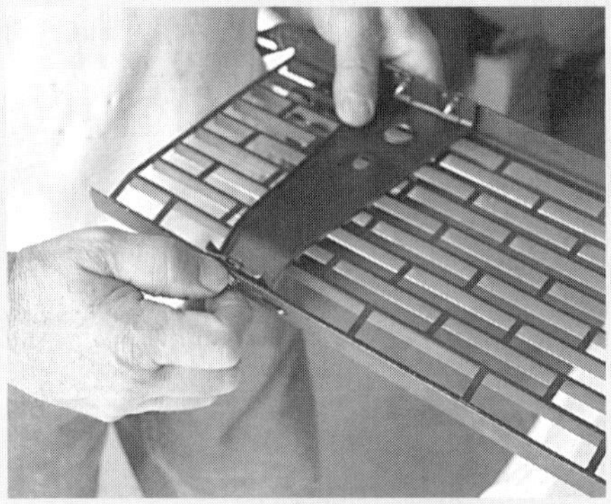

11. Install the foglight brackets at each end of the grille using the supplied screws. Note that the holes for the '66 brackets are already in the grille edges, two at the top and one at the bottom. Don't tighten the screws at this point because the brackets will likely require adjusting.

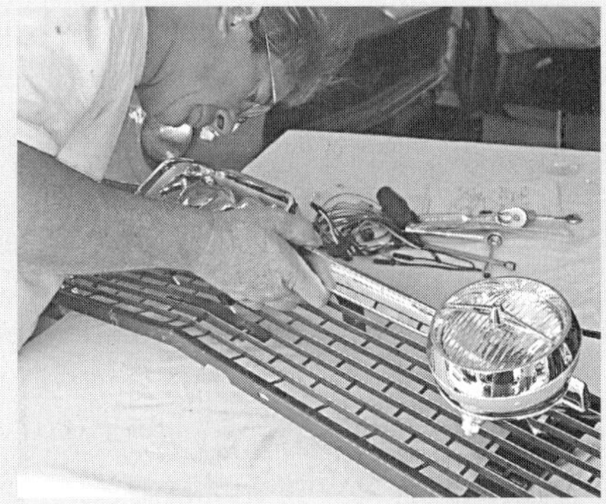

12. Route the foglight wiring through the large opening in each bracket, then mount the foglight assembly to the grille using the four studs in the corral and one stud on each lightbar. Now you can tighten the bracket screws.

13. Reinstall the grille with the foglight assembly attached.

14. In factory '65–'66 GTs, a pair of 3⁄4-inch holes were cut in the radiator support for the foglight wiring to pass through. Most radiator supports have dimples to locate the positions of the openings. You can use a hole-saw or step-type drill bit to make the holes. A plastic grommet is supplied to prevent chafing the wiring.

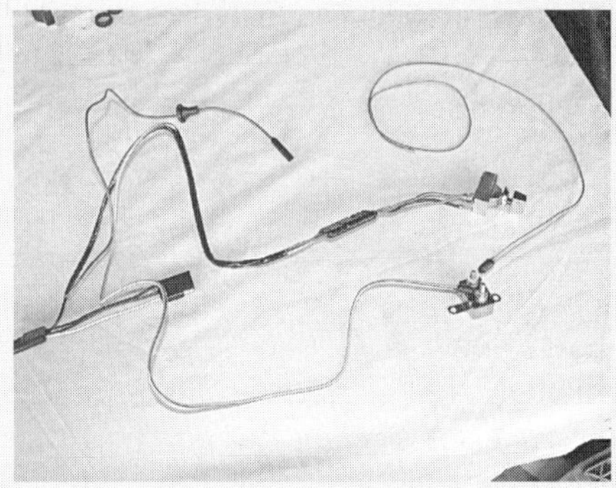

15. The wiring for the Virginia Classic Mustang foglight kit is identical to OEM. Per the factory installation, the harness connects into the headlight-switch harness so the rear taillights illuminate when the foglights are switched on. The circuit breaker mounts on the firewall.

16. Route the wiring along the radiator support and driver-side engine compartment.

17. On factory GTs, the foglight switch was located on the extreme left side of the instrument panel, near the courtesy-light switch in the door jamb. However, on a non-factory–type installation, the switch can be located anywhere, including underneath the instrument panel.

With the new reproduction Rally-Pac from CJ Pony Parts, you can add the factory's sporty tach and clock combo to your '65-'66 Mustang.

Chapter 42

How to Install a Rally-Pac

Text and Photos by Mark Houlahan

There's always one last item that makes a project perfect. The handblown glass Christmas-tree topper, the fancy ribbon around a gift box, or even the cherry on top of a sundae are examples that come to mind. For *Mustang Monthly's* Project '66, the one item we wanted to top off our restoration was a Rally-Pac. The Rally-Pac Ford offered in the '60s became a much-coveted option through the years, and having one in your restored Mustang was icing on the cake. Though our project car has been completed and on the road for more than a year, we've added a few items as they became available, such as the newly reproduced console trim that allowed us to finally install a console. Now we can do the same with a Rally-Pac.

Until recently, the problem has been finding a decent one in restorable condition, sort of like finding someone under 18 at a Barry Manilow concert. But things have begun to shape up for us Mustang enthusiasts (although I can't say the same for Barry). The highly sought-after Rally-Pac is now available as an accurate reproduction for '64 1/2 to '66 Mustangs (both low- and high-profile models) in 6000- and 8000-rpm tachometer versions. A brand-new exterior-case casting (wrinkle-finished like the original) is the foundation for all new internals as well. An accurate electronic air-core-movement tachometer and a quartz clock assembly with all new lenses, chrome trim, attaching hardware, and wiring are part of the kit. The tachometer and clock faces are silk-screened like the originals, though they are now backlit for easier viewing at night.

Installation isn't difficult (just some minor wiring and turning a screwdriver or ratchet gets the job done) and you can add the classy look of a Rally-Pac to your Mustang in about an hour. It comes only in black or white, so if you want to color-match the Rally-Pac to your interior (only correct for the low-profile model in cars with the Interior Décor option), we suggest picking the white model and buying a can of interior spray paint so you can color-match it yourself. Our Interior Décor color is black, so we didn't need to refinish the exterior housing. If you decide to color match yours, use light coats of paint to prevent losing the wrinkle-finish look.

We picked up our '66 Rally-Pac from CJ Pony Parts (PN RP5, $434.95). CJ also carries the '65 high profile-model.

Sources

CJ Pony Parts
www.cjponyparts.com

Scott Drake Mustang Parts
www.scottdrake.net

1. The all-new '66 low-profile reproduction Rally-Pac is made by Scott Drake and is a high-quality reproduction. The cast housing is identical to the original and the Nordskog-built gauges within the housing look exactly like they should. We heard that one of the toughest parts of the reproduction process was building a clock that had the "ticking" second hand like the original, but Scott Drake did.

2. Ford used the interior illumination circuit to power the Rally-Pac's clock. Remove the driver-side door-sill scuff plate and driver-side kick panel to access the cowl side panel to route your wiring to the door-jamb switch.

3. The rest of the wiring under the dash attaches to the dash-light circuit, a good ground, and the stud on the back of the ignition switch. It's possible to find these connections lying under the dash, but at our age the thought of removing six Phillips head screws and setting the gauge cluster aside sounded easier on our backs. A small battery-operated screwdriver, like this Craftsman 3.6-volt Brite Driver, helps locate the screws you need to remove and replace.

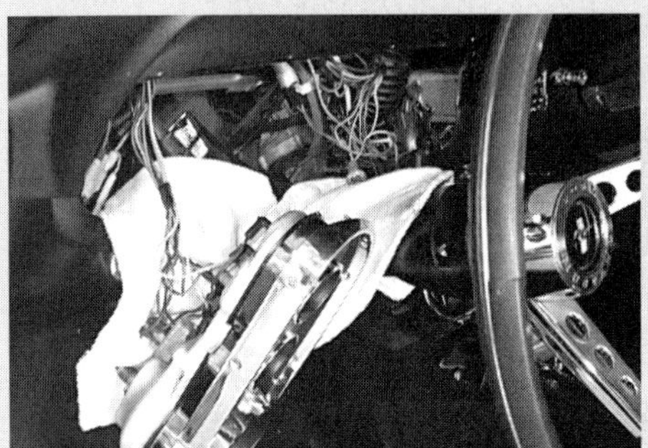

4. You can probably snake your hand behind the gauge cluster to make the connections, but if you carefully label the wiring going to the back of the cluster, it's easier to disconnect the wiring from one side of the cluster and let it hang out of the way. You'll have to unscrew the speedometer cable to do this. Notice the towel over the steering column to protect the paint.

5. With room to work, start with the underdash harness installation. The first part of the harness deals with the clock power and connects in series with the doorjamb switch. Make sure you've disconnected the battery before unplugging the doorjamb switch. The green wire with the green end is the original wire to which you will connect the clock feed.

6. The remaining underdash wiring connects near the ignition switch, so carefully route the harness in that direction. The blue wire from the Rally-Pac underdash harness will connect to the dash illumination circuit. This is a red plug with three female bullet terminals by the ignition switch. Connect the blue wire to one of the three available spots in the connector.

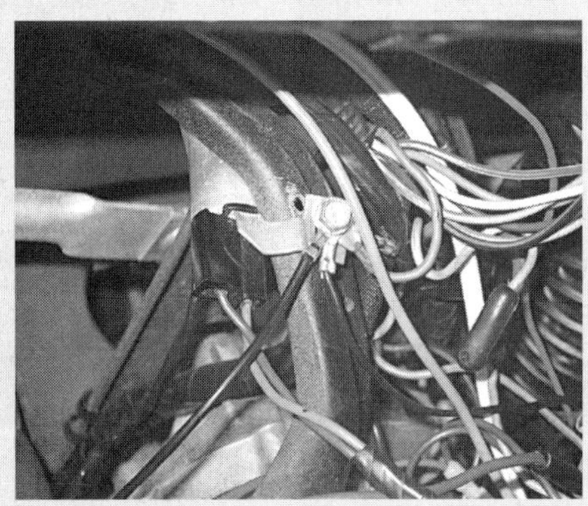

7. The black ground wire (with the small ring terminal soldered to it) is the main ground lead for the Rally-Pac and needs to be connected to a clean, secure ground path. There are plenty of dash screws you can use, but we simply added it to the several other ground wires found at the right dash brace. The red wire with the green connector is not used in the V-8 application.

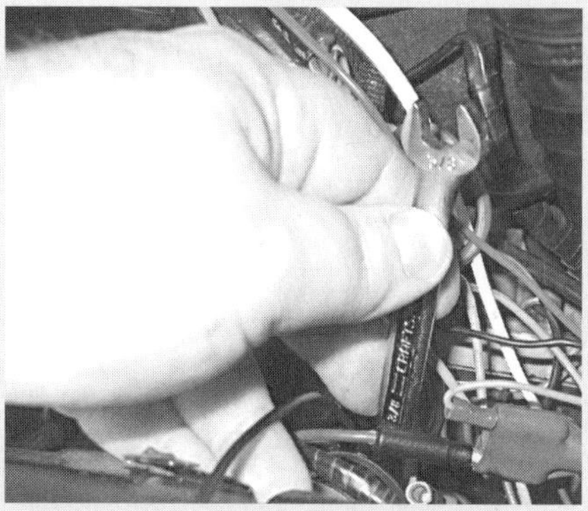

8. Two wires come with the main Rally-Pac harness that you can disconnect and install separately of the Rally-Pac. One is the power feed from the ignition switch and the second is the tachometer signal wire. The short black wire connects to the stud on the back of the ignition switch. Temporarily remove the retaining nut, then slide the ring terminal of the short black wire over the stud and re-secure the nut. Now you can reinstall the gauge cluster and kick-panel trim.

9. The long black wire is the tachometer feed and needs to pass through the firewall. The instructions suggest the accelerator-pedal-rod seal, but we found enough gap in the A/C-hose grommet to pass the wire through at this point. You can also drill a small, discreet hole to pass the wire through.

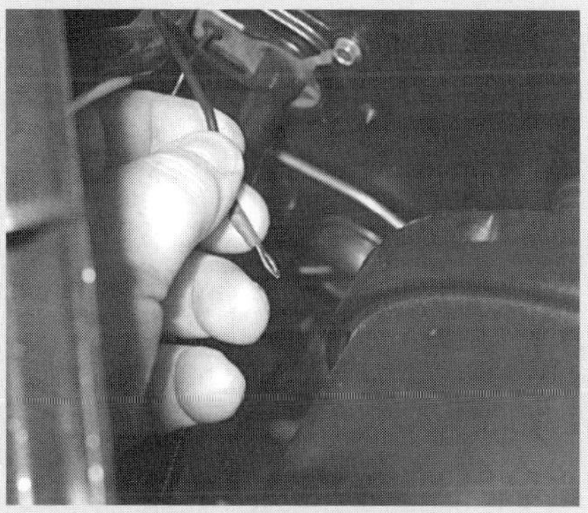

10. Once you've gotten the wire through the firewall, it has to be routed to the ignition coil. We routed the wire along the intake at the valve-cover edge right along with the A/C clutch wire. The wire is black, so it conceals easily. If you want to take it a step further, cut the electrical tape off the gauge-feed harness on the engine and rewrap it with the tachometer wire included.

11. Once you reach the coil, the ring terminal is slipped over the negative-side coil terminal, marked "DIST" on the yellow top coil shown here. Aftermarket coils have a "–" that indicates the negative side of the coil. We used a small nut to retain the wire, then reinstalled the distributor wire over it.

12. As you can see, with a bit of housekeeping the tachometer wire can be hidden. This wire is the only deviation from the original design (that picked up its tach signal in line with the ignition switch under the dash).

13. With the air-filter assembly back on, the wire looks totally hidden. Unless someone with knowledge of a Mustang's ignition circuit looks carefully, no one will know there's one extra black wire under the hood.

14. Back inside the car, route the underdash harness, tach signal wire, and ignition feed wire to a common location at the base of the steering column.

15. It's time to mount the Rally-Pac to the steering-column tube. Carefully place the Rally-Pac on the steering column and bend the wire harness to mimic the L shape on the driver side of the Rally-Pac mounting bracket. Start the retaining bolts for the Rally-Pac bracket on this side first, ensuring the wire harness is within the L of the bracket.

16. Once the Rally-Pac is secured to the steering column, the wire-harness guide channel can be installed. One end of the channel simply slides into place behind the L of the Rally-Pac mounting bracket. The opposite end has to be secured to the steering column with the supplied Phillips head screw. Depending upon the location of the bracket, you might have to lower the steering column to get a drill bit into the right place. If you're uncomfortable drilling into a curved surface, you can always use double-sided foam tape to secure the end of the wire channel.

17. Finally, connect the Rally-Pac harness to the underdash harness, tach feed wire, and ignition switch wire per the instructions. Everything is color-coded, green to green, and so on, but there is one illumination wire that has a black end instead of a red one. Double-check the actual wire color, not the colored end, and you'll be fine.

18. The completed installation looks like the car came with it 39 years ago! The last things to do are fire up the car and check the operation of the tach, set the clock, and check the gauge illumination. For those who have never had a Rally-Pac, the white illumination is correct on the low-profile models.

19. But if it bothers you (it does us a little), you can purchase green bulb covers from any vendor that carries aftermarket gauges. Simply remove the two screws at the rear of the Rally-Pac for each gauge, slide out the gauge, pop out the bulb and socket assembly from the rear of the gauge, then slide a bulb cover over the bulb. You can also use a green permanent marker to color the bulbs (shown here) if you can't find the correct-size bulb covers for the low-profile '66 model.

20. For owners of '64 1/2 Mustangs and '65 models with the standard (non-GT or Décor Interior Group) instrument panel, CJ Pony Parts also offers the early, or low-profile, version of the Rally-Pac.

INTERIOR

Chapter 43

How to Replace a '71–'73 Dashpad

Text and Photos by Jim Smart

Just Dashes' new '71–'73 dashpad will help get your "big" Mustang interior restoration on the right track.

Pity the enthusiast who is into '71–'73 Mustangs. You want to curse the '65–'66 Mustang crowd, which has everything it's ever wanted for a full-scale restoration. The '65s and '66s were the first to get interior restoration components such as seat upholstery, carpet, instrument bezels, and dashpads. It has taken many years for '71–'73 Mustang restorers to get what they've wanted and needed for a factory original restoration. These folks have had to make do with what they could find.

You can call it the food-chain theory of manufacturing and marketing. Mustang-parts reproducers first go where the numbers are because it gives them the financial horsepower to develop products for the minorities later. This is true with any type of product, be it automotive restoration parts, clothing, furniture, or refrigerators. The most popular stuff gets covered first.

The '71–'73 Mustang dashpad has been a long time coming, and it's now available from Just Dashes, which can take virtually any upholstered dashpad, console, or door panel and make it like new again. The company's '71–'73 Mustang dashpad isn't a reskinned piece; it's brand-new, ready for quick and easy installation in your Mustang. Our friend, Marlon Mitchell of Fly-Ford Racing, helped with our '71 Mustang dashpad installation. You'll find that '71–'73 Mustang dashpad replacement is easier than '65–'70 because it can be done in an hour.

Sources

Fly-Ford Racing
www.fly-ford.com

Just Dashes
www.justdashes.com

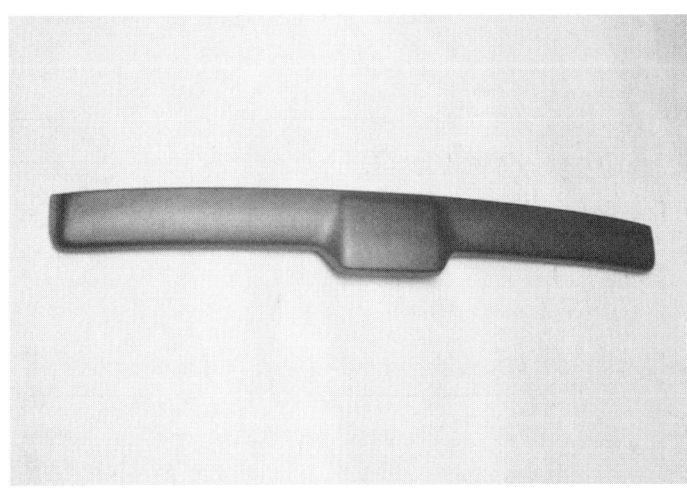
1. This is the '71–'73 Mustang dashpad from Just Dashes. We like this quality piece because it's identical to original equipment in every way.

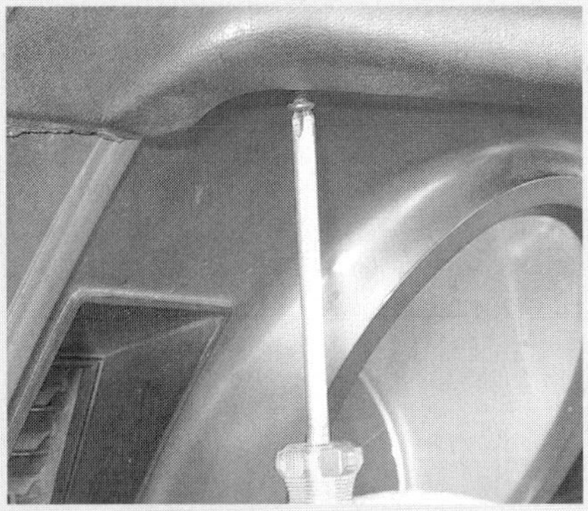

2. The top of the '71–'73 dashboard can be removed, which makes dashpad removal easy. Begin removal with six screws across the front of the top of the dashboard. There are three screws above the instrument panel.

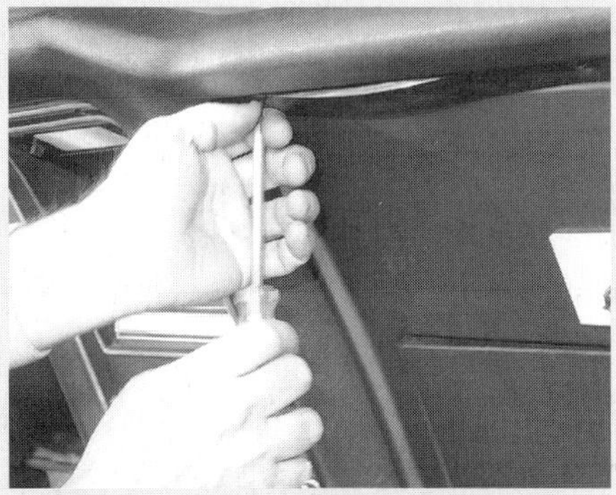

3. There are three more screws to be removed across the top of the passenger side. These screws anchor the top of the dashboard.

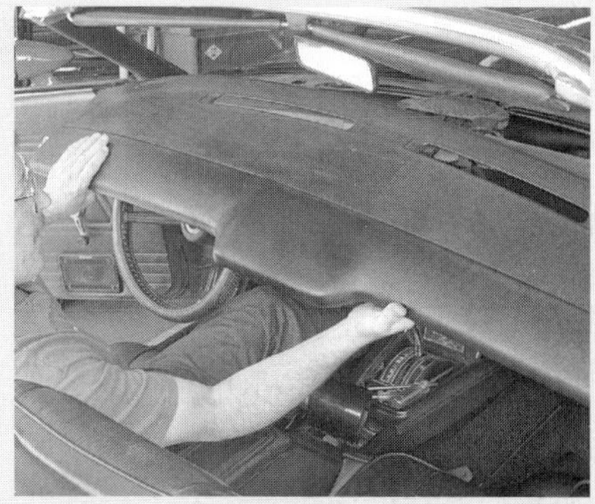

4. Marlon Mitchell of Fly-Ford Racing lifts the top of the dashboard and pulls out the assembly. Marlon loves to build Mustangs. He's also the most trusted front-end alignment guy we know in Southern California.

5. The dashpad is anchored to the dashboard crown with five tinnerman nuts. Remove the nuts with an 11⁄32-inch socket.

6. Separate the dashpad and crown as shown. Install the new dashpad the same way.

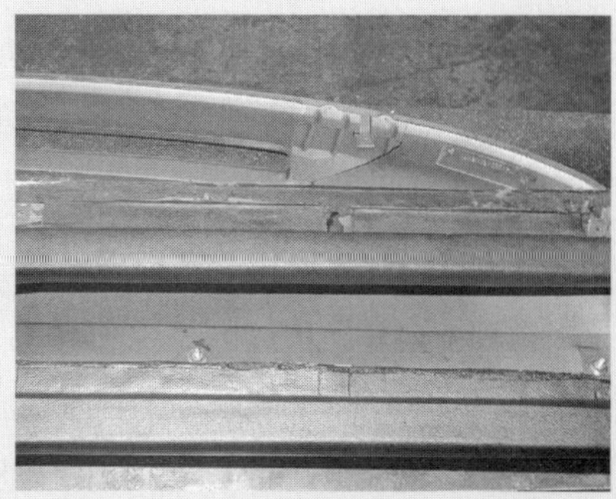

7. Each side of the dashpad has two locations for nut clips. These locations have to be drilled with a 3⁄8-inch bit to make way for the nut clip. In the middle of each side, material must be removed to make way for the nut clips attached to the dashboard.

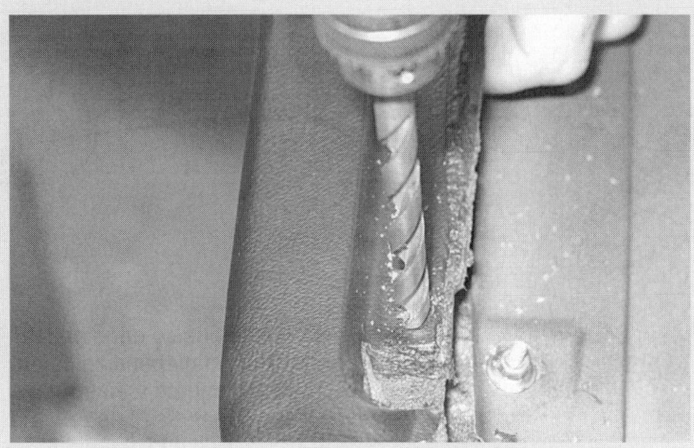

8. Two locations on each side of the dashpad get pilot holes first, then are drilled with a 3⁄8-inch bit as shown. This is for the nut clips. The 3⁄8-inch hole provides room for adjustment.

9. Nut clips are transferred from the old dashpad to the Just Dashes dashpad.

10. In the middle of each side, material must be removed to make way for the dashboard nut clips. Look at your Mustang's original dashpad, and you'll see where material needs to be removed. You can use a jigsaw; Marlon uses a nibbler from The Eastwood Company. This little guy buzzes its way through material efficiently.

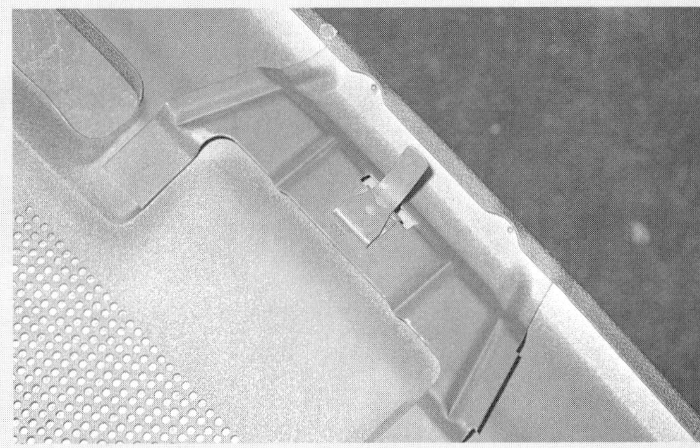

11. The dashboard crown is retained with screws and three retainer clips that slip into three square openings in the top of the dashboard. Be careful with these clips. They can break, leaving you looking for a parts car.

12. The new dashpad and crown are installed with ease by simply slipping the clips into the square openings and pushing the assembly in place. Remember, your Mustang's vehicle identification number (VIN) is riveted to this assembly. If you replace the crown, check your state motor-vehicle laws for VIN removal and transfer. In most cases, it must be witnessed by a law enforcement officer or your motor vehicle administration.

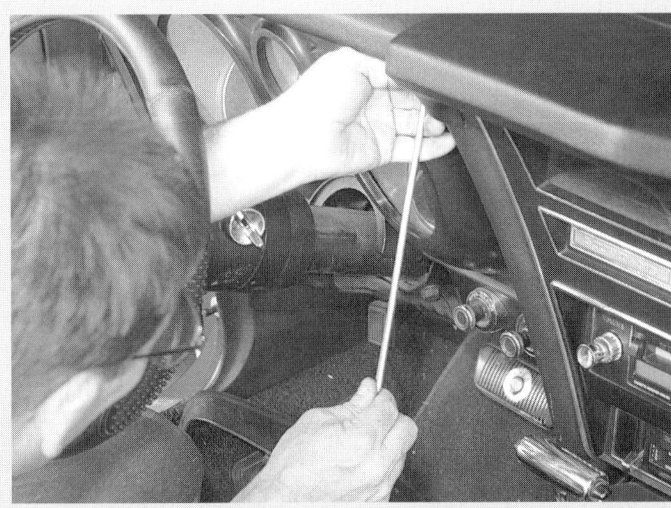

13. Reinstall the screws that retain the dashboard crown and dashpad, then you can admire your new dashpad from Just Dashes. For '71–'73 Mustang owners, a long-awaited day has arrived. It took us less than an hour to perform the swap.

Chapter 44

Installing a Perfect Fit A/C System

Text and Photos by Donald Farr

Classic Auto Air's Perfect Fit air-conditioning systems are perfect for '65–'73 street Mustangs in the summer.

My wife has a saying, usually regarding her tile work or cabinet painting, that proclaims, "Perfection is boring." But in the case of Classic Auto Air's Perfect Fit Series air-conditioning systems for vintage Mustangs, we can modify that to say, "Perfection is easy." With the Perfect Fit kits for '64 1/2 to '73 Mustangs, it's easy to add air conditioning to first-generation Mustangs.

Classic Auto Air (CAA) designed the Perfect Fit Series as a complete heat/cool and defrost climate-control system that replaces the old heater-defrost box with a brand-new unit, incorporating a large A/C coil, for a factory-air look. Available in '64 1/2–'66, '67–'68, '69–'70, and '71–'73 Mustang versions, the Perfect Fit systems utilize the factory heater-control panel and use stock vent locations, except for '64 1/2–'66, where vents are mounted under the instrument panel, one in each corner and a dual unit in the center. Under the hood, a state-of-the-art Sanden compressor requires less horsepower and runs smoother when compared to the '60s Tecumseh or York compressors. CAA's kit replaces the factory fan switch with an updated switch that includes three speeds plus "Off." Other features include 134a refrigerant compatibility, dehumidified defrost, a high-output OEM-style blower, large, separate A/C and heat coils just like factory systems, OEM-quality control cables, and a nylon extruded door to eliminate rust-out with a heavy-duty shaft assembly.

Best of all, the CAA Perfect Fit systems come with thorough installation instructions. The one for our '68 installation tallied 22 pages with diagrams, helpful templates, and illustrations.

Source
Classic Auto Air
813/251-2356

Of course, CAA got into the business over 25 years ago by restoring and refurbishing original Mustang A/C equipment, which it still offers today for Mustang owners who want the original factory equipment and appearance.

To check out the installation and operation of a Perfect Fit system, we followed along as CAA installers, Doug Gresian and Fritz Adolf added A/C to Russ Aljoe's "restomod in progress" '68 hardtop. CAA recommends disassembling and preparing the Mustang first, then installing the Perfect Fit components. With the system installed, the car will need to be delivered to a certified air-conditioning shop for evacuating and charging with 134a refrigerant. Any leaks will turn up during the vacuuming.

As we said previously, the instructions for installing the Classic Auto Air Perfect Fit A/C system are well-detailed, with plenty of tips (like using refrigerant oil on the fittings) to help make the installation go smoothly.

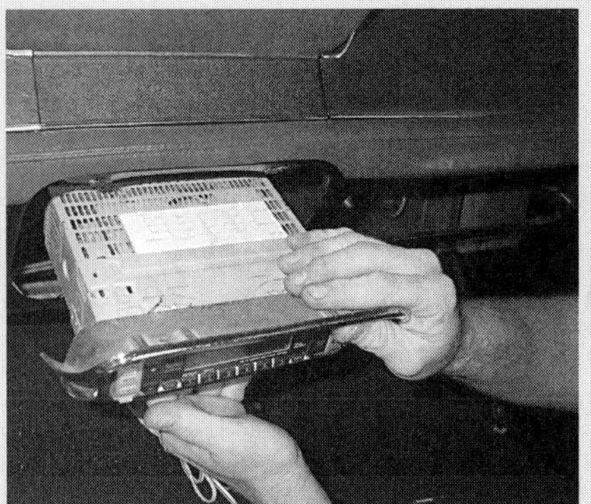

1. The first order of business on a '67–'68 Mustang is to remove the radio, ashtray, and glovebox for better access behind the instrument panel. If the car is equipped with a console, it must be removed as well.

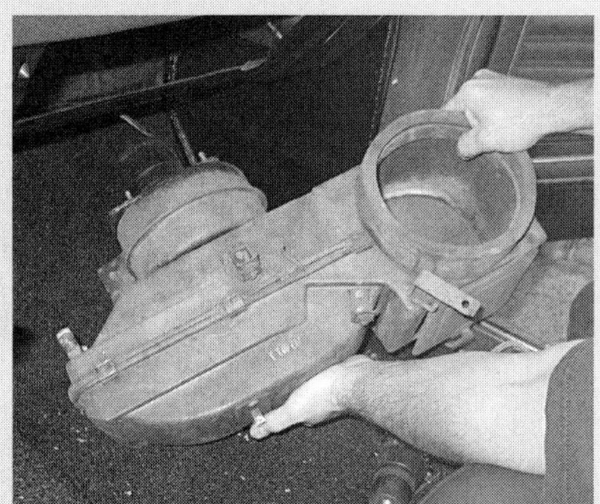

2. At the original heater assembly, disconnect the three control cables and electrical harness and remove the factory defrost ducting, then locate and remove the mounting screws under the instrument panel. After draining the radiator, cut or remove the existing heater hoses at the firewall. Finally, unbolt the original heater box from the firewall and remove.

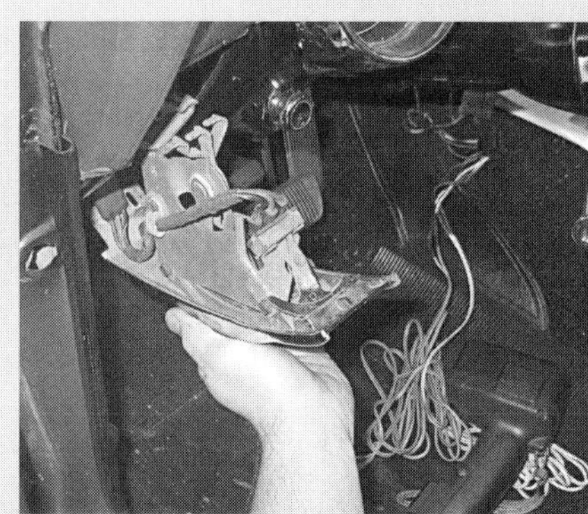

3. Remove the original heater-control assembly, which is located on the driver side of the instrument panel on '67–'68 Mustangs. Remove the control cable, then move the assembly to the work bench to replace the original blower switch.

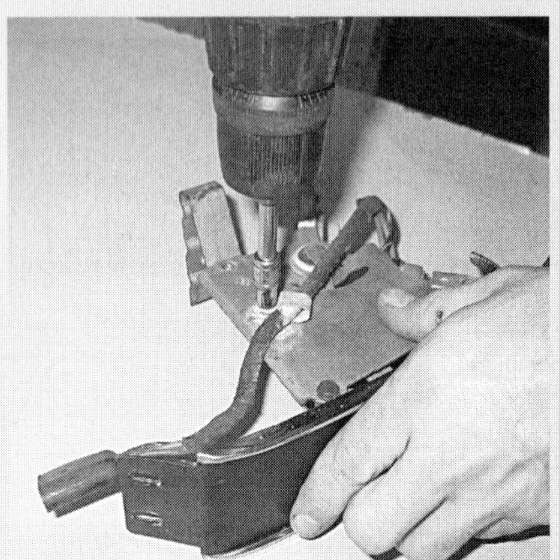

4. Unbolt the original blower switch and wiring harness.

5. This three-position, plus "Off," switch replaces the original blower switch.

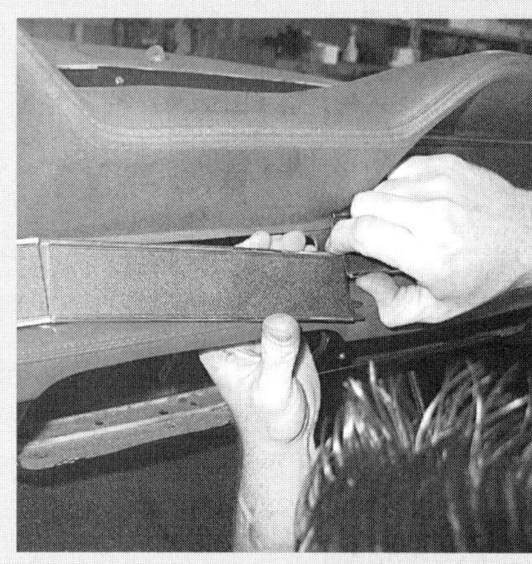

6. Remove the plastic center trim bezel. It should be retained from the rear with a nut.

7. On '67–'68 Mustangs, the center opening in the instrument panel must be trimmed to provide room for the CAA center vents and hoses. CAA installer Doug Gresian uses an electric saw, but he says the job can be done with tin snips, although they will likely distort the metal. Fortunately, sloppy work will be hidden by the center vents. Notice the tape around the opening that helps prevent accidentally marring the dashpads.

8. It's a good idea to test-fit the center vent panel. That way, you can trim more if needed.

9. After pinpointing the location based on the measurements supplied in the instructions, use a 2-inch round hole saw to drill a hole into the passenger-side panel for the vent. A 2-inch snap-in hose adaptor fits into the opening at this point.

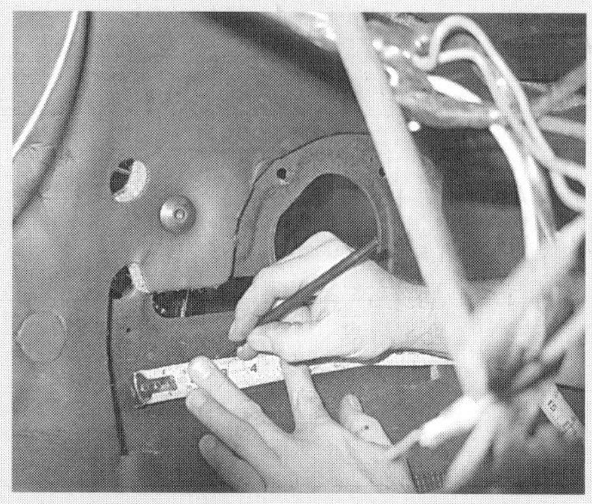

10. You'll also need to drill an opening in the firewall for the drain tube. CAA provides measurements and a template for the hole. Before drilling, check the engine side of the firewall for possible brake-line interference. You may need to bend the line out of the way.

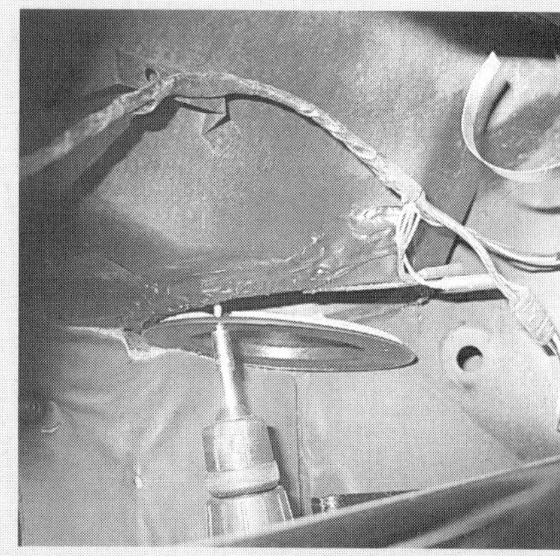

11. A block-off plate is included in the kit to cover the original fresh-air inlet in the cowl.

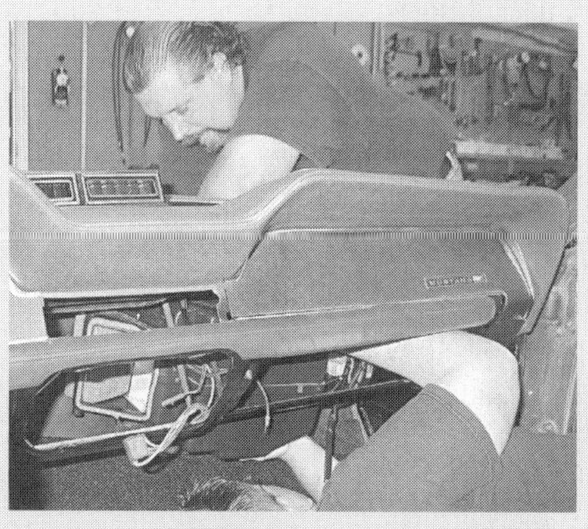

12. After installing the pair of mounting J-clips under the dash, the condenser/heater-core box can be slipped into place. It helps to have an extra pair of hands on the other side of the firewall to position the heater hose and refrigerant connections through the original heater-box opening. Several screws secure the box in place, as noted in the instructions. You'll notice that the heater hose connections now come through the firewall in a different location from the original. You can plug the original firewall holes. The instructions provide complete directions for modifying the original fan wires for the new blower.

13. The defrost duct assembly installs on the condenser/heater-core box with a pair of screws.

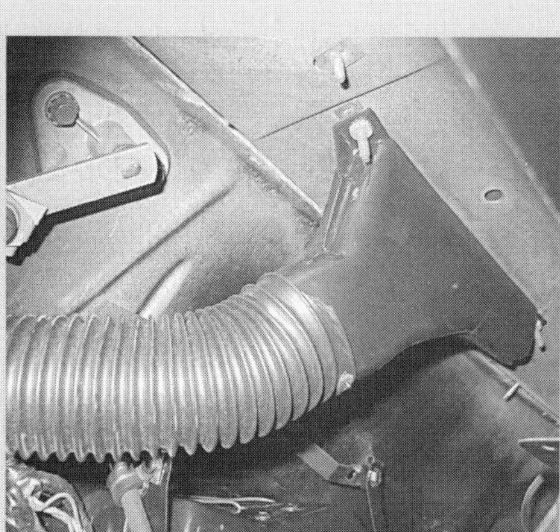

14. After mounting the defrost ducts under the instrument panel, the duct hosing can be run from the defroster-duct assembly. When measuring the supplied hose, be sure to stretch it out before measuring and cutting.

15. Moving to the engine compartment, remove the fan and shroud. If the Mustang was originally equipped with a four-blade fan, CAA recommends upgrading to a five-blade, which is an available option.

16. The CAA compressor-mounting-bracket bolts to the water pump and head. Correct-length bolts, to accommodate the thickness of the bracket, are included, along with an idler pulley for the A/C belt.

17. The original crankshaft pulley is replaced by CAA's version with an extra groove for the A/C belt.

18. The Perfect Fit A/C system comes with a modern Sanden compressor. A much-improved design over the original Mustang compressors, the Sanden is more efficient, uses less horsepower, and eliminates vibrations.

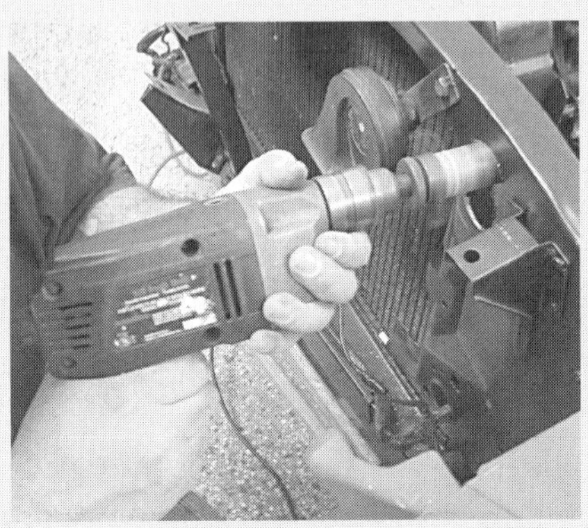

19. To prepare for the condenser and dryer installation, you'll need to disassemble the front of the car, removing the bumper, grille, and latch mechanism. With everything out of the way, locate the pair of precut punch-outs on the driver side of the radiator support and open up the punch-outs using a 1 3⁄8-inch hole saw. A pair of drill openings, one on top of the other, will usually do the trick for making the oval opening; then file down the raw edges and coat with paint for rust protection.

20. Install the condenser/dryer assembly and route the discharge hose and liquid tube through the radiator-support opening. A cutoff switch is included to prevent damage to the compressor should the system lose pressure.

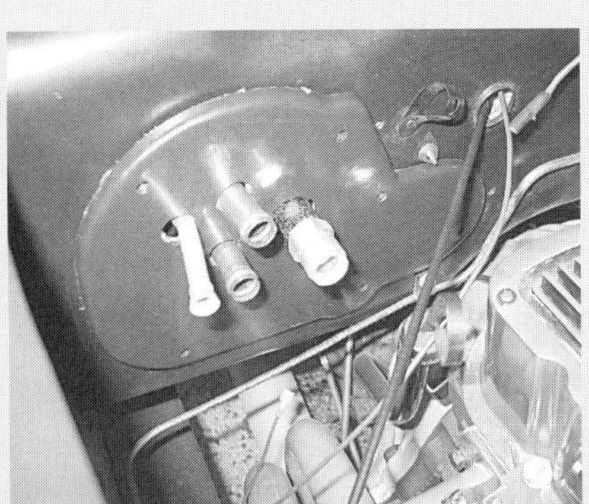

21. The Perfect Fit system comes with this cover to help seal the firewall opening for the heater hose and refrigerant connections.

22. Route the supplied hoses from the refrigerant connection at the firewall to the compressor.

23. Back inside, connect the Perfect Fit wiring harness to the new switch on the control panel and route the wiring under the instrument panel. Connectors and a diagram are included.

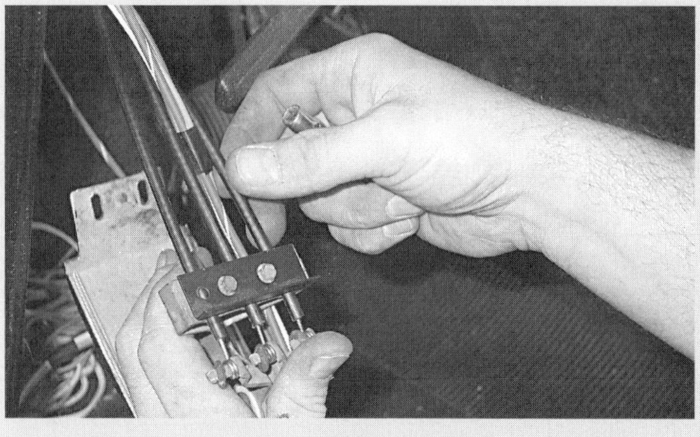

24. The control cables—for defrost, heater door, and water valve—attach to the control panel. The longest cable goes through the firewall to the water valve, which controls the amount of hot water reaching the heater core.

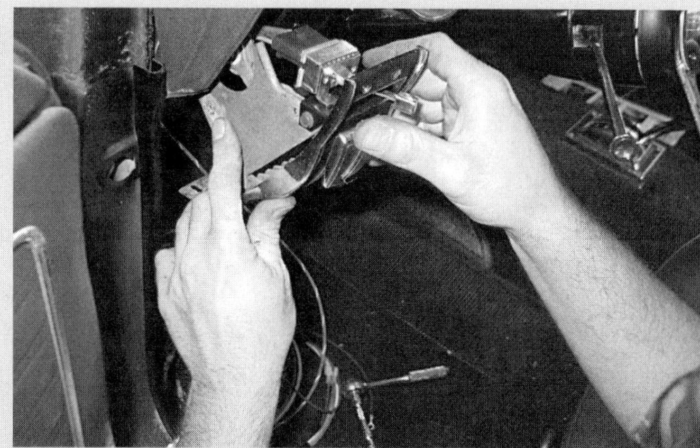

25. With the wiring and cables attached, the control panel can be reinstalled.

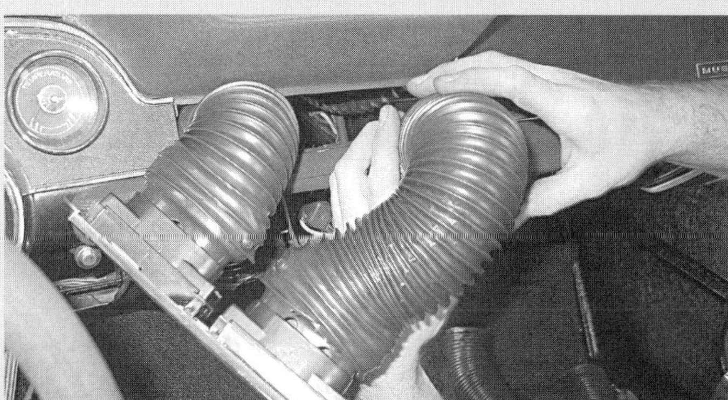

26. The vent hoses attach to the vents—two at the center and one at each corner. At the center, attach the hoses to the vent and feed them through the instrument-panel opening. They attach to a faceplate at the condenser/heater-core box.

27. The outside vents mount at each corner: the passenger side (shown) in the lower panel and the driver side underneath the control panel.

28. The Perfect Fit system provides a factory look in the interior.

29. Finally, after everything is buttoned up, the car can be delivered to a certified air-conditioning shop for evacuating and charging. When the system was fully operational on our '68 Mustang, the air coming out of the vents registered 42 degrees with the windows down.

Chapter 45

How to Install an Overhead Console

Text and Photos by Jim Smart

Thanks to a new reproduction from Scott Drake Mustang Parts, owners of '67–'68 hardtops and fastbacks can add this popular factory option.

When Ford redesigned the Mustang in 1967, it took an already great idea and made it better by sculpting and honing the classic lines to accentuate the positives. Ford also gave the Mustang more options for 1967: big-block power, a sizzling exterior trim package, a redesigned Interior Décor Group in brushed aluminum, a louvered hood, full instrumentation with a tachometer, and more. One stylish extra was the overhead console, a rarely ordered option everyone liked to search for in the salvage yards 25 years ago. It was a groovy way to dress up a '67–'68 Mustang hardtop or fastback interior.

Scott Drake Mustang Parts understands what Mustang enthusiasts like because Scott is one himself. What's more, quite a few of his associates are also enthusiasts. The '67–'68 Mustang overhead console for hardtops and fastbacks is one of Drake's latest reproductions. Available in traditional Mustang colors, the '67–'68 overhead console is easy to install, and adds a touch of elegance to any '67–'68 Mustang interior. Although this option was originally designed for '67–'68 Mustangs, we're convinced it can be installed in a '65-'66 Mustang with some perseverance.

Before we get started on our fastback overhead-console installation, we're happy to report you don't have to replace the headliner for this Scott Drake reproduction. A couple of small slits in the headliner will get the job done, but be sure to use caution when you cut it.

Sources

Mustangs Etc.
www.mustangsetc.com

Scott Drake Mustang Parts
www.scottdrake.com

This is the reproduction '67–'68 Mustang overhead console from Scott Drake Mustang Parts. The brackets are available separately because cars that originally came with overhead consoles may already have them.

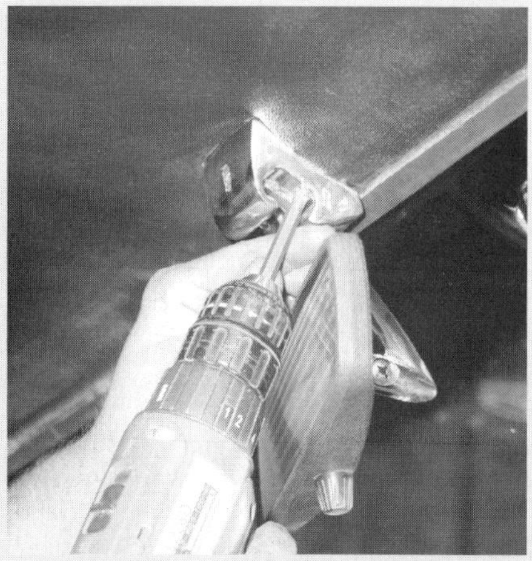

1. To start the installation, swing the sunvisors out of your way and remove the rearview mirror.

2. A test-fit is required to locate the mounting holes that are predrilled into the roof structure for the forward console bracket. Use a razor blade to make a small slit in the headliner midship. The slit should be 2 inches aft of the rearview-mirror mounting location because there is approximately 2 inches of visible headliner between the console and rearview mirror. The bracket slips through the opening in the headliner.

3. With the bracket secured with a single screw in the middle, carefully drill two 1/8-inch holes, one on each side. These holes are for the bracket retaining screws.

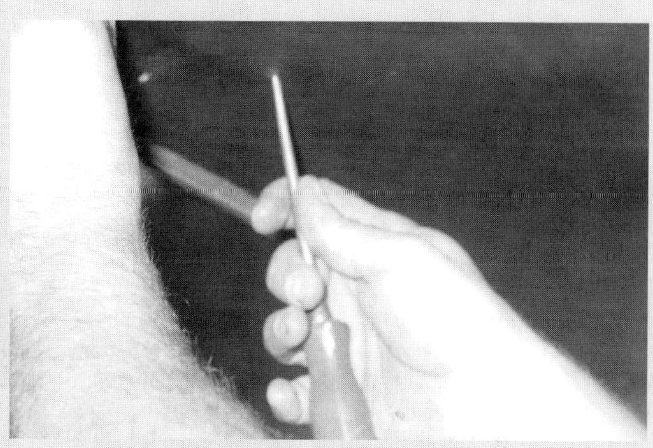

4. For fastbacks, remove the rear "targa" band across the back of the cabin. For hardtops, skip this step.

5. Install the rear console bracket on the outside of the headliner. Predrilled console-bracket mounting holes are already behind the headliner; all you have to do is find them with an awl.

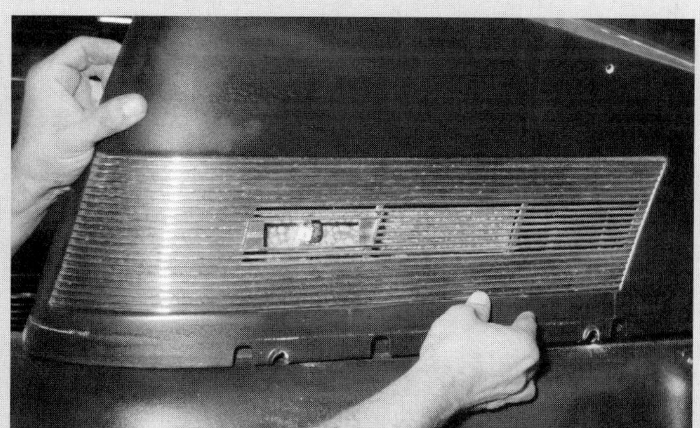

6. Power is needed for the console's map lights. With hardtops, get the power from the dome-light lead. It's "hot" all the time. When the doors are opened or the headlamp switch is turned to courtesy-light mode, the circuit is grounded, which gives us light. With fastbacks, power for the overhead console comes from the courtesy lights, but it's in a different location. Remove the air-extractor panel to access the power.

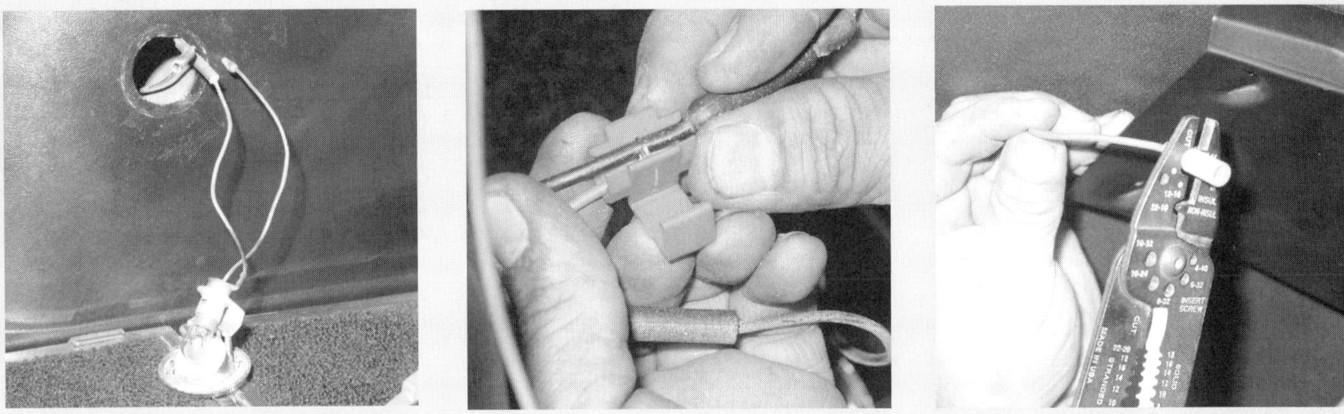

7. Source your map-light power from the courtesy-light lead as shown using a connector like this. Route the wire across the headliner to the rear-console bracket like this. Make sure the wire isn't pinched anywhere and install a female bullet connector.

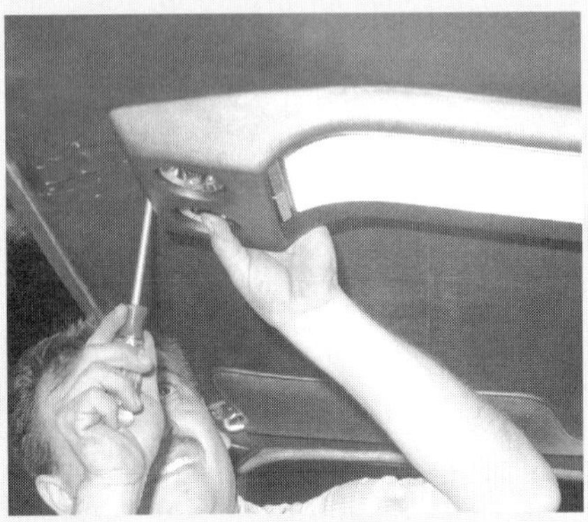

8. Gilbert Bravo of Mustangs Etc. installed our Scott Drake overhead console. Any kind of overhead work is challenging, and getting the new console to fit is likely the hardest part. You'll need a helper in the back seat.

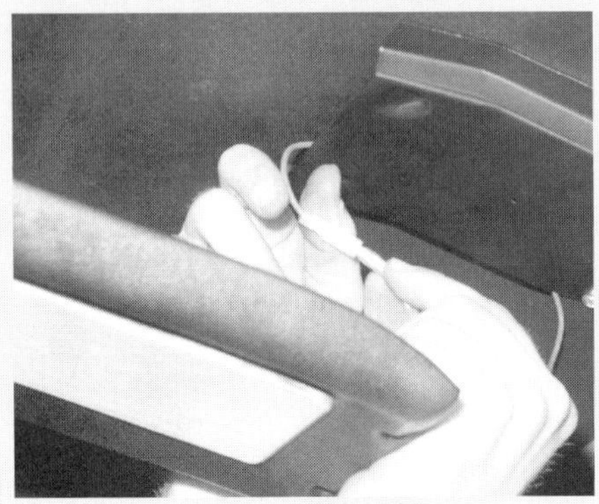

9. Map-light power comes from this female bullet connector tied to the courtesy-lamp circuit (black wire with a blue stripe).

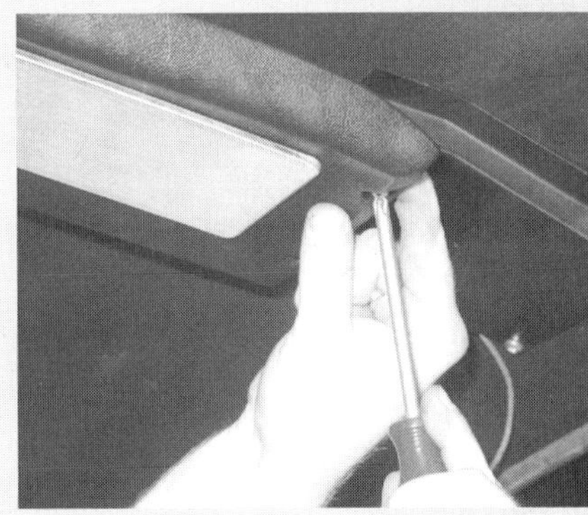

10. Two rear console screws are tightened into the bracket.

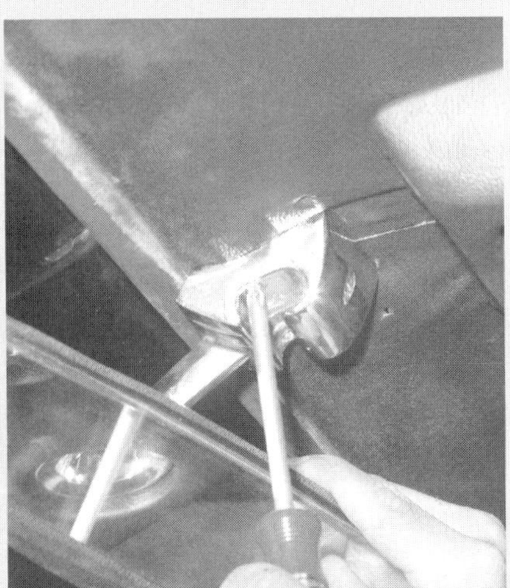

11. Reinstall the rearview mirror.

12. Reinstall the quarter-trim air-extractor panel (fastbacks only).

13. Reinstall the targa band (fastbacks only).

Working on a vintage Mustang's interior is often more involved than meets the eye. Here are some helpful hints to make tasks easier.

Chapter 46

Pro Interior Restoration Tips

Text by Miles Cook
Photos by Miles Cook and Jim Smart

The interior in a typical vintage Mustang might appear like a relatively simple entity, with seats, door panels, carpeting, a headliner, and a dashboard. But things can get complicated in a hurry when it comes time to resurrect that tattered old cabin and make it look new again.

Even in the early '65–'66 cars, which seemingly have the most rudimentary interiors of all Mustangs, there's still much to be learned about getting the inside back into presentable shape. Our goal is to teach you about vintage Mustang interiors and, in the process, make it easier for you to complete your project, regardless of the condition of your car's interior.

To cover all the bases, we're going to imagine a Mustang interior that needs nearly everything. If your car's interior isn't that bad, consider yourself lucky. If your car falls into our category, we're confident our helpful hints will make restoring a vintage Mustang interior easier.

While much of the following information pertains to restoring an interior to factory condition, we'll also look at some of the many upgrades available these days that provide meaningful advances. But first we'll look at the scores of interior resto tips we dug up at an experienced local upholstery shop.

Tips from a Professional

In an effort to discuss something other than buying new interior parts, we went to see one of our buddies who's spent his entire professional life working on automotive interiors, with plenty of vintage Mustangs included. Mike Ambrose of Ambrose Custom Interiors is the consummate specialist, and we can't think of an interior issue he doesn't know something about. We can only imagine how many cars he's worked on over the years, and he can fabricate a complete custom street-rod interior from scratch. Restoring a regular interior in a popular production car like a vintage Mustang is something he can do in his sleep.

When you set out to restore the interior in an old Mustang (or any vintage car), Mike emphasized performing the various tasks in a specific order. You'll see how this plays out as we go along.

Thinking of the process as a step-by-step procedure, Mike began by suggesting you "take the interior apart yourself." If you'll be doing the restoration project, familiarize yourself by

Carefully trim carpet for a smooth look upon final installation.

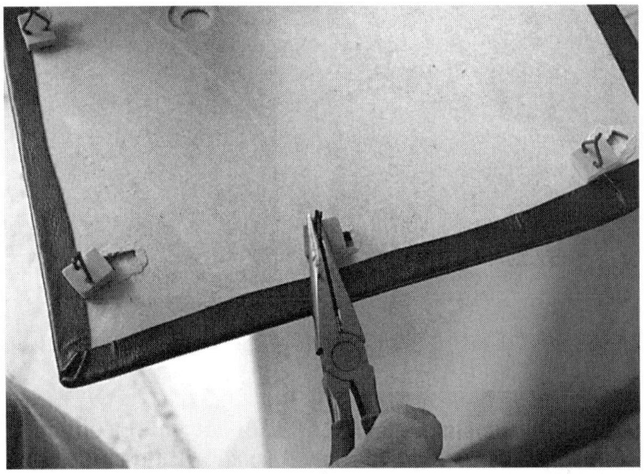

Door-panel clips are much stronger than the door panels they hold in place. Be especially careful if a new or valuable original panel needs to be removed.

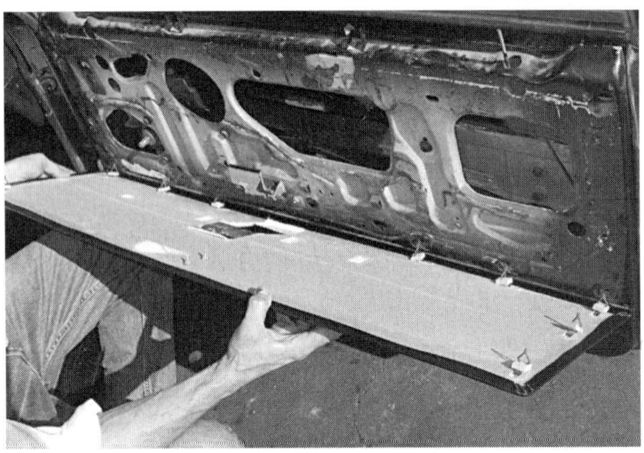

Carefully determine placement of panel clips when installing new or original panels.

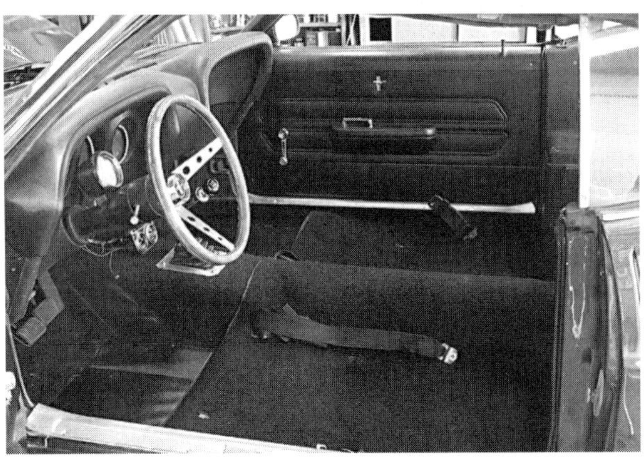

With this car's interior almost completed, seats are one of the last items to install during the course of an interior restoration.

doing all the disassembly work so you know what's what. Don't just rip and tear. Slowly and carefully remove the seats, carpet, door panels, sill plates, headliner, and whatever else you intend to replace, in a careful and deliberate manner.

According to Mike, it's important to invest in a digital camera. Even if you don't have a computer, a small and inexpensive camera can shoot hundreds of photos and you can have them printed after the interior is apart. "Shoot the interior coming apart as much as possible," Mike said. "It's impossible to take too many photos." If you have a small flash card, shoot fairly low-resolution pictures. For the purpose of putting everything back together again,

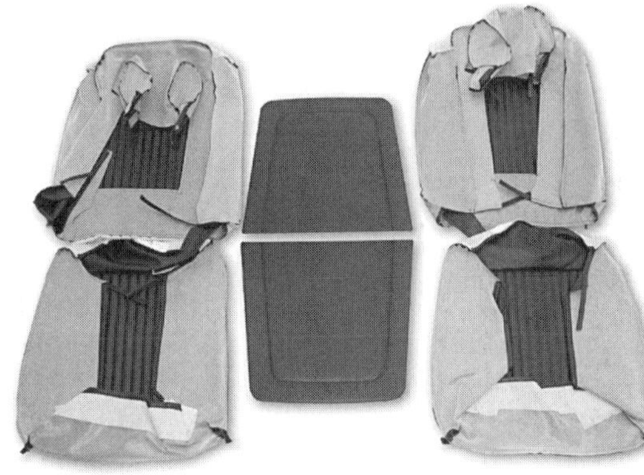

New upholstery is about the most visible element of a freshly restored interior. Proper installation is critical for it to look nice.

it's better to have 200 low-res photos than 50 high-res shots taking up the whole compact flash card. But if you have a computer, you can dump the photos into the computer as you go along.

The next step was to "check for rust in the floors once the carpet is out." If any repairs are needed with the interior of the car stripped bare, now is the time to do it.

We're assuming a full resto process for our imaginary interior, which would include a new headliner. If the old headliner has been removed, Mike stressed to remove and discard the old fiberglass insulation and replace it with a modern automotive sound-deadening material, such as Scosche Industries' Accumat. According to Mike, "Get rid of that nasty fiberglass headliner insulation material as soon as the old liner comes out.

As mentioned earlier, put a new interior back together in specific steps. Put the new headliner in first. Both the front windshield and rear window will need to be removed to install the new headliner correctly, so remove the glass first if you're doing a headliner. "There's simply no other way to do it right," said Mike.

Now we're down to the bare shell of an interior, with everything out from carpet to headliner. What's the order of installing new interior pieces? According to Mike, it should go like this: headliner, any interior wiring, carpet, doors (panel assembly and installation), kick panels (with built-in speakers if you want; Mike suggests this is a good time to upgrade to a set of speakers from Custom Autosound), seatbelts, seats (reupholstered if necessary), and rocker-sill plates. Reinstalling the windshield and rear glass (and fixing any leaks the car might have in these areas) would be the final step.

With the basics out of the way, we discussed tips

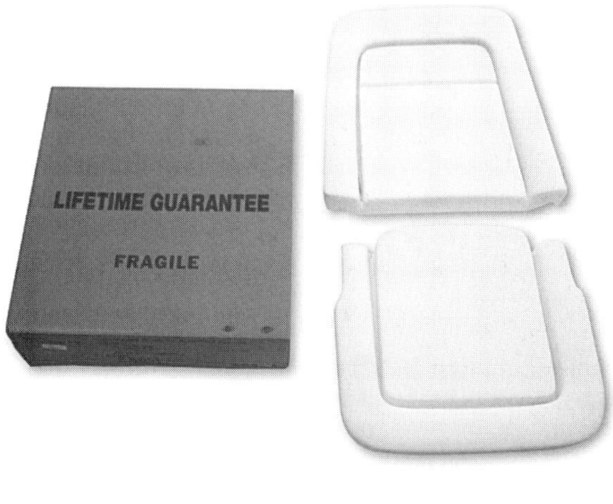

Don't skimp on seat buns. They're a must with new upholstery for the finished seats."

Vintage Mustang seat frames are simply old. Thoroughly inspect them for weld breaks to ensure they're reusable

A seat's original listing wire combined with new burlap is essential for a proper seat restoration. It's standard procedure at Ambrose Custom Interiors.

for installing specific components, and the headliner took up a good chunk of time. Before working with it, leave the headliner out in the warm sun. In cold weather, use a heat gun in the garage, but be careful not to damage the fabric by holding the gun too close. Mike mentioned the two bow retainers in the back of the headliner. He said, "Be sure these are reused. They're color-coded so you'll know where they're supposed to go." These bows often need to be cleaned up because they're either rusty or have adhesive material on them from the factory. As we said earlier, use good sound insulation material between the headliner and roof, as a vintage Mustang will be noisy inside with nothing up there.

After the headliner is installed, the carpeting is next. "Run all wires (stereo and so on) down the sills before the installation of carpeting, so you can access them later if need be without having to remove the carpet," said Mike. Another tip is to install and leave the seatbelt bolts in place under the carpet as it's installed so you can locate the holes after the carpet is in place. carefully cut the carpet around the bolts with a carpet knife. You'll be able to reinstall the seatbelts, and the holes will be right there since you cut out enough carpet to remove the bolts and put them back in with the seatbelts attached. This same trick also applies to the sunvisor screws when installing a new headliner. The fabric can be carefully cut around the screws once the liner is completely installed.

A new package tray with some sound insulation would be next. "Use the same stuff here as you would in other areas of the car," said Mike. While on the subject of package trays, we showed Mike our trusty old '69 Mustang project car that, among other things, still has a leaky rear window. This has slowly taken its toll on the car's package tray that was replaced about two years ago. "To repair the leak properly, the window has to be removed," he told us.

With a new headliner, package tray, and carpeting installed, we turned our discussion to door panels. Mike likes the idea of installing new weather-shield material between the doors and panels. Unless the existing material in your car (sometimes known as "vapor barrier") is in really good shape, we agree. "Use a good-quality plastic weather-shield sort of material," Mike suggested.

New door panels and new (or original) retaining clips can be a dangerous combination if extra care isn't taken with their installation. Be sure the door-panel retaining clips are lined up with the holes in the door before pressing the new panels into place to avoid damage to the fragile cardboard panel. Another trouble spot is panel removal. Mike uses a special Mac pliers-type tool (PN BT326) to get the clips out of the door while avoiding damage to the panels. The clips are much stronger than the

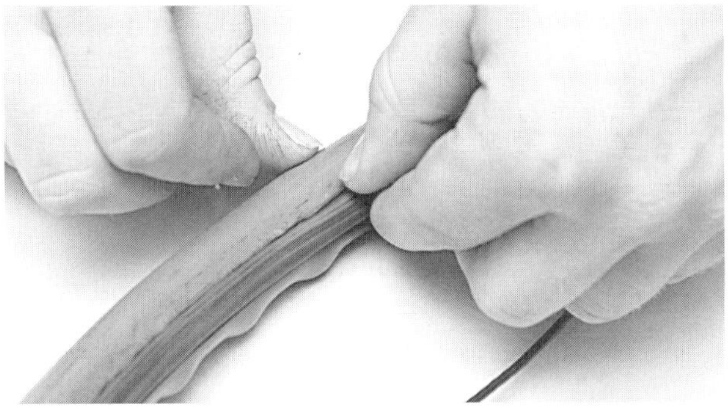

Optional on Mustangs from '69 to '73, the Rim-Blow wheel lets the driver sound the horn simply by squeezing the steering-wheel rim. As you can imagine, it can be a finicky piece of equipment and after more than 30 years, a Rim-Blow wheel will likely need a new switch. National Parts Depot offers the switch. (www.npdlink.com)

Flaming River offers a tilt steering column alone and in conjunction with its rack-and-pinion steering setups, although that's beyond the scope of our discussion. But the tilt column is both a comfort and a safety update because it's collapsible. Best of all, you can run with a stock steering wheel should that be your cup of tea. (www.flamingriver.com)

Though not directly related to a car's interior, the door glass has some effect while you're sitting in the car. With that in mind, we think swapping '69 glue-in glass for the more solid '70 bolt-in setup is a cool upgrade. Repop '70 door glass is available from Muscle City Glass.

cardboard in the door panel, so avoid ripping the panel off while leaving the clips in place in the door.

The final piece of the interior puzzle is the seats and re-covering them if necessary. Mike mentioned plenty of tips on this subject. Before putting new upholstery on a seat, it's taken down to the bare frame. Mike said to thoroughly inspect the seat frame for weld breaks. Repair as needed or get a replacement frame. After 30–40 years, Mike said he's seen his fair share of broken Mustang seat frames.

No big secret here: Better sound equipment such as the Custom Autosound head unit, 10-disc CD changer, and speakers shown here are staple resto ideas for a vintage Mustang that will be driven on a regular basis. We've covered this subject several times over the years.

Reassembling the seat is straightforward stuff for a good upholstery shop like Ambrose Custom Interiors, but if you want to tackle this job yourself, here are a few pointers. If you're installing new upholstery, you must also get new seat buns. "The new covers won't fit over old seat buns," said Mike. "They'll always be wrinkled and it just won't look right." He also recommended two pieces of new burlap (for a double thickness) under the seat buns and pointed out that the original seat listing-wire needs to be reinstalled over the new burlap and secured with new hog rings. Mike says there's only one size ring that's used in vintage automotive seats.

Use new hardware for the seat where possible. Mike advises assembling seats slowly and carefully, and concentrating on doing the job right rather than fast. As for reinstalling the seats, it's easier to start the nut under the car with a socket than with your bare hand. "Secure the nut in place in the socket with some masking tape around the nut to get it started on the seat stud," Mike said. "It's easier than trying to get it started with your fingers." Good advice, because if the nut falls in between the inner and outer floor panels underneath the car, you won't be able to retrieve it. Mike also stressed reusing (or getting new) rubber caps under the car that cover the seat-nut access holes to keep moisture out, just as the factory intended.

Looking again at procedural order, the seats are installed last after the seatbelts. Finally, Mike completes an interior restoration project by installing new (or existing) rocker-sill plates as the last component inside the car.

Upgrades

Here are several upgrades that can be added to any vintage Mustang interior, even one that's considered fully finished. Of course, they can also be added to a cabin that's undergoing a full restoration at the same time.

Custom Leather Upholstery. Leather upholstery that mimics original-style seat-cover patterns is now available. Shown here are real leather covers with original-type '65-'66 Interior Décor Group running horses. This is similar to what came in '65-'66 Mustangs with Pony interiors. These covers are available from JMC Motorsports in various colors (www.jmcmotorsports.com) and can be installed by a competent interior upholstery shop.

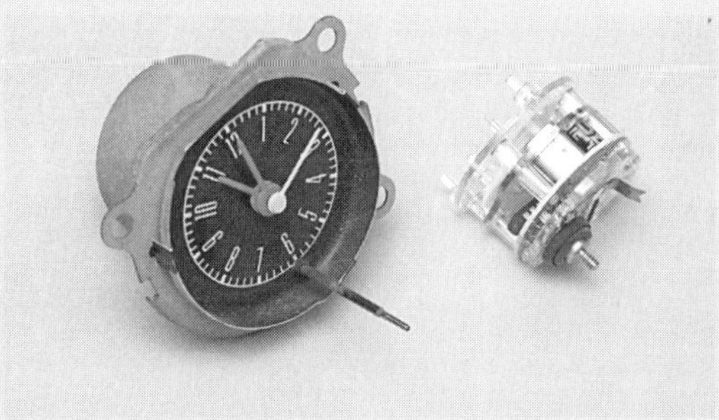

Quartz Clock Movement. Although this option has been around for many years, it's always worth mentioning. A smooth quartz clock movement is a great way to top off a fully decked out and restored Mustang interior. Quartz movements are available for just about any '65–'73 clock (including '65–'66 Rally-Pacs), with the only real exception being the square clocks used in '69–'70 standard interiors.

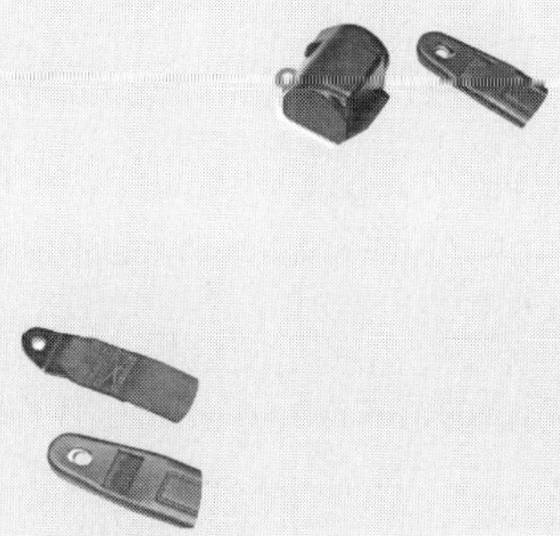

Three-Point Seatbelts. A major safety update for any vintage Mustang, three-point seatbelts are worthy of consideration in a car that is driven daily, even one with shoulder belts that require you to close two buckles for full protection. Custom Accessories makes these retrofit belts for the front and rear seats and they're available from Virginia Classic Mustang (www.virginiaclassicmustang.com).

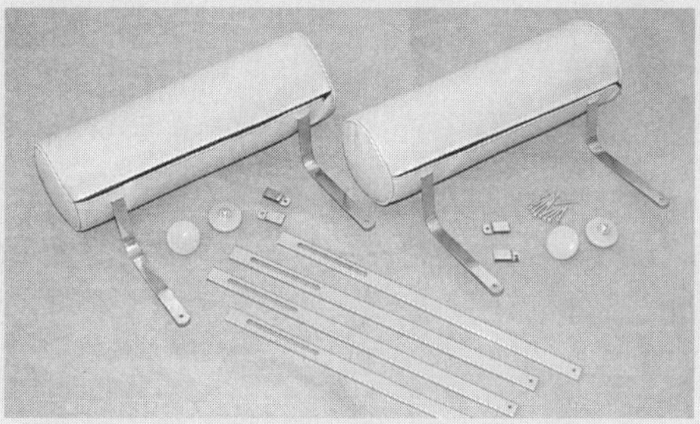

Although these headrests were available as a dealer option, not many are around today. Luckily, reproduction versions are offered, and these are a great safety upgrade even for an interior that otherwise needs nothing. SafeCode offers the headrests (www.mustsangproject.com).

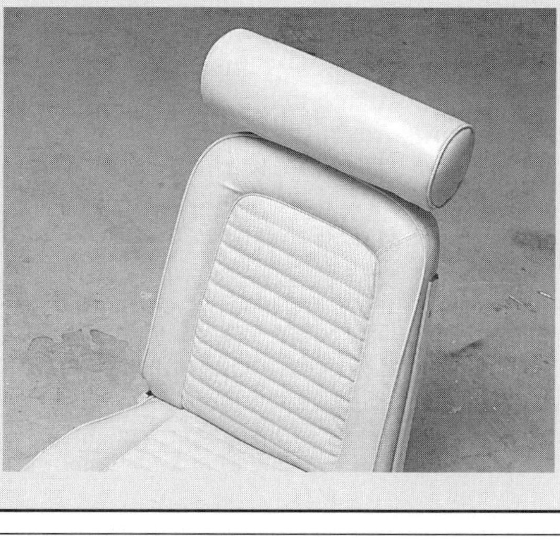

Chapter 47

How to Install a New Headliner

Text and Photos by Jim Smart

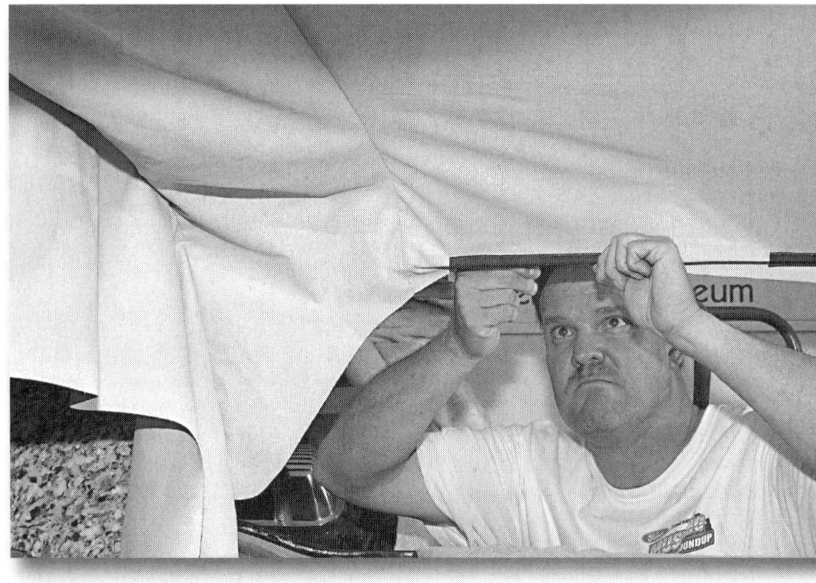

Installing new moonskin will help give your Mustang a factory-fresh look.

This is likely one of the most dreaded classic Mustang tasks—besides bodywork and paint, of course. But headliner replacement doesn't have to be a difficult job if you start with the right mindset. And, when you are finished, you will have saved the cost of hiring an upholstery shop and will feel a whole lot better about your own talents. This is a large part of what restoration is about—tackling something you've never done before, which improves self-confidence.

So what does it take to install your own headliner? To find out, we called Bill Faull at Mustangs Plus, and ordered a white TMI Products' headliner for Gary Mattson's '66 Mustang hardtop as the beginning of a full-scale interior restoration.

Headliner installation requires a warning of what's involved. You have to work hard and be unwilling to accept sloppy craftsmanship. When you have wrinkles, you have to go the extra mile to ensure they go away. The rule of thumb is to begin headliner installation at the middle and work your way fore and aft, then left and right. The first thing to do is make sure all the bows are in the correct slots. Ford gives you a couple of options in the roof structure for headliner bows, but take note of where the bows are before you remove the old headliner. Take pictures and keep them on hand before beginning the installation.

The windshield and back window must be removed to replace the headliner. Before you remove the glass, get acquainted with a windshield shop that has plenty of practice with classic-car rubber-gasket windows. Most shops today understand modern glue-in windows, but have little experience with the rubber gasket type.

Sources

Mustangs Etc.
www.mustangsetc.com

TMI Products
www.tmiproducts.com

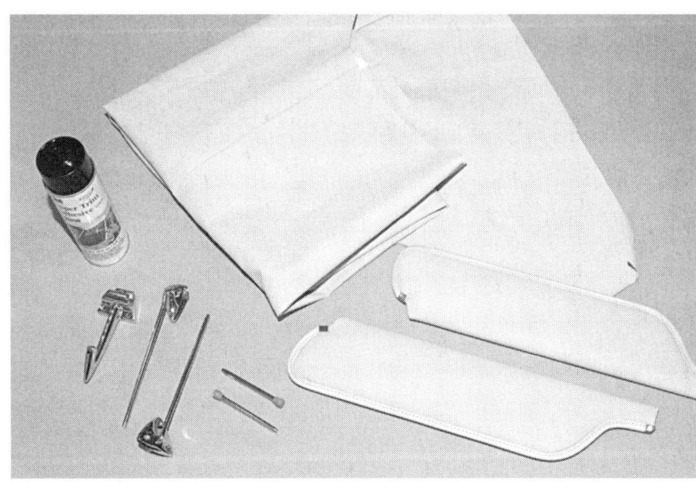

Here's everything needed to replace the headliner. We're also installing new sunvisors and brackets, as well as a new rearview mirror and wind lace.

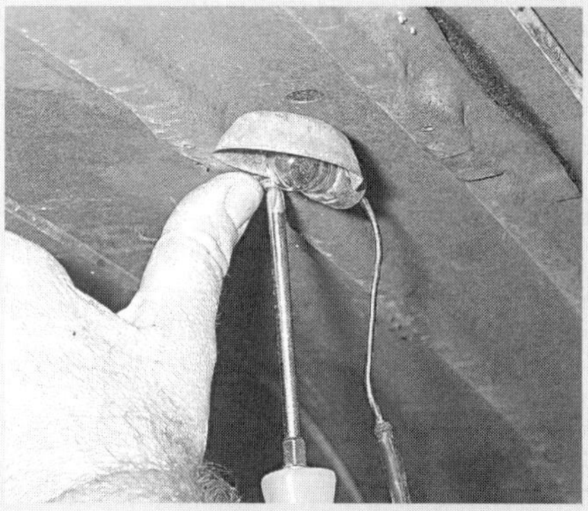

1. We'll never understand why Ford didn't install an overhead courtesy light in '65–'66 hardtops. The overhead roof support has a provision for one; all the screw holes are there. We went to Mustangs Etc. and picked up a used courtesy-light package from a '68 hardtop. The courtesy lamp installs easily at the overhead location on the roof support. Route the lead toward the driver side along the support, keeping the butt connector in the valley of the support. Otherwise, it will show in the headliner.

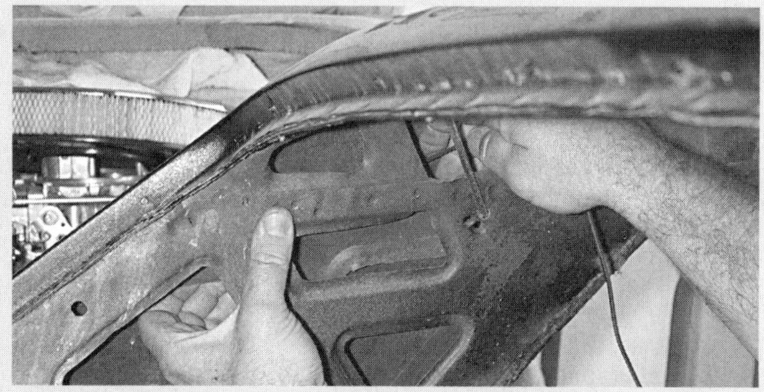

2. Gary routes the courtesy-lamp lead back along the driver's-side roof rail down the inside of the sail pillar to the trunk area. Power comes from the main wiring loom to the taillight lead pigtail on the driver's side. This pigtail is for fastback models and isn't used on the hardtop.

3. Gary checks the headliner bows for proper positioning before he gets to work. Note Gary has installed new roof insulation. Because foam insulation was in short supply the day of our installation, Gary used carpet padding. For best results, we suggest foam insulation with an aluminum-foil backing between the foam and the roof.

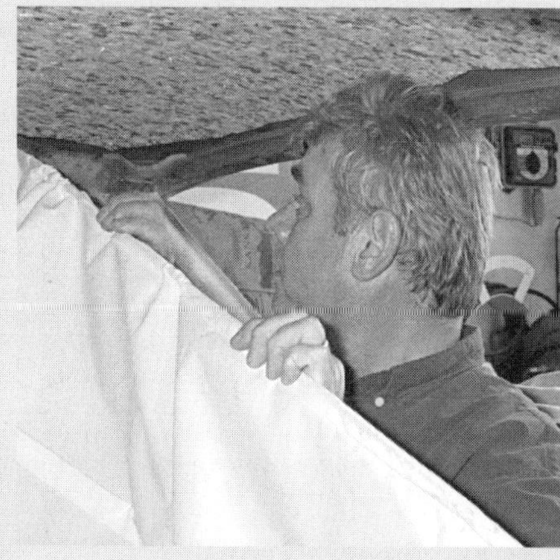

4. Bows are slipped into each sleeve from front to rear. Bunch the headliner toward the middle of the bows in order to install them.

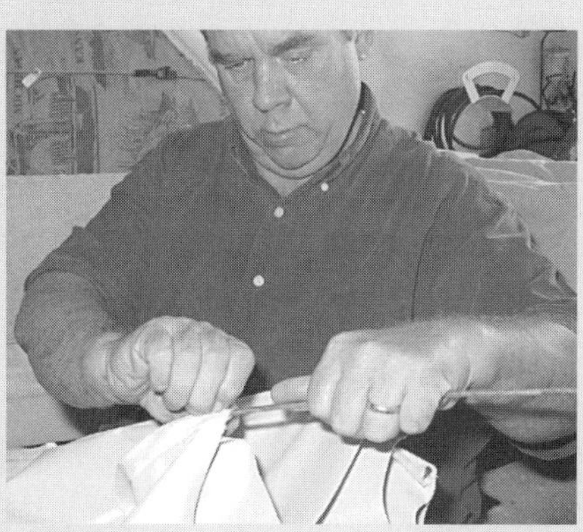

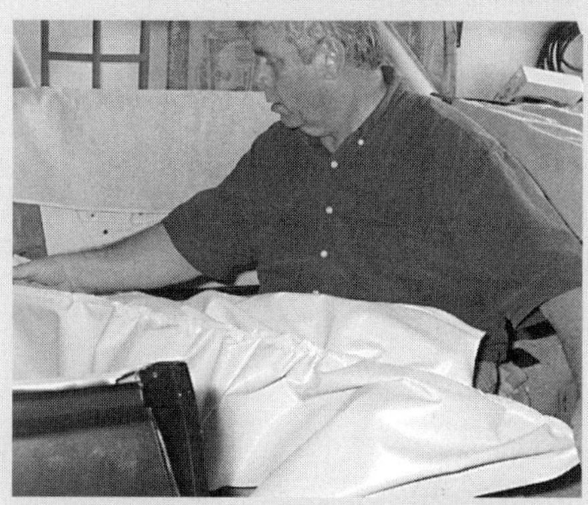

5. With the installation of each bow, Gary bunches the headliner toward the middle, which makes installation a snap. Expect the bows to roll over from the weight of the headliner. You will have to roll the bows upright against the insulation, then begin your work. Sometimes, you will have to remove bows already installed in order to install others. It is a process you will have to work through to determine what works. Bow positioning depends on your Mustang and how the headliner was cut and stitched.

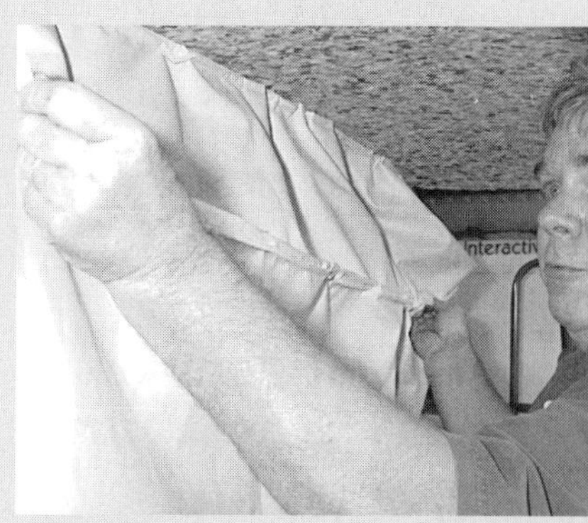

6. Once positioned, each bow and panel takes close inspection to ascertain proper bow positioning in the roof rails.

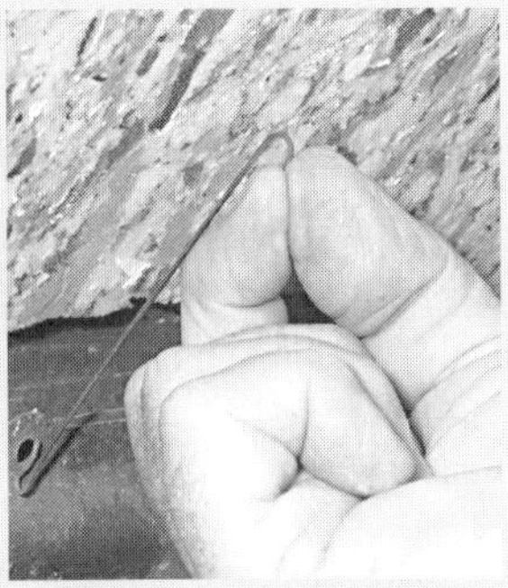

7. These headliner bow-retaining wires attach at the rear roof rail and No. 4 bow. They keep the last bow properly positioned. These retainers are available from Scott Drake Mustang Parts dealers.

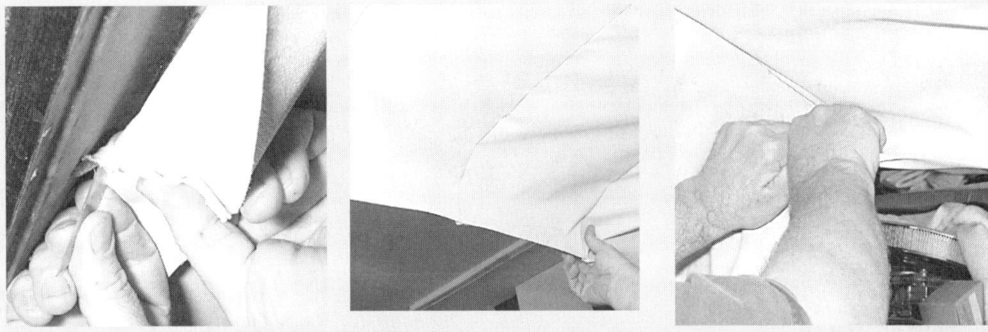

8. Getting the headliner where you need it for lockdown is challenging and demands patience. Get impatient and you will ruin the headliner and have to order another one. Use a razor blade and gently cut each sleeve, working each headliner panel toward the outside. Work each panel until all wrinkles are gone. Go too far and you will create arcs in each panel.

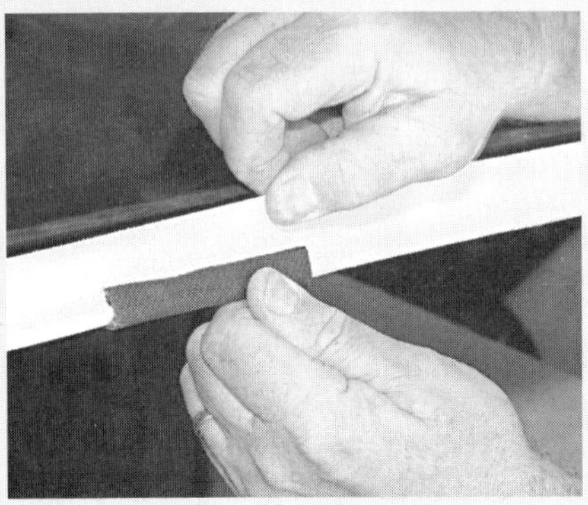

9. We suggest a dry run with the headliner before actually gluing it to the body. You can do this using pieces of wind lace from your old headliner. Cut them into 3 to 4-inch sections to use as temporary retainers. This enables you to work the headliner before making a permanent decision.

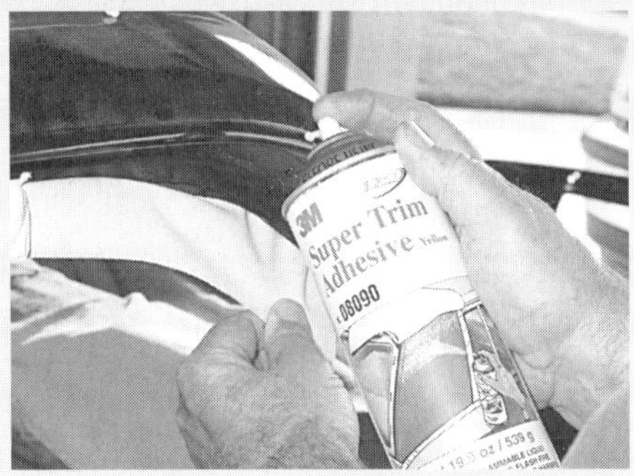

10. We never use anything but 3M Super Trim Adhesive on our headliner installations. Spray both surfaces with a modest amount of adhesive and allow it to tack-off (become sticky), which takes about five minutes. Next, pull the headliner firmly toward the front of the vehicle until all wrinkles are gone, then secure the panel, beginning at the middle and working outward. Secure the headliner using pieces of wind lace. Do this in front first, working down the sides to the back.

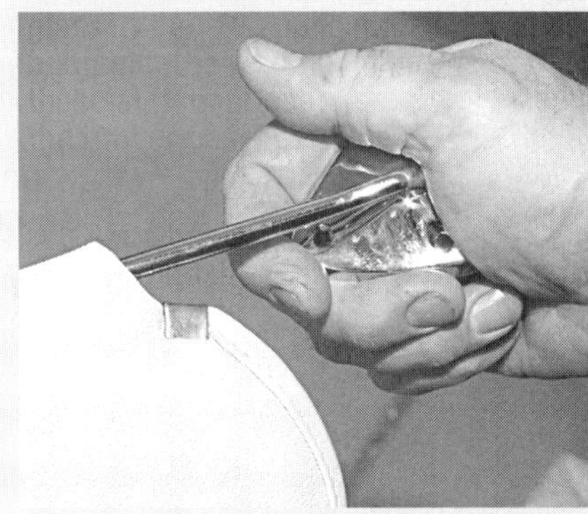

11. Sunvisors can be as challenging as the headliner, especially inserting the brackets and pins. Lubricate them with WD-40 or a pinch of white grease. Insert them with a twisting action, working them into the sunvisor. Positioning is tricky because they need to be attached to the vehicle to get that right.

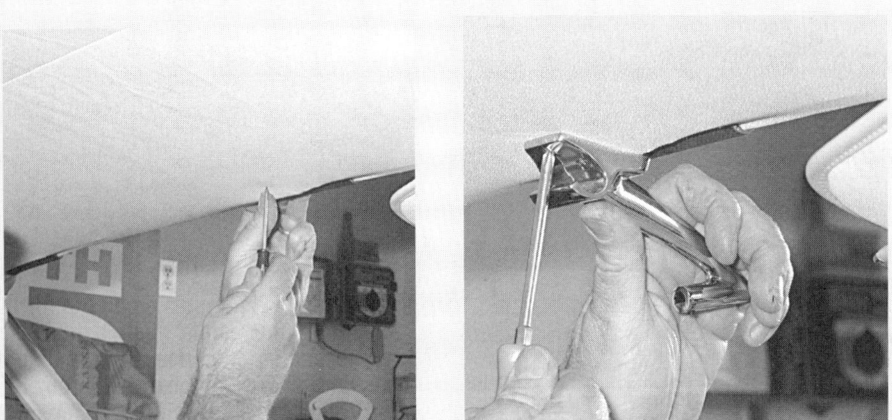

12. Gary finds the screw holes for the rear-view mirror and the sunvisor brackets. These become part of the headliner anchoring in front. Do not install the mirror or brackets until you are positive about headliner positioning.

13. Remember, your strategy needs to be front first, then down both sides, checking for proper positioning as you go, and securing the headliner with adhesive and wind lace strips.

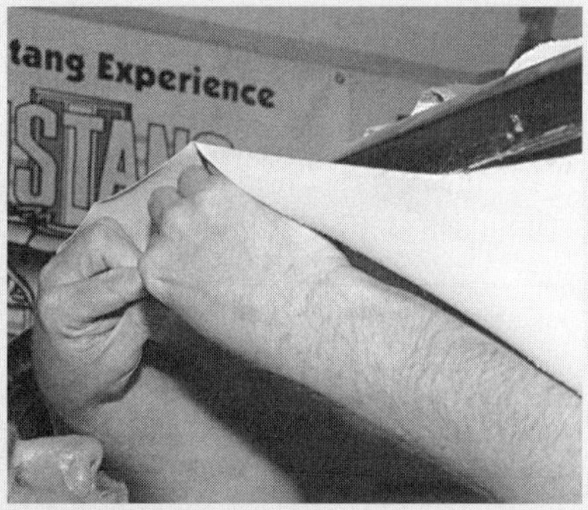

14. Once all points north of the sail pillars are secure, yank and secure the headliner at the rear roof rail. Pull firmly and secure at the center first, working outward toward the sail pillars. Remember, wrinkles are unacceptable.

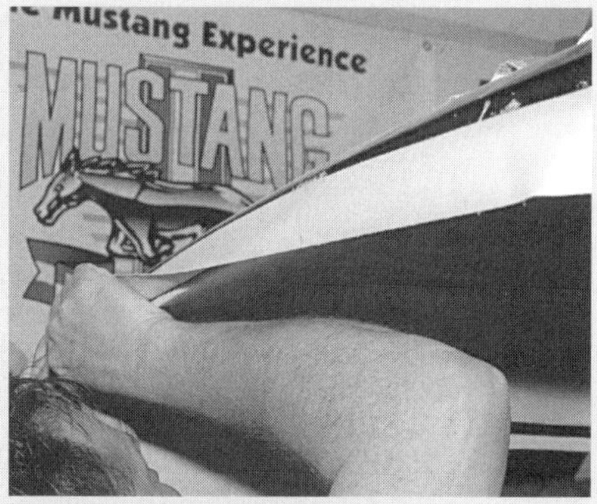

15. As the rear window headliner attachment points are secured, work outward to the sail pillars and lock down the headliner. Again, check for wrinkles before gluing the headliner.

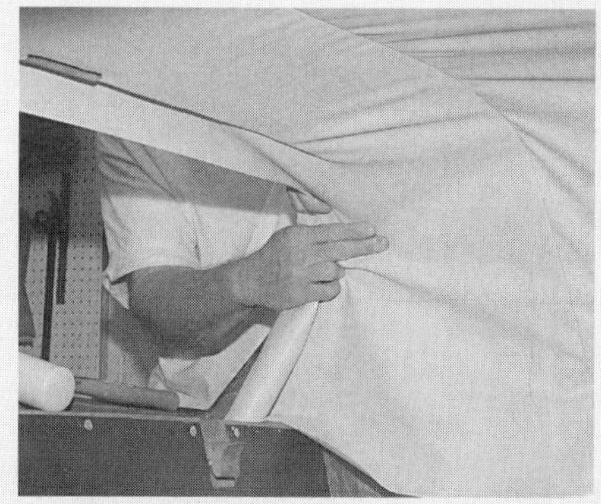

16. Each sail pillar has to be worked carefully because of the potential for wrinkles. Pull fore and aft, watch for wrinkles, and secure the sail-pillar segment. Minor wrinkles can be worked out with a heat gun.

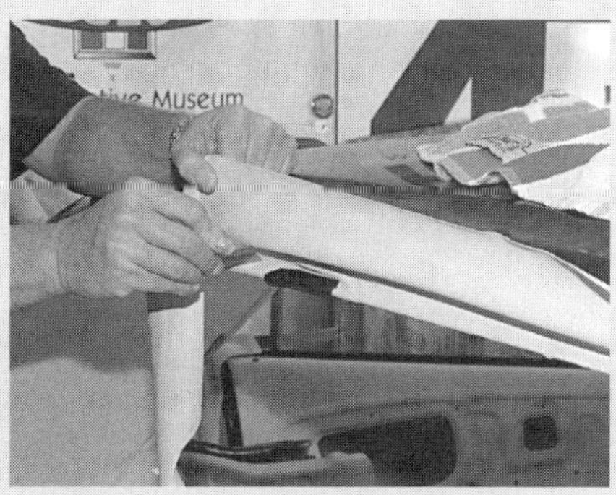

17. When you are confident the headliner is secure and wrinkle-free, trim the excess around the perimeter with a razor blade.

18. Minor wrinkles can be worked out with a heat gun. Remember, your headliner is made of vinyl plastic. It will melt if you use too much heat or get the heat gun too close to the vinyl. Keep a safe distance of at least six inches and work the vinyl. Watch the wrinkles fade away. You can also use a hairdryer.

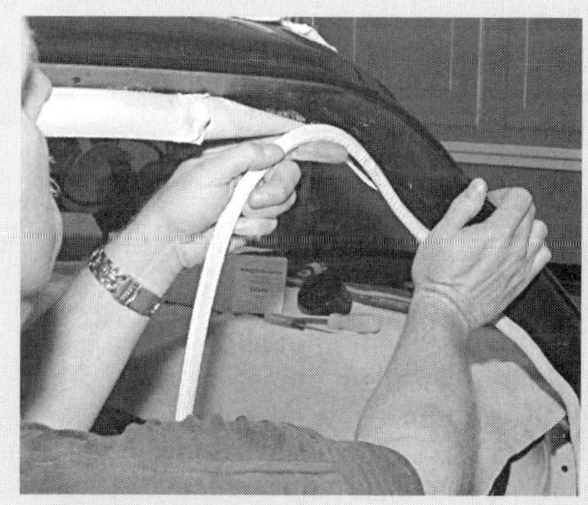

19. When the trim adhesive is set, remove the temporary wind lace and install the new wind lace. You will need at least 16 feet, but play it safe and order 20 feet.

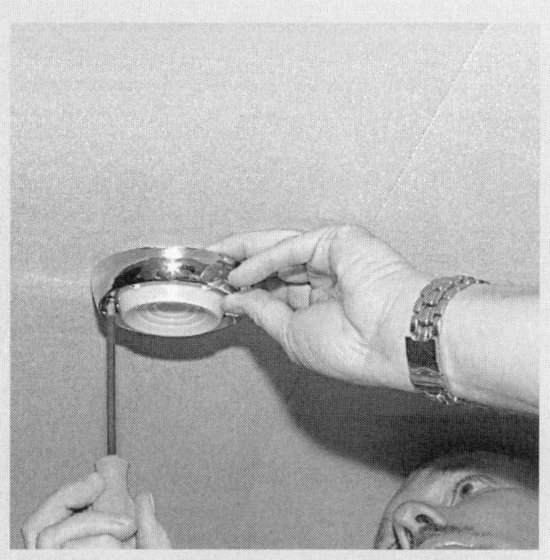

20. Finally, for our dome-light addition, the lens and bezel are installed. Don't forget the lens/bezel screws when you order your courtesy light from Mustangs Etc.

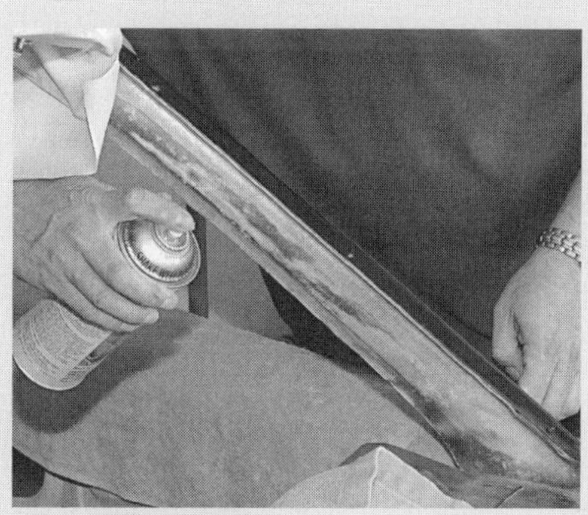

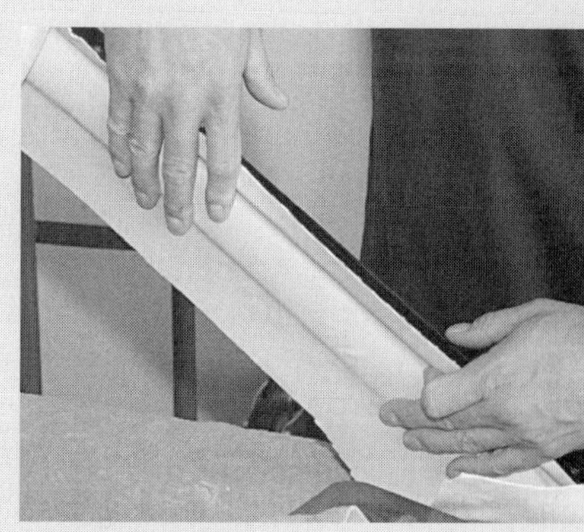

21. In 1965–'66, and later in 1968, Ford didn't cover the A-pillars with anything but paint. Go one better by covering your A-pillars with vinyl or leather in the interior color, like we did with Gary's Mustang restomod. A good rule of thumb is to spray 3M Super Trim Adhesive on the metal first, then lightly spray the vinyl just prior to installation. Too much trim adhesive can make the vinyl bubble. Apply a light dusting of trim adhesive and lay it down as shown. Work out any wrinkles with a heat gun.

Chapter 48

How to Replace a '65–'66 Dashpad

Text and Photos by Miles Cook

Installed and ready to go in our '66 convertible project car, a new dashpad is a nice upgrade for any vintage Mustang. The pads are quite different for '65–'66, '67–'68, '69–'70, and '71–'73 Mustangs, so installation procedures vary. This time we're concentrating on the earliest generation.

The dashpad in a vintage Mustang is a rather high-profile interior component that's often one of the first pieces requiring replacement. Constant exposure to the hot sun is the main reason, and the result is usually cracks and/or surface warp.

The latter happened to the pad in Debbie Starr's '66 convertible, likely due to the fact that the car was her only means of getting to and from her job as an elementary school teacher for more than five years. Driving the car daily meant it sat in the school parking lot every day under the hot sun. In fact, on hot days, you could easily fry an egg inside, especially since this car has a black interior.

Now that Debbie has another vehicle to drive, her Mustang has become more of a hobbyhorse of sorts. Now she can concentrate on a few restoration upgrades and other general improvements knowing the car will lead an easier life.

We decided to begin with a new dashpad. It turns out Mustang Country International was also able to spend some time on a few other things while installing the new pad. This included a new pair of defroster ducts as well as pointing out the variations between '65 and '66 pads. Check out the sidebar at the end of the installation instructions to learn more about the differences between the two years.

Source
Mustang Country International
www.mustangcountryintl.com

Mustang Country carries a full selection of pads for all '65-'73 Mustangs. A black '66 pad is one of the more popular year/color combinations, so we were ready to begin installation upon arrival.

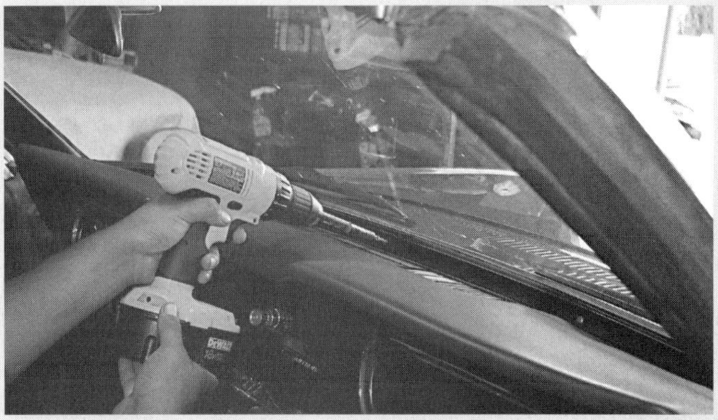

1. We began by removing the line of screws located directly below the windshield.

2. Once the screws are out, the trim pieces can be removed. These are set aside to be reused with the new pad. If needed, you can repaint the trim strips for a better appearance.

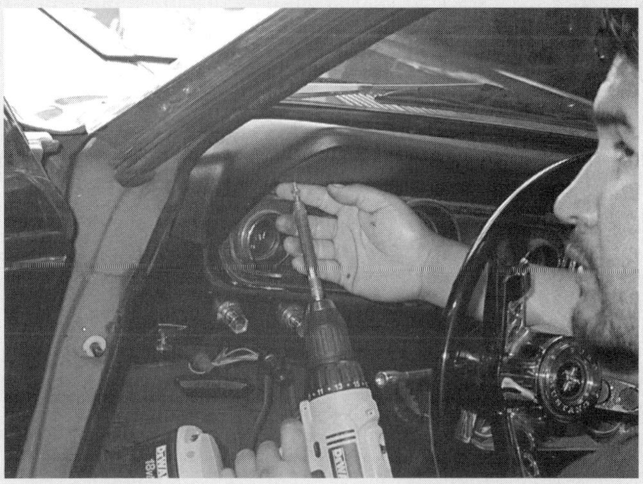

3. There are four screws above the gauge cluster and four above the glovebox door. Remove these eight screws to prepare for taking the pad out of the car.

4. Next, remove the screws that hold the speaker grille, and remove it from the top of the pad.

5. The dash speaker—in this case, what's left of a dual aftermarket setup—needs to be removed as well. This typically involves a few screws and unplugging the speaker wire from the radio. Next, remove the pair of defroster ducts, which attach to the top of the dashpad with special clips.

6. With the speaker and defroster ducts removed, access is gained to two nuts that need to be removed from inside the dashpad. In this car, one of the aftermarket speakers was sitting inside the dash, and we were able to reconnect it after putting everything together again.

7. There are a total of eight screws (four above the gauge cluster and four above the glovebox door) and four nuts (two at the center and two at each end) that hold a '66 dashpad in place. Once these 12 fasteners are removed and the dashpad trim pieces under the windshield are removed, the pad can come out. The two nuts at either end of the dash aren't visible when the pad is still in the car. Reach inside the dash with a 1/4-inch–drive ratchet and socket to access them.

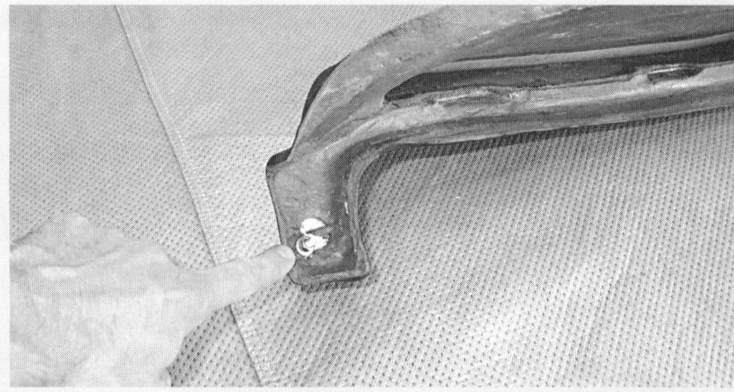

8. This is the new pad with its outer nuts still in place. Here you can see where they go once the pad is in place. After the pad is in position, these nuts are installed and secured by threading them behind and up inside the dash. Although they are the hardest to access, it can be done by feel.

9. Another pair of nuts hold the pad in place at the center. As shown, these are easily accessed from the area where the speaker grille resides. These nuts aren't used on '65 pads.

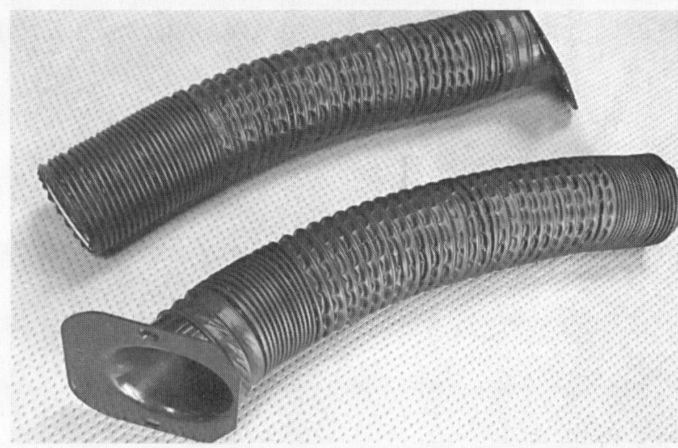

10. Because the defroster ducts in our car were damaged, we replaced them with a set of reproduction ducts from Scott Drake Enterprises; they are also available from Mustang Country.

11. Mustang Country likes to trim some of the foam away from the edge of the pad to allow it to fit snugly against the windshield base. Be careful, however, as you can easily end up going through to the top of the pad if you remove too much.

12. Now we're ready to install the new pad. At this point, the new piece is basically a drop-in replacement for the old one.

13. With the pad in place, install the new defroster ducts next. They thread into place and attach to the heater box just like the old ones.

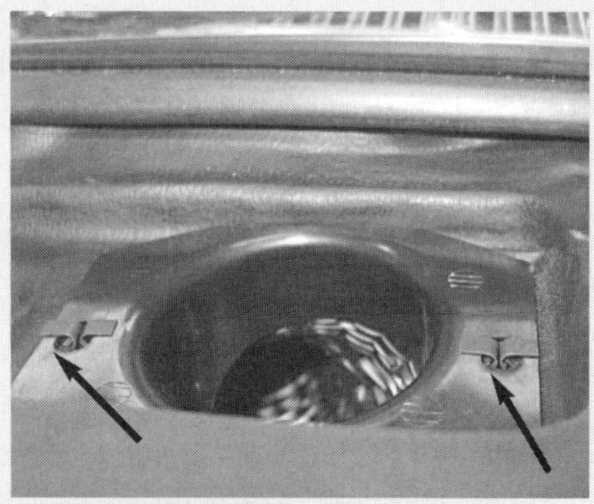

14. These specialized clips hold the ducts in place on top of the pad.

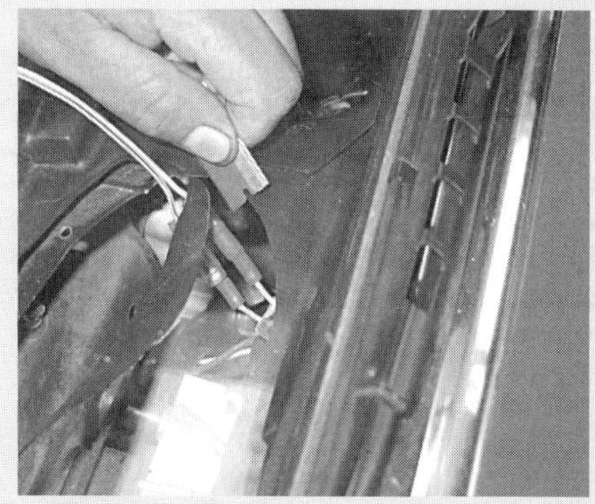

15. Some slight trimming may be required; it will vary with each car. Here, a small amount of pad vinyl is trimmed away to make room for the speaker and speaker grille.

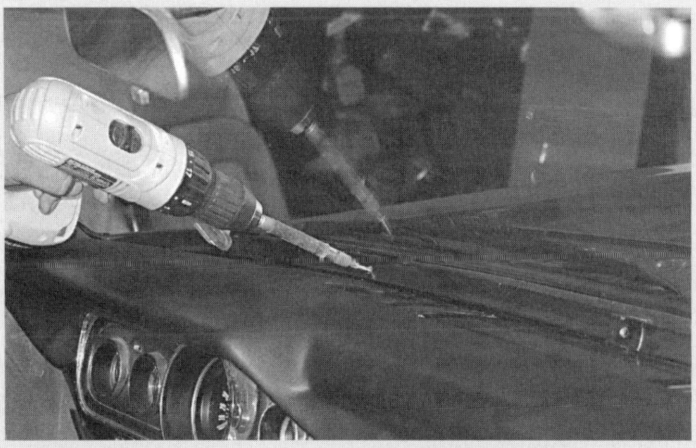

16. With the pad in place, the only task that remains is reinstalling the fasteners. The two inside nuts shown earlier are already back in place, as are the speaker and speaker grille. Here the windshield trim pieces are reinstalled.

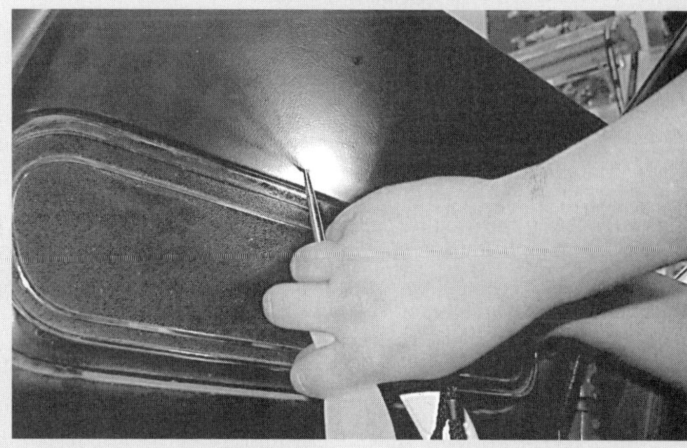

17. With the four nuts installed, eight screws remain to complete the installation. New holes need to be created in the new pad to secure the pad to the dashboard. This is relatively easy, as shown here, using a small punch and a flashlight to locate the threads that receive the screws.

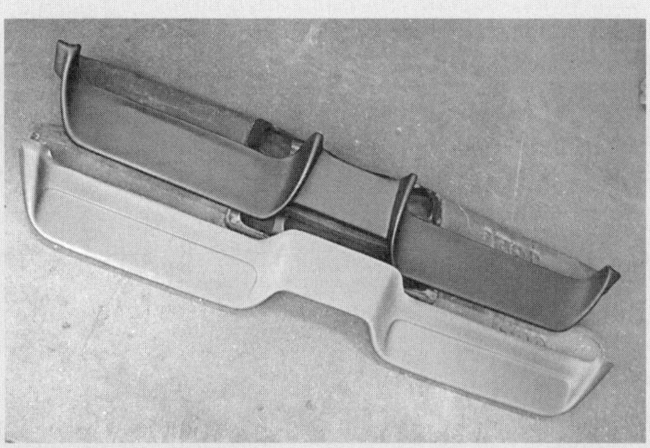

18. Although installation of a '65 pad is similar to the '66, there are some differences. The two nuts in the middle of the '66 pad aren't there for the '65 pads. However, the '65s have a pair of chrome strips that run along the front edge of the pad where it meets the gauge cluster and glovebox door. Also, the blue '65 pad has a different shape compared to the black '66 pad. The '66 pad also has a pair of "wings" in the center that the '65 pad lacks.

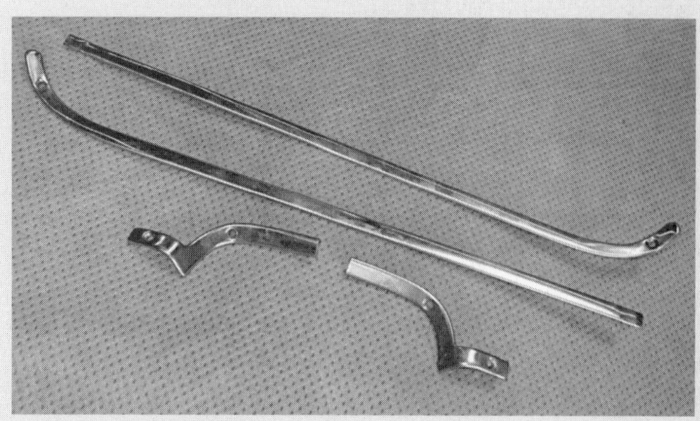

19. This is a set of dashpad strips for a '65 Mustang. The long and short strip join together to make a complete strip for each side of the pad, one above the gauge cluster and the other above the glovebox door.